Differential-Algebraic Equations Forum

W0114796

Differential-Algebraic Equations Forum

The series "Differential-Algebraic Equations Forum" is concerned with analytical, algebraic, control theoretic and numerical aspects of differential algebraic equations (DAEs) as well as their applications in science and engineering. It is aimed to contain survey and mathematically rigorous articles, research monographs and textbooks. Proposals are assigned to an Associate Editor, who recommends publication on the basis of a detailed and careful evaluation by at least two referees. The appraisals will be based on the substance and quality of the exposition.

More information about this series at
http://www.springer.com/series/11221

Achim Ilchmann • Timo Reis

Editors

Surveys in Differential-Algebraic Equations II

 Springer

Editors

Achim Ilchmann
Institut für Mathematik
Technische Universität Ilmenau
Ilmenau, Germany

Timo Reis
Fachbereich Mathematik
Universität Hamburg
Hamburg, Germany

ISBN 978-3-319-11049-3   ISBN 978-3-319-11050-9 (eBook)
DOI 10.1007/978-3-319-11050-9
Springer Cham Heidelberg New York Dordrecht London

Library of Congress Control Number: 2013935149

Mathematics Subject Classification (2010): 34A08, 65L80, 93B05, 93D09

Printed on acid-free paper

Springer is part of Springer Science+Business Media (www.springer.com)

Preface

We are pleased to present the second of the three volumes of survey articles in various fields of differential-algebraic equations (DAEs). In the preface to Volume I, we tried to give a historical sketch of the development of DAEs, a systematic classification of the subject within applied mathematics and between engineering and mathematics, and a list of the different and broad topics studied in DAEs.

In this volume we complete the list of survey articles in the sense that they are of theoretical interest and equally relevant to applications. In particular, this volume contains articles in which DAEs are regarded from the viewpoint of classical disciplines of mathematics (algebra, functional analysis), control aspects (observers, optimal control) and concrete applications of DAEs in chemical engineering.

We hope that this issue will also contribute to complete the picture of the latest developments in DAEs. This collection of survey articles may indicate that DAEs are no longer just a collection of results to the same topic but a field in its own right under the rank of ordinary differential equations.

Ilmenau, Germany
Hamburg, Germany
June 2014

Achim Ilchmann
Timo Reis

Contents

Linear Differential Algebraic Equations and Observers

Karen S. Bobinyec and Stephen L. Campbell

Abstract Observers play an important role in the control of linear systems. Given the importance of DAE models, it is natural that there has been considerable interest in designing and using observers for DAEs. The first part of this paper surveys some aspects of DAEs and observers that lay the foundation for the second part which discusses a recent general approach to observer design for linear DAEs using completions. This approach also holds great promise for nonlinear DAEs.

Keywords DAE • Linear systems • Observer

Mathematics Subject Classification (2010) 34H15, 34A09, 93C05

1 Introduction

Dynamical systems called observers play an important role in numerous areas of engineering. Many physical systems are modeled by systems of differential algebraic equations (simply referred to here as DAEs). Thus, it is not surprising that over the last two decades there has been considerable interest in observers for DAEs and in DAEs as observers. If the observer itself is a DAE we shall call it a DAE observer. This survey focuses on new design techniques for observing DAEs and discusses how these results provide a starting point for answering observer-related questions. We also point out the wide variety of ways in which DAE and observer theory intertwine. Our desire for this paper is that it be widely accessible, or readable for such groups as control engineers and numerical analysts. Therefore, in order to place these results in context, the first part of this paper mentions observer theory in the framework of controllability, observability, stabilization, and various other control topics. However, an extensive survey of these interconnections is not included because it would result in a series of volumes duplicating many excellent

K.S. Bobinyec • S.L. Campbell (✉)
Department of Mathematics, North Carolina State University, Raleigh, NC, USA
e-mail: ksbobiny@ncsu.edu; slc@ncsu.edu

© Springer International Publishing Switzerland 2015
A. Ilchmann, T. Reis (eds.), *Surveys in Differential-Algebraic Equations II*,
Differential-Algebraic Equations Forum, DOI 10.1007/978-3-319-11050-9_1

books and papers already available. Much of our later work included in the second part of this paper has yet to appear in journal publications.

Section 2 provides background on three topics. Section 2.1 briefly reviews what observers are for linear ordinary differential equations (ODEs). Control engineers can skip this first section, but later sections may be hard to follow for readers unfamiliar with the material in Sect. 2.1. Section 2.2 highlights some needed facts about DAEs, but we assume most readers have a basic understanding of this topic, so the presentation is brief. Finally, Sect. 2.3 discusses completions of DAEs and introduces the three completions discussed in this paper. Throughout Sect. 2, proofs and all but the most necessary definitions are omitted.

Section 3 focuses on some of the many ways DAEs and observers have previously appeared in the literature. Although we cover a wide array of contributions in order to show the richness of this topic, this collection is only a sampling. This survey section has a considerable number of references.

Sections 4, 5, 6, and 7 provide an example of current research involving DAEs and observers. Section 4 develops several types of observers based on completions of DAEs. Section 5 presents some theoretical results relating completions to control properties. In some sense, the linear time-invariant case is known from the linear time-varying case. However, as is typical in systems theory, the algorithms and results in the linear time-invariant case are cleaner and of considerable interest themselves and the linear time-varying results can be more complicated and technical. Accordingly, we first discuss the linear time-invariant case and then the linear time-varying case. Computational examples are included in Sects. 6 and 7.

Whether a given expression is constant or time-varying will be made clear by subheadings and statements as to whether we are in the linear time-varying or linear time-invariant case.

In this survey, results that are now standard and appear in textbooks are given without citation. Additionally, although the discrete time case has received a lot of attention and is important in many applications, we consider only continuous time systems.

2 Background

We first provide some background on observers, DAEs, and completions of DAEs.

2.1 Observers

The starting point for literally thousands of engineering papers and books is

$$x' = Ax + Bu \tag{2.1a}$$

$$y = Cx + Du, \tag{2.1b}$$

where x' denotes the time derivative. Quantities A, B, C, D are constant matrices in the linear time-invariant case and time dependent matrices in the linear time-varying case. x, y, u are vector valued functions of time. x is the state, y is the output of the system available for measurement or other use, and u is the control or input. Throughout this paper x is n dimensional, y is m dimensional, and u is p dimensional. All vectors and matrices may be complex unless otherwise stated. We will only show the dependence on t when it exists if it is not clear where a function is being evaluated or it helps to emphasize that the problem is time varying. Typically n is much larger than either m or p. Matrix D may be present in a particular application but is often dropped from theoretical discussions. In real-world applications, a lot of modeling and reformulation often occurs before (2.1) results. Other terms sometimes present in (2.1) are discussed later.

Our emphasis is on estimating x from y. The control or input u is considered known unless we state otherwise, for example, in the comments on estimating disturbances.

We focus on smooth solutions x and sufficiently smooth inputs u. Impulsive and non-smooth solutions have also been historically of interest, but we do not consider them here. One could write a whole additional survey paper on impulsive systems.

2.1.1 Linear Time-Invariant Systems

We first focus on the linear time-invariant case. The controllability matrix $\mathcal{C}$ and the observability matrix $\mathcal{O}$ are defined by

$$\mathcal{C} = \begin{bmatrix} B & AB & \cdots & A^{n-1}B \end{bmatrix}, \quad \mathcal{O} = \begin{bmatrix} C \\ CA \\ \vdots \\ CA^{n-1} \end{bmatrix}. \tag{2.2}$$

The range of $\mathcal{C}$, denoted $R(\mathcal{C})$, and the nullspace of $\mathcal{O}$, denoted $N(\mathcal{O})$, are both A-invariant subspaces. $R(\mathcal{C})$ is the controllable subspace and $N(\mathcal{O})$ is the unobservable subspace. If $R(\mathcal{C})$ is all of C^n, the system is controllable. If $N(\mathcal{O})$ is zero dimensional, the system is observable. Subspaces M_1, M_2, and M_3 are defined so that $R(\mathcal{C}) = R(\mathcal{C}) \cap N(\mathcal{O}) \oplus M_1$, $N(\mathcal{O}) = R(\mathcal{C}) \cap N(\mathcal{O}) \oplus M_2$, and $C^n = R(\mathcal{C}) \cap N(\mathcal{O}) \oplus M_1 \oplus M_2 \oplus M_3$. Performing a similarity transformation relative to the decomposition $M_1 \oplus R(\mathcal{C}) \cap N(\mathcal{O}) \oplus M_3 \oplus M_2$ produces the Kalman decomposition (here x is a new state variable)

$$x' = \begin{bmatrix} A_{11} & 0 & A_{13} & 0 \\ A_{21} & A_{22} & A_{23} & A_{24} \\ 0 & 0 & A_{33} & 0 \\ 0 & 0 & A_{43} & A_{44} \end{bmatrix} x + \begin{bmatrix} B_1 \\ B_2 \\ 0 \\ 0 \end{bmatrix} u \tag{2.3a}$$

$$y = \begin{bmatrix} C_1 & 0 & C_2 & 0 \end{bmatrix} x + Du. \tag{2.3b}$$

This form is fundamental in understanding some of our later discussions. There are instances when some of these block rows and columns are missing.

The eigenvalues of A, which determine what types of functions appear in various variables, are the union of each A_{ii}'s eigenvalues. The eigenvalues of A_{22} and A_{44} are described as unobservable since they do not affect the output y, while the eigenvalues of A_{11} and A_{33} are observable. The eigenvalues of A_{11} and A_{22} are identified as controllable because they can be altered by state feedback, while the eigenvalues of A_{33} and A_{44} are uncontrollable. If the control in (2.1) is implanted as a feedback,

$$u = Fx + v, \tag{2.4}$$

then (2.1a) becomes

$$x' = (A + BF)x + Bv. \tag{2.5}$$

This feedback can simultaneously change the controllable eigenvalues of A to any values the user wants but cannot alter the uncontrollable eigenvalues. Matrix F is called either the gain or feedback matrix.

Feedback is very important in engineering applications. Unfortunately we usually know y and not x, so if we want to use a full state feedback like (2.4), we need to estimate x. One way to estimate the state is with observers. An observer is a dynamical system that takes y and returns an estimate $\hat{x}$. This estimate asymptotically converges to x independent of the initial conditions $x(0)$ and $\hat{x}(0)$. Of course, the convergence must be rapid enough to give a useful answer in time for the estimate to be used but slow enough to keep from responding to every little perturbation. Thus, we often care about the rate the estimation error $x - \hat{x}$ converges to zero. Some uses of observers are highlighted in Sect. 3.

The starting point for many discussions about observers is a Luenberger observer

$$\hat{x}' = A\hat{x} + Bu + L(y - \hat{y}) \tag{2.6a}$$

$$\hat{y} = C\hat{x} + Du. \tag{2.6b}$$

We know y, our measured output, and u, the control we are applying. Note that (2.1) is an actual physical system while (2.6) exists in a computer or on a chip. If $e = x - \hat{x}$ is the estimation error, then

$$e' = (A - LC)e. \tag{2.7}$$

Matrix L is chosen to make the observable eigenvalues of A any value the user wants, but the unobservable eigenvalues of A remain eigenvalues of $A - LC$ for any L. Therefore, in order for the estimation error to go to zero, the unobservable eigenvalues of A need to have negative real parts. However, we cannot control the rate at which the unobservable terms go to zero. If $u = F\hat{x} + v$ is used instead of (2.4), then the combined system of (2.1) and (2.6) has both the desired eigenvalues of $A + BF$ and the eigenvalues of $A - LC$ introduced by the observer.

2.1.2 Linear Time-Varying Systems

In the linear time-varying case we still have (2.6) but the details are more technical since eigenvalues no longer tell the full story without a lot of extra assumptions. Arguments based on Lyapunov functions start to replace those based on eigenvalues. Also, the derivation of a Kalman canonical form like (2.3) is not the same in the linear time-varying case. For nonlinear systems, the use of linearizations may produce local observers. Some of these issues are addressed later.

Results on linear time-varying observers date back to at least [80, 81, 113, 114] to name a few. Many papers on this topic will say to take an object X with property Y and then show that they can build an observer. Careful discussions of how to find object X in the first place are often either a special case for a special problem, not given, or highly technical and computationally difficult. In our examples we go into some detail on how to find object X.

2.2 Differential Algebraic Equations

A number of books on DAEs, their analysis, their numerical solution, and their applications have been published including [23, 86, 88, 94, 119]. In this section we focus on material needed for the linear case and for discussing observers. We are interested in differentiable solutions.

A linear DAE

$$E(t)x'(t) + F(t)x(t) = B(t)u(t) \qquad (2.8)$$

is solvable if, for any given sufficiently smooth B and any sufficiently smooth u, the system's solution is uniquely determined by a consistent initial condition [23]. These consistent initial conditions must satisfy all constraints. Due to possibly hidden constraints, consistent initial conditions may not be obvious and the set of consistent initial conditions will depend on the particular B and u. Research on methods for finding consistent initial conditions includes [24, 44, 93, 96], but the examples to which we apply our observer construction approach in Sects. 6 and 7 have consistent initial conditions that can be found by hand.

We assume the linear DAEs are solvable, so methods such as those in [35] or [103] are not required for checking the solvability of a system. We also assume the coefficient matrices of the linear DAEs are smooth, meaning at least as many derivatives exist as are needed. The solution manifold, or the manifold of consistent initial conditions in [27], is characterized by the constraints of the DAE, some of which may be implicit, and is where the solutions of a DAE live for a consistent initial condition [33]. The solution manifold depends on u.

Other phrases in the literature synonymous with DAEs are descriptor systems [64], general or generalized state space systems [61], implicit systems[6], semi-state systems [109], and singular systems [97]. This variety in terminology exists because of separate initial investigations into these systems by different disciplines. Examples of applications involving DAEs include chemistry [25, 141]; circuit theory [108, 121]; constrained mechanical systems [15, 69]; gas networks [118] (discrete time and observer consideration); power systems [75]; and robotic motion [109]. Article [28] also provides an example overview on higher index DAEs.

2.2.1   Linear Time-Invariant DAEs

There are a number of ways to approach the theory of linear time-invariant DAEs. In some approaches control variables are just viewed as an additional unknown algebraic variable. However, if the control is specified or assumed given, the approaches simplify to a consideration of the Weierstraß canonical form. That is, if we have the linear time-invariant DAE

$$Ex' + Fx = Bu \tag{2.9}$$

with constant E, F, B and u taken as an input, then the pencil $sE + F$ is called regular when a value of s exists for which the pencil is invertible. Those values of s for which $sE + F$ is singular are called the matrix pencil eigenvalues. The condition of being regular is equivalent to (2.9) having a solution for every B and for every u and the solution being uniquely determined given a consistent initial condition. In some situations the variables x, u are combined and one wants to know for what choices of control the system is regular. In other cases (2.9) is part of an optimization problem and what is important to the optimization software is whether there is a choice of control that in some sense gives a good problem. These questions are important, and the interested reader is referred to [40, 45, 62]. Here we assume that u is a known control or input.

Under the assumption of regularity, there exist invertible transformations P, Q so that if we multiply (2.9) by P and let $x = Qz$, (2.9) becomes the quasi-Weierstraß (or Kronecker) form [70]

$$\begin{bmatrix} I & 0 \\ 0 & N \end{bmatrix} z' + \begin{bmatrix} H & 0 \\ 0 & I \end{bmatrix} z = \begin{bmatrix} B_1 \\ B_2 \end{bmatrix} u \tag{2.10}$$

with $N^k = 0$ and $N^{k-1} \neq 0$ (if $N = 0, k = 1$). k is called the index of the DAE. These transformations decompose the DAE into two systems. The first, $z_1' + Hz_1 =$

$B_1 u$, is an ordinary differential equation. The second, $N z_2' + z_2 = B_2 u$, has only the one solution

$$z_2 = \sum_{i=0}^{k-1} (-N)^i B_2 u^{(i)}, \tag{2.11}$$

where $u^{(i)}$ denotes the ith derivative. If H in (2.10) is $\ell \times \ell$, the solutions of (2.9) live on a ℓ dimensional manifold referred to as the solution manifold. The $d = n - \ell$ constraints given by (2.11) define this manifold.

2.2.2 Linear Time-Varying DAEs

The development of the theory for linear time-varying DAEs is quite different and, in many ways, is closer to the general nonlinear case. In particular, matrices and Jacobians can have changing ranks and additional numerical calculation is needed at each time step with numerical integrators. Also, eigenvalues may not tell the full story on stability. This similarity provides a way to test ideas for the nonlinear case and is one reason why the linear time-varying case is so important besides just the fact that linear time-varying systems occur. It is already clear from (2.11) in the linear time-invariant case that some type of differentiation is needed to describe the solutions of the DAE. Since taking the derivative of numerically computed quantities is usually avoided, we base all differentiations on the original equations. After differentiating the linear time-varying DAE (2.9) k times, we get the derivative array equations

$$\mathcal{E}(t) w(t) + \mathcal{F}(t) x(t) = \mathcal{B}(t) v(t), \tag{2.12}$$

where

$$\mathcal{E} = \begin{bmatrix} E & 0 & 0 & \cdots \\ E' + F & E & 0 & \ddots \\ E'' + 2F' & 2E' + F & E & \ddots \\ \vdots & \vdots & \vdots & \ddots \end{bmatrix}, \quad \mathcal{B} = \begin{bmatrix} B & 0 & 0 & \cdots \\ B' & B & 0 & \ddots \\ B'' & 2B' & B & \ddots \\ \vdots & \vdots & \vdots & \ddots \end{bmatrix},$$

$$\mathcal{F} = \begin{bmatrix} F \\ F' \\ \vdots \\ F^{(k)} \end{bmatrix}, \quad w = \begin{bmatrix} x' \\ x'' \\ \vdots \\ x^{(k+1)} \end{bmatrix}, \quad v = \begin{bmatrix} u \\ u' \\ \vdots \\ u^{(k)} \end{bmatrix}.$$

Here $\mathcal{E}, \mathcal{B}$ are $(k+1) \times (k+1)$ block matrices. Having assumed matrices E, F, and B from the linear DAE are smooth, matrices $\mathcal{E}, \mathcal{F}$, and $\mathcal{B}$ are smooth as well.

While the formulas have $u^{(k)}$ in them, in practice many of these higher derivatives do not actually appear in the observer. For example, in the index three DAEs from constrained mechanics the input u does not appear in either the algebraic state constraint nor in the velocity constraint. As a consequence the higher derivatives of u do not appear. For example, in the example of Sect. 6, if u_2 is zero which is often the case, then the higher derivatives u_1'', u_1''' of the other control do not appear.

Note that if a linear DAE is solvable, then matrix $\mathcal{E}$ from the derivative array (2.12) has constant rank even though matrix E in (2.8) may not [43].

We assume k is large enough so the following three assumptions hold:

A1. $[\mathcal{E}, \mathcal{F}]$ is full row rank for all t.
A2. $\mathcal{E}$ has constant rank.
A3. The first n columns of $\mathcal{E}$ are linearly independent and are linearly independent of $\mathcal{E}$'s remaining columns. Equivalently, if $\mathcal{E}v = 0$ for some vector v, then the first n entries of v are zero.

Property A3 is sometimes called 1-fullness. For the linear time-invariant case, assumptions A1–A3 holding for some k are equivalent to the pencil being regular and the smallest k for which the assumptions hold is the index. However, these assumptions holding for some k are more general in the linear time-varying case and are a basis for developing a nonlinear DAE theory. Note also [88].

For real analytic linear time-varying DAEs, A1–A3 are equivalent to solvability. They are almost equivalent to solvability for systems which are just sufficiently differentiable. See [94] for a treatment with reduced smoothness in some components. Thus for many systems the assumptions are not restrictive.

Definition 2.1 For linear time-varying systems we define the index of the DAE to be the smallest k such that A1–A3 hold [35].

An approach based on differentiation originally due to Fliess and coworkers has some similarities to what is presented here but differs in several ways [65–67]. Their approach's mathematical background is in differential algebra; that is, the authors use a noncommutative algebra with a differentiation operator. Another difference with the approach of this paper is that we only differentiate equations but they, at least formally, differentiate outputs. Differentiation of real outputs can be carried out by using smoothing operators provided the noise has certain properties [120]. Since good surveys on this alternative approach exist, we refer the interested reader to the cited references and their bibliographies.

2.3   Completions of DAEs

Completions of DAEs play an important role in the latter parts of this paper. Many simulation packages require an ODE model, so the ability to automatically generate an ODE from a DAE motivates some of the interest in completions. A completion of an index k linear DAE (2.8) is a linear ODE

$$\tilde{x}'(t) = \tilde{A}(t)\tilde{x}(t) + \sum_{i=0}^{k-1} \tilde{B}_i(t)u^{(i)}(t) = \tilde{A}(t)\tilde{x}(t) + \tilde{B}(t)v(t) \qquad (2.13)$$

designed so the solutions of (2.13) contain those of (2.8). The tilde in (2.13) indicates that the matrices are computed quantities whose values can depend on how the completion is done. For an n-dimensional state vector x, a linear DAE is defined on the less than n-dimensional solution manifold. Equation (2.13) is described as a completion since it is defined on a complete n-dimensional space that has the solution manifold as a submanifold [34, 116]. Matrices $\tilde{A}$, $\tilde{B}$ are usually computed numerically at each needed time t and should be smooth up to numerical error. The solutions of (2.13) that are not solutions of the original DAE are called the additional dynamics. Theory in [34] identifies a general form relating all possible completions of a linear DAE.

It is important to note that the matrix $\mathcal{E}$ in (2.12) is neither full column nor full row rank. Thus there are many ways that one could numerically derive (2.13) from (2.12). In the linear time-varying case it is essential that the result be smooth in t. We focus on three completions of linear DAEs: the least squares completion (LSC), the stabilized least squares completion (SLSC), and the alternative stabilized completion (ASC). Each of these will result in smooth coefficients of the completion and each have their advantages.

An early article on the LSC is [34]. The LSC ensures the smoothness of coefficients since the Moore–Penrose inverse of $\mathcal{E}$ is smooth if $\mathcal{E}$ is constant rank which it often is. This completion finds the unique minimum norm least squares solution for w of the derivative array equations. A noted benefit in [26] of working from the derivative array equations is the derivatives are of the original linear DAE and not of numerically computed values. A concern of [39] revisited in [116, 117] is that consistent initial conditions may not be enough to keep the numerical solution of the LSC from drifting off the solution manifold. The desire is to affect the additional dynamics so the completion's solutions satisfy the constraints of the linear DAE and converge to the solution manifold if perturbed.

The processes proposed in [116] of finding the LSC and an alternative least squares completion are each modified in [117] with stabilized differentiation to produce the SLSC and the ASC, respectively. From Sect. 2.3.4, the additional dynamics of the LSC are not asymptotically stable and for systems of index greater than two they are unstable. Stabilized differentiation has been used since [8] to stabilize the additional dynamics that come from explicit constraints. The SLSC and the ASC extend that idea to general linear time-varying fully implicit DAEs.

Section 2.3.1 details the LSC; Sect. 2.3.2 describes how the process in Sect. 2.3.1 is modified to produce the SLSC; and Sect. 2.3.3 develops the ASC. For these completions' numerical algorithms, the coefficient matrices $\mathcal{E}, \mathcal{F}$, and $\mathcal{B}$ in (2.12) can be computed by symbolic or automatic differentiation [37, 46, 48]. Advantages and disadvantages of each type of completion will be given along with their development and carefully discussed in Sect. 2.3.4.

2.3.1 Least Squares Completion

With the LSC, the vector w in (2.12) is solved for in the minimum norm least squares sense. If $\mathcal{E}^\dagger$ is the Moore–Penrose pseudoinverse [41] of $\mathcal{E}$, then the minimum norm least squares solution for w is

$$\overline{w} = \begin{bmatrix} x' \\ * \end{bmatrix} = -\mathcal{E}^\dagger \mathcal{F} x + \mathcal{E}^\dagger \mathcal{B} v. \tag{2.14}$$

Matrix $\tilde{A}$ is taken as the first n rows of $-\mathcal{E}^\dagger \mathcal{F}$ and matrix $\tilde{B}$ is taken as the first n rows of $\mathcal{E}^\dagger \mathcal{B}$. Both matrices $\tilde{A}$ and $\tilde{B}$ are smooth because they are defined from smooth matrices. $\mathcal{E}^\dagger$ is smooth since $\mathcal{E}$ is smooth and has constant rank from the solvable assumption A1–A3. Readers interested in the technical details are referred to [27, 30, 34, 35, 43, 63].

If $\mathcal{G}$ is a maximal rank left annihilator of $\mathcal{E}$, then $\mathcal{G}(\mathcal{F} x - \mathcal{B} v) = 0$ describes the solution manifold. This description is used later and we will see that sometimes $\mathcal{G}$ need not be a smooth function of t.

2.3.2 Stabilized Least Squares Completion

While the LSC completion is the easiest to compute and analyze it has the disadvantage that the additional dynamics are almost never asymptotically stable. As a way to try and overcome this difficulty the SLSC was developed and analyzed extensively in [115].

For the stabilized least squares completion, the k derivatives in the derivative array equations come from differentiating the linear DAE (2.9) using differential polynomial $\frac{d}{dt} + \lambda$, resulting in

$$\mathcal{E}_\lambda = \begin{bmatrix} E & 0 & 0 \cdots \\ E' + F + \lambda E & E & 0 \ddots \\ E'' + 2F' + 2\lambda E' + 2\lambda F + \lambda^2 E & 2E' + F + 2\lambda E & E \ddots \\ \vdots & \vdots & \vdots \ddots \end{bmatrix},$$

$$\mathcal{B}_\lambda = \begin{bmatrix} B & 0 & 0 \cdots \\ B' + \lambda B & B & 0 \ddots \\ B'' + 2\lambda B' + \lambda^2 B & 2B' + 2\lambda B & B \ddots \\ \vdots & \vdots & \vdots \ddots \end{bmatrix},$$

$$\mathcal{F}_\lambda = \begin{bmatrix} F \\ F' + \lambda F \\ F'' + 2\lambda F' + \lambda^2 F \\ \vdots \end{bmatrix}, \quad w = \begin{bmatrix} x' \\ x'' \\ \vdots \\ x^{(k+1)} \end{bmatrix}, \quad v = \begin{bmatrix} u \\ u' \\ \vdots \\ u^{(k)} \end{bmatrix}.$$

$\frac{d}{dt} + \lambda$ represents stabilized differentiation and works to stabilize the additional dynamics of the completion. Stabilization parameter λ is discussed further in Sect. 2.3.4. The derivative array equations are now

$$\mathcal{E}_\lambda w + \mathcal{F}_\lambda x = \mathcal{B}_\lambda v \tag{2.15}$$

and (2.15) is then solved in the minimum norm least squares sense. Note that $\mathcal{E}_\lambda = P(\lambda)\mathcal{E}, \mathcal{B}_\lambda = P(\lambda)\mathcal{B}, \mathcal{F}_\lambda = P(\lambda)\mathcal{F}$, where $P(\lambda)$ is a block lower triangular matrix with identity on the principal diagonal. Thus, solving (2.15) in the least squares sense is the same as solving (2.12) in the weighted least squares sense. We drop the λ subscript if it is clear which completion is being discussed. Usually λ is taken as real and nonnegative.

2.3.3 Alternative Stabilized Completion

The third completion we shall discuss is the ASC. The previous two completions are in some sense the most natural given earlier work in the area and can often be used to construct useful observers. However, in working a number of problems we have found that the observers based on the ASC sometimes perform better even if they are more computationally intensive. There are two reasons for this. One is that for LTI DAEs the SLSC with a parameter λ produces Jordan blocks of the same size as the index whereas the ASC produces simple eigenvalues. This affects the additional dynamics and has the consequence that tracking error with SLSC sometimes looks like $p(t)e^{-\lambda t}$ with a polynomial $p(t)$ while the tracking error with ASC looks like $e^{-\lambda t}$. Thus while both observers may converge at the same rate mathematically, in practice one sometimes sees that ASC observers go more monotonically and are sooner within tolerance of zero while the SLSC observers will sometimes move larger before tending to zero. The second reason concerns the LTV case. By the way it is constructed the additional dynamics of the ASC have the property that the error in the equations defining the solution manifold when using the additional dynamics of the completion goes to zero at the specified rate. Given information on the conditioning of these equations then gives us needed information on the stability behavior of the additional dynamics. On the other hand with the SLSC on a LTV system we lose that direct connection with the stabilization parameter and the asymptotic behavior of the additional dynamics. The theory requires λ be chosen large enough but it is sometimes hard to figure out exactly how large to choose λ. This is discussed in [115]. These advantages of the ASC are illustrated with several of our computational examples in Sects. 6 and 7.

The process for finding the ASC originated with a method from [87] that constructs a canonical form for a linear DAE. Articles [89, 90, 92] expand on [87], extracting an index 1 linear DAE from a higher index one without altering the system's solutions. This extraction is motivated by the availability of numerical methods for solving index 1 but not higher index linear DAEs. Reference [88] provides a comprehensive overview of the material in [87, 89, 90, 92].

This extraction technique [116] generates a system, the alternative least squares completion, for which x' is defined explicitly. Article [117] extends the linear time-invariant focus of [116] to linear time-varying systems and introduces stabilized differentiation into the process of finding the alternative least squares completion. The resulting alternative stabilized LSC is now referred to as the ASC [17–21]. Dissertation [115] includes additional details about the methods in [116, 117].

The first step for finding the ASC of an index k linear DAE is to construct (2.12), the derivative array equations using differential polynomial $\frac{d}{dt}$. From this point on, the derivations differ for the linear time-invariant and time-varying cases, so their developments are presented separately.

ASC: Linear Time-Invariant Case

For a linear time-invariant DAE, the coefficient matrices of (2.12) simplify to

$$
\mathcal{E} = \begin{bmatrix} E & 0 & 0 & \cdots \\ F & E & 0 & \ddots \\ 0 & F & E & \ddots \\ \vdots & \vdots & \vdots & \ddots \end{bmatrix}, \quad
\mathcal{F} = \begin{bmatrix} F \\ 0 \\ 0 \\ \vdots \end{bmatrix}, \quad
\mathcal{B} = \begin{bmatrix} B & 0 & 0 & \cdots \\ 0 & B & 0 & \ddots \\ 0 & 0 & B & \ddots \\ \vdots & \vdots & \vdots & \cdots \end{bmatrix}.
$$

Following the construction of the derivative array equations, the first k block rows of matrices $\mathcal{E}$, $\mathcal{F}$, and $f_v = \mathcal{B}v$ are denoted by $\tilde{\mathcal{E}}$, $\tilde{\mathcal{F}}$, and $\tilde{f}_v$, respectively.

The process continues with the calculation of three orthonormal bases to define matrices Z_2^T, T_2, and Z_1^T. Matrix Z_2^T is determined such that $Z_2^T \tilde{\mathcal{E}} = 0$. In block column notation $Z_2^T = \begin{bmatrix} Z_{2,0}^T \cdots Z_{2,k-1}^T \end{bmatrix}$, where each of the k blocks has n columns. Next matrix T_2 is determined such that $Z_2^T \tilde{\mathcal{F}} T_2 = 0$. Finally, matrix Z_1 is defined such that its columns form an orthonormal basis for $R(ET_2)$. By construction, matrix $\begin{bmatrix} Z_1^T E \\ Z_{2,0}^T F \end{bmatrix}$ is nonsingular.

Matrices Z_2^T and Z_1^T are used to build the $(k+1)n \times (k+1)n$ matrix

$$
\Gamma = \begin{bmatrix} Z_1^T & 0_{\ell \times kn} \\ Z_2^T & 0_{d \times n} \\ 0_{d \times n} & Z_2^T \\ \hline & Z_3^T \end{bmatrix},
$$

where d is the rank of Z_2^T, $\ell = n - d$ and Z_3^T is a $kn \times (k+1)n$ matrix selected so that Γ is nonsingular. Then

$$
\Gamma \mathcal{E} = \begin{bmatrix} Z_1^T E & 0_{\ell \times n} & \dots & 0_{\ell \times n} & 0_{\ell \times n} \\ 0_{d \times n} & 0_{d \times n} & \dots & 0_{d \times n} & 0_{d \times n} \\ Z_{2,0}^T F & 0_{d \times n} & \dots & 0_{d \times n} & 0_{d \times n} \\ * & * & \dots & * & * \end{bmatrix}, \quad \Gamma \mathcal{F} = \begin{bmatrix} Z_1^T F \\ Z_{2,0}^T F \\ 0_{d \times n} \\ * \end{bmatrix},
$$

$$
\Gamma f_v = \begin{bmatrix} Z_1^T f \\ Z_{2,0}^T f + Z_{2,1}^T f' + \dots + Z_{2,k-1}^T f^{(k-1)} \\ Z_{2,0}^T f' + \dots + Z_{2,k-2}^T f^{(k-1)} + Z_{2,k-1}^T f^{(k)} \\ * \end{bmatrix} = \begin{bmatrix} Z_1^T f \\ Z_2^T \tilde{f}_v \\ Z_2^T \tilde{f}_v' \\ * \end{bmatrix}.
$$

The augmented form of $\Gamma \mathcal{E} w = -\Gamma \mathcal{F} x + \Gamma f_v$ is

$$
[\Gamma \mathcal{E} \| - \Gamma \mathcal{F} | \Gamma f_v] = \begin{bmatrix} Z_1^T E & 0 & \cdots & 0 & -Z_1^T F & Z_1^T f \\ 0 & 0 & \cdots & 0 & -Z_{2,0}^T F & Z_2^T \tilde{f}_v \\ Z_{2,0}^T F & 0 & \cdots & 0 & 0 & Z_2^T \tilde{f}_v' \\ * & * & \cdots & * & * & * \end{bmatrix}. \tag{2.16}
$$

The third row in (2.16) equals the first derivative of the second row. If stabilized differentiation is applied to the second row of (2.16) instead, the augmented form becomes

$$
\begin{bmatrix} Z_1^T E & 0 & \cdots & 0 & -Z_1^T F & Z_1^T f \\ 0 & 0 & \cdots & 0 & -Z_{2,0}^T F & Z_2^T \tilde{f}_v \\ Z_{2,0}^T F & 0 & \cdots & 0 & -\lambda Z_{2,0}^T F & Z_2^T \tilde{f}_v' + \lambda Z_2^T \tilde{f}_v \\ * & * & \cdots & * & * & * \end{bmatrix}. \tag{2.17}
$$

The first and third rows of (2.17) can be considered together to form the system

$$
\begin{bmatrix} Z_1^T E \\ Z_{2,0}^T F \end{bmatrix} \tilde{x}' = -\begin{bmatrix} Z_1^T F \\ \lambda Z_{2,0}^T F \end{bmatrix} \tilde{x} + \begin{bmatrix} Z_1^T f \\ Z_2^T \tilde{f}_v' + \lambda Z_2^T \tilde{f}_v \end{bmatrix}, \tag{2.18}
$$

from which $\tilde{x}'$ can be solved for explicitly,

$$
\tilde{x}' = -\begin{bmatrix} Z_1^T E \\ Z_{2,0}^T F \end{bmatrix}^{-1} \begin{bmatrix} Z_1^T F \\ \lambda Z_{2,0}^T F \end{bmatrix} \tilde{x} + \begin{bmatrix} Z_1^T E \\ Z_{2,0}^T F \end{bmatrix}^{-1} \begin{bmatrix} Z_1^T f \\ Z_2^T \left(\tilde{f}_v \right)' + \lambda Z_2^T \tilde{f}_v \end{bmatrix}.
$$

$$
\tag{2.19}
$$

Equation (2.19) is the ASC with

$$\tilde{A} = - \begin{bmatrix} Z_1^T E \\ Z_{2,0}^T F \end{bmatrix}^{-1} \begin{bmatrix} Z_1^T F \\ \lambda Z_{2,0}^T F \end{bmatrix}$$

and $\tilde{B}$ equal to

$$\begin{bmatrix} Z_1^T E \\ Z_{2,0}^T F \end{bmatrix}^{-1} \begin{bmatrix} Z_1^T B & 0 & \cdots & 0 & 0 \\ \lambda Z_{2,0}^T B & \left(Z_{2,0}^T + \lambda Z_{2,1}^T\right) B & \cdots & \left(Z_{2,k-2}^T + \lambda Z_{2,k-1}^T\right) B & Z_{2,k-1}^T B \end{bmatrix}.$$

ASC: Linear Time-Varying Case

For the linear time-invariant case, the ASC was derived from a subsystem of the derivative array equations that did not include the kth derivative of the linear DAE. The information provided by an additional differentiation is superfluous in the linear time-invariant case but necessary in the linear time-varying case as we now show. Note that all matrices now depend on t.

Following the construction of the derivative array equations, a matrix $Z_2^T(t)$ is defined such that the columns of $Z_2(t)$ form an orthonormal basis for $N\left(\mathcal{E}^T(t)\right)$. Multiplying (2.12) on the left by $Z_2^T(t)$ and then moving all remaining terms to the right-hand side results in

$$0 = -Z_2^T(t)\mathcal{F}(t)x(t) + Z_2^T(t)\mathcal{B}(t)v(t). \tag{2.20}$$

Applying stabilized differentiation once to (2.20) produces

$$0 = \left(\frac{d}{dt} + \lambda\right)\left(-Z_2^T \mathcal{F} x + Z_2^T \mathcal{B} v\right)$$
$$= -Z_2'^T \mathcal{F} x - Z_2^T \mathcal{F}' x - Z_2^T \mathcal{F} x' - \lambda Z_2^T \mathcal{F} x + \left(Z_2^T \mathcal{B} v\right)' + \lambda Z_2^T \mathcal{B} v$$

from which

$$Z_2^T \mathcal{F} x' = -\left(\left(Z_2^T \mathcal{F}\right)' + \lambda Z_2^T \mathcal{F}\right) x + \left(\left(Z_2^T \mathcal{B} v\right)' + \lambda Z_2^T \mathcal{B} v\right). \tag{2.21}$$

Next matrices T_2 and $Z_{1,0}^T$ are determined so the columns of T_2 are an orthonormal basis for $N\left(Z_2^T \mathcal{F}\right)$ and the columns of $Z_{1,0}$ form an orthonormal basis for $R(ET_2)$. By construction, matrix $\begin{bmatrix} Z_{1,0}^T E \\ Z_2^T \mathcal{F} \end{bmatrix}$ is nonsingular. Therefore,

$$Z_{1,0}^T E x' = -Z_{1,0}^T F x + Z_{1,0}^T B u$$

considered together with (2.21) produces system

$$
\begin{bmatrix} Z_{1,0}^T E \\ Z_2^T \mathcal{F} \end{bmatrix} \tilde{x}' = - \begin{bmatrix} Z_{1,0}^T F \\ (Z_2^T \mathcal{F})' + \lambda Z_2^T \mathcal{F} \end{bmatrix} \tilde{x} + \begin{bmatrix} Z_{1,0}^T Bu \\ (Z_2^T Bv)' + \lambda Z_2^T Bv \end{bmatrix}
$$

whence $\tilde{x}'$ can be solved for explicitly. The ASC is

$$
\tilde{x}' = - \begin{bmatrix} Z_{1,0}^T E \\ Z_2^T \mathcal{F} \end{bmatrix}^{-1} \left(\begin{bmatrix} Z_{1,0}^T F \\ (Z_2^T \mathcal{F})' + \lambda Z_2^T \mathcal{F} \end{bmatrix} \tilde{x} - \begin{bmatrix} Z_{1,0}^T Bu \\ (Z_2^T Bv)' + \lambda Z_2^T Bv \end{bmatrix} \right).
$$

$$(2.22)$$

One could go farther to get $\tilde{x}' = \tilde{A}\tilde{x} + \tilde{B}v$, but isolating $\tilde{B}$ is complicated and requires additional differentiation.

The process of finding the ASC includes the computation of three orthonormal bases to define matrices Z_2^T, T_2, and $Z_{1,0}^T$ and the computation of the first derivative of Z_2^T. A method presented in [91, 122] finds a smooth decomposition of a linear time-varying matrix by performing an orthogonal transformation on a non-smooth decomposition. Article [91] also describes a process for determining the first derivatives of the smooth decomposition factors. One key to the smoothness of the computed completions is that Z_2^T and $Z_{1,0}^T$ can be multiplied on the left by an invertible matrix without changing the completion [88].

2.3.4 Completion Stability

The original motivation behind stabilizing the LSC and developing the ASC was to improve the stability of the additional dynamics so a completion's numerical solution would converge to the solution manifold if a numerical perturbation caused drift. The implementation in [117] of stabilized differentiation with differential polynomial $\frac{d}{dt} + \lambda$ during completion development was influenced by the work in [8]. Stabilization parameter λ is generally defined as a constant nonnegative scalar.

Linear Time-Invariant Case

The effects from selecting such a λ are known for linear time-invariant completions due to the relationship between the eigenvalues and the stability of a linear time-invariant ODE [5].

By design, the matrix pencil eigenvalues of a linear time-invariant DAE are also eigenvalues of matrix $\tilde{A}$. A completion's other eigenvalues come from its additional dynamics and are referred to as the additional dynamics eigenvalues.If the additional dynamics eigenvalues have a negative real part, the completion's numerical solution will converge to the solution manifold due to the asymptotically stable additional dynamics. For a constant scalar λ, the additional dynamics eigenvalues of the stabilized and the alternative least squares completion equal $-\lambda$. Therefore, the desired negative real part results when $\lambda > 0$ [115, 117]. However, the Jordan form of the two stabilized completions is usually different.

If $\lambda = 0$, the SLSC reduces to the LSC and its additional dynamics eigenvalues equal 0. The Jordan canonical form of $\tilde{A}$ from the LSC reveals the 0 eigenvalues are contained in a $k \times k$ Jordan block. For an index 1 linear time-invariant DAE, the additional dynamics of its LSC are stable. Thus, the analytical solution of the completion that starts near the solution manifold will stay near the solution manifold. For a higher index system, if a perturbation occurs, the resulting drift grows as linear combinations formed with terms t^j, $0 \leq j \leq k - 1$, and the additional dynamics affect the completion's numerical solution. Matrix $\tilde{A}$ from the alternative least squares completion (the ASC when $\lambda = 0$) has a different Jordan canonical form. Instead of a $k \times k$ Jordan block, there is a 1×1 Jordan block for each 0 eigenvalue. No matter the index, the additional dynamics are stable since the only polynomial terms that appear in the linear combinations are constants. When $\lambda > 0$, the linear combinations formed with terms $t^j e^{-\lambda t}$ for the SLSC and with terms $e^{-\lambda t}$ for the ASC are asymptotically stable [115–117].

Linear Time-Varying Case

A completion's additional dynamics are not as well-understood in the time-varying case and just having $Re(\lambda) > 0$ may not be enough to stabilize the additional dynamics. Stabilization parameter λ has to be taken sufficiently large and in some cases λ needs to be time-varying [115, 117]. Theory is discussed in [115] for linear time-varying stabilization parameters and remains to be formalized for stabilizing with a matrix instead of a scalar [18].

We note that when reading [20], the $\lambda \left(Z_2^T \mathcal{F} x - Z_2^T \mathcal{B} v \right)$ term in (7b) should be subtracted and not added since λ is defined as nonnegative in [20].

2.3.5 Nonlinear Systems

The focus of this survey is on linear DAEs and observers and includes a description of an observer construction approach based on completions and the derivative array. However, this approach has great potential for extension to the nonlinear case due to the availability of a substantial body of work on completions of nonlinear DAEs [30, 38]. This is commented on more fully in Sect. 8.

3 Prior Work on DAE and ODE Observers

The state of a physical system is not always available for monitoring or feedback. Measurements may be hindered by cost or sensor inaccessibility, for example. But in some applications, knowing the state is important. In [102], Luenberger proposed a method (2.6) for estimating the state so the estimate in lieu of a true value could be used in a feedback control law. Since then observers have been designed to estimate a system's unknown parameters [68] or inputs [60]; applied to fault estimation, detection, and isolation [84, 85, 98, 141, 147] and disturbance detection [99]; implemented for tracking [139]; and included in power systems [54] and circuit theory [104]. The work on determining if a fault has occurred and identifying the fault and its size is important for making sure a problem in a modern industrial system can be realized before damage or inefficiency hinders normal operations.

Initial observer design focused on physical systems described by linear time-invariant ODEs or difference equations. As research progressed, the systems became more general, incorporating time-varying coefficients and nonlinear structure. Other modifications allowed for systems with delays [74] and constraints on the control [57]. Research has also been extended to constructing observers for physical systems described by DAEs, the focus of this survey. In some studies the observers are ODEs and in some studies the observers are DAEs. Note that the term "DAE observer" or "ODE observer" refer to the observer equation. These terms are equivalent to "DAE based observer" and "ODE based observer." The system whose state is being estimated is given after the word "observer." Thus there are four possible combinations of DAE/ODE observer of a DAE/ODE. For example, ODE based observers of higher index DAEs are used for fault detection in [125] and fault identification in [124]. The fault detection and fault isolation problem for descriptor systems is also considered in [53].

There are numerous papers on observers for linear time-invariant DAEs. Article [107] designs full-order and reduced-order observers for systems in descriptor standard form, a form dependent on the matrix pencil. Another form in [140, 142] from which to construct a reduced-order observer is the generalized staircase form. The authors express concern, though, about the computational efficiency of the algorithms required to transform a linear time-invariant DAE into this special form. The observers in [77] are derived from Luenberger observer theory but the systems to be observed should be in normal form. Switched DAEs are considered in [136].

The construction of the full-order and reduced-order observers in [59] requires the computation of more matrices than the one gain matrix commonly found in the design of observers for linear time-invariant ODEs. The observers in [61, 64] incorporate singular value decompositions, while article [127] focuses on generalized inverses. Articles [56, 145] include a proportional-integral (PI) control in the observer for reducing the index of the DAE under consideration. A PI feedback is not only proportional to a quantity but also includes an integral based on that quantity.

Another approach [16] for constructing an observer for a linear time-invariant DAE results in another DAE and not an ODE. Its construction is based on eigenvalue placement. In this approach the first derivative of the observer's state vector has the same singular coefficient matrix as the system to be observed. Feedback is used to make the error equation an index 1 DAE. One of the requirements for this observer to asymptotically estimate the state is dependent on the regularity of a modified matrix pencil. The author of [130] also considers DAE observers for linear time-invariant DAEs. Dependences on eigenvalue placement, matrix pencils, and special forms make it difficult to generalize these observers for estimating linear time-varying and nonlinear DAEs.

An observer in [112] applies a reduction algorithm to a linear time-invariant DAE in order to reveal a system that can be observed by an ODE. If a similar procedure were implemented for a linear time-varying DAE, an appropriate linear time-varying ODE observer would need to be selected. Even though the eigenvalues of a linear time-varying ODE do not provide any indication of the system's stability [146], eigenvalue placement techniques are used to design the observers in [50, 95, 100]. The author of [95] writes to improve rather than promote this construction technique since eigenvalue placement is implemented for linear time-varying engineering applications. The observers for linear time-varying ODEs in [111] are also designed with an eigenvalue placement technique. This article's approach assumes the system to be observed is lexicographic, which means the system is observable and has an observability matrix with the same rows linearly independent for all time. Article [110] extends the research in [111] to reduced-order observers, and a lexicographic assumption can also be found in [129]. In response to this restriction, the authors of [51] propose forming an augmented system that is lexicographic when the original one is not. Section 4.1 describes why computing, and thereby constructing an observer dependent on, the observability matrix is avoided for our time-varying observers. An observable assumption is also limiting, commented on by the authors of [76]. Other attempts at generalizing linear time-invariant observer designs for linear time-varying ODEs incorporate canonical forms [126, 128].

Our observer construction approach for estimating the state of a linear DAE constructs an observer for an ODE utilizing a Luenberger observer mentioned in Sect. 2.1 and further described in Sect. 4. Previous work in [12, 13, 43] constructs a similar observer for linear time-varying DAEs, but our design takes advantage of recent developments in stabilized completions. The procedure from [12, 13, 43] generates the projection onto the solution manifold as the solution of an ODE and then generates estimates on the solution manifold.

Observers for nonlinear DAEs are considered in [7, 22, 58, 148, 150]. Frequently the DAEs are index 1, while in [7] both the DAE and its observer are index 1. Another nonlinear study in [82] considers the input/output (I/O) of a nonlinear higher index DAE. The authors transform the DAE to an ODE on a manifold and then linearize the I/O behavior using feedback. The observer they propose can be used to globally stabilize the system. A nonlinear system is also considered in [143]. There it is shown that by considering a DAE observer of higher index, it is possible to get an observer for which the error dynamics are linear and, hence, the convergence rate is controlled more easily. The results in [143] are one illustration of the advantages that sometimes exist for DAE observers.

Another paper showing an advantage of DAE observers is [71]. The authors consider systems with disturbances such as

$$x' = Ax + Bu + Md_1 \tag{3.1a}$$

$$y = Cx + Du + Nd_2 \tag{3.1b}$$

where d_1 and d_2 are input and output disturbances, respectively. Article [71] shows how to use a DAE observer combined with some proportional and integral terms to not only estimate the state but to also estimate the disturbances. The "integral terms" mean the equations defining the observer and its output have extra integrators. The authors use the proportional part of the gain to get stability of the error equation and the integral terms to reduce the effect of the noise.

Fundamental to working with observers are control-theoretic properties like stability, observability, and controllability. The book [55] includes one of the first fairly complete basic discussions on these issues for linear time-invariant DAEs. Examinations of these control properties for DAEs in [42, 47, 137, 138] work from the derivative array. An alternative approach based on a linear time-varying version of the Kronecker form [70] is in [144]. Stabilization of index 1 linear time-varying DAEs using a Lyapunov approach is found in [1]. A survey on controllability of linear DAEs is given in [11]. In a behavioral point of view one looks at the set of signals that satisfies the equations and does not necessarily distinguish the states from the controls at the start. The authors of [78] discuss controllability and observability of DAEs from a behavioral point of view for systems with real analytic coefficients.

Common structural assumptions on a system's index [29] or on the system being in Hessenberg form [31, 32, 52] may simplify the processes for solving linear DAEs. For example, observers designed for mechanical systems often exploit a Hessenberg-like structure. Another advantage is that there are positive results for linearized approximations of nonlinear DAEs with Hessenberg structure. These linearization results can be important in designing local observers for nonlinear DAEs. Our observer construction approach given in the next section constructs observers for linear DAEs using completions, which requires no assumptions on a system's structure and addresses the constraints concern expressed for DAEs by using a stabilization.

4   Completion-Based Observers for DAEs

Completions allow for the construction of observers for DAEs using design techniques associated with ODEs. We shall also see a modified ODE observer that takes advantage of a DAE's solution manifold. The development of the observers being considered here assumes a completion of a DAE with output equation has already been found:

$$\tilde{x}'(t) = \tilde{A}(t)\tilde{x}(t) + \tilde{B}(t)v(t) \tag{4.1a}$$

$$\tilde{y}(t) = C(t)\tilde{x}(t). \tag{4.1b}$$

Additionally, $\tilde{A}$, $\tilde{B}$, and C should be smooth matrix functions and are assumed bounded. Unless otherwise noted, C has full row rank m.

Section 4.1 makes some preliminary comments on observability and detectability. In designing a completion-based observer, there are two separate design considerations. One is what type of completion to use. The second is what type of observer we are going to construct from the type of completion. We shall present three types of observers. The full-order observer in Sect. 4.2 and the reduced-order observer in Sect. 4.3 are classical observers. The maximally reduced observer in Sect. 4.4 is a new observer that is unique to DAEs.

Having presented the three types of observers to be constructed, Sect. 4.5 then begins the discussion of how the completion process interacts with the observer construction and why we need to consider different types of completions. This discussion is continued in Sect. 5

4.1 Observability and Detectability

Not every system of ODEs can be observed. In order for the system to be observable, the information provided through the output needs to make the state distinguishable from the zero solution. For linear time-invariant ODEs, observability is just that $N(\mathcal{O}) = \{0\}$ where $\mathcal{O}$ is defined in (2.2).

With linear time-varying ODEs the situation becomes more complicated. As with the linear time-invariant case we define what is unobservable since that forms a subspace. The state can be observable at one time and not at another. Let $\Phi(t, t_0)$ be the fundamental solution matrix of $x' = Ax$, that is $\Phi' = A\Phi$, $\Phi(t_0, t_0) = I$. Then a state x_0 is unobservable at time t_0 if $C(t)\Phi(t, t_0)x_0 = 0$ for all $t \geq t_0$. The unobservable at t_0 vectors form a subspace. If that subspace is only the zero vector we say the system is observable at time t_0. The system is observable if it is observable for all t_0 from some time on. Extensive discussions can be found in linear systems texts such as [5]. Here we will just discuss some of the ideas we use in working our examples.

It is possible to check the observability of a linear time-varying ODE using the observability matrix [131]. However, for (4.1) the $n - 1$ derivatives of $\tilde{y}$ needed to construct $\mathcal{O}$ require knowing the first $n - 2$ derivatives of $\tilde{A}$ and the first $n - 1$ derivatives of C. Unless matrix $\tilde{A}$ can be defined symbolically in terms of quantities with known derivatives, differentiating a computed matrix may introduce numerical error into the observability matrix and possibly produce an incorrect rank. This is especially true if higher derivatives are needed. Derivatives of $\tilde{A}$ can be computed using larger derivative arrays, but that approach is not presently considered.

One alternative method for checking the observability of a linear time-varying ODE solves a Lyapunov differential equation

$$\frac{\partial W_o}{\partial t} = -\tilde{A}^T W_o - W_o \tilde{A} - C^T C \tag{4.2}$$

by integrating backwards in time with $W_o(t_f, t_f) = 0$ [49]. The observability Gramian $W_o(t_0, t_f)$ is defined

$$W_o(t_0, t_f) = \int_{t_0}^{t_f} \Phi(\tau, t_0)^T C(\tau)^T C(\tau) \Phi(\tau, t_0) \, d\tau, \tag{4.3}$$

where $\Phi(t, t_0)$ is the state transition matrix ($\Phi'(t, t_0) = \tilde{A}\Phi(t, t_0)$ and $\Phi(t_0, t_0) = \Phi(t_0) = I$). The method implemented for the linear time-varying example in Sect. 7 integrates forward in time with $W_o(t_0, t_0)$ and solves the system

$$\frac{d}{dt} W_o = \Phi^T C^T C \Phi \tag{4.4a}$$

$$\frac{d}{dt} \Phi = \tilde{A}\Phi. \tag{4.4b}$$

If there exists a time t_f such that rank $(W_o(t_0, t_f)) = n$ for $t_0 < t_f < \infty$, then system (4.1) is observable and $(\tilde{A}, C)$ is its observable pair [5]. If system (4.1) is not observable, a similarity transformation can be applied to identify the standard form for the unobservable system. The transformation matrix should be smooth and bounded and have a bounded inverse [135]. From [134, 135], if the unobservable system in (2.3) is uniformly asymptotically stable, then (4.1) is detectable.

4.2 The Full-Order Observer (FO)

The full-order observer

$$\hat{x}' = \tilde{A}\hat{x} + \tilde{B}v + L(\tilde{y} - \hat{y}) \tag{4.5a}$$

$$\hat{y} = C\hat{x} \tag{4.5b}$$

is an example of a Luenberger observer because of its use of known inputs and outputs from the physical system to be observed and because of its assignment of observable eigenvalues to the left half-plane in the linear time-invariant case [101, 102]. Observer (4.5) uses a correction term to compare the plant output $\tilde{y}$ with the observer output $\hat{y}$ (the output resulting from the observer calculated state). If data exists for the plant output, then these results from the physical system can be substituted into (4.5a) for $\tilde{y}$.

The gain matrix L in (4.5) needs to be chosen appropriately. A measure of the observer's state estimate is the estimation error $e = \tilde{x} - \hat{x}$, which satisfies

$$e'(t) = (\tilde{A}(t) - L(t)C(t)) e(t). \tag{4.6}$$

The methods chosen for constructing the gain matrix L differ for the linear time-invariant and linear time-varying cases.

4.2.1   FO Gain: Linear Time-Invariant Case

In the linear time-invariant case, gain matrix L is designed to assign the observable eigenvalues of $\tilde{A}$ to the left half-plane. The algorithm from [5] used in our examples is the classical construction, which utilizes the controller form of the linear time-invariant system's observable pair.

4.2.2   FO Gain: Linear Time-Varying Case

The method chosen for constructing the gain matrix in the linear time-varying case comes from the development presented in [80, 113, 114]. From the theory of Lyapunov, if $V_L = e^T P_L e$, if $-V'_L = e^T Q_L e$, for real symmetric, positive definite P_L, Q_L, and if $(\tilde{A}, C)$ is uniformly completely observable, then the observer (4.5) will asymptotically estimate the state.

The problem is how do we find L, P_L, Q_L which satisfy these conditions? We shall now show one way to approach this. Pick a Q_L. Let M be a real symmetric positive definite matrix which will be chosen later. Let P_L be defined by

$$P'_L = -P_L \tilde{A} - \tilde{A}^T P_L - Q_L + C^T M C. \tag{4.7}$$

Let $L = \frac{1}{2} P_L^{-1} C^T M$. Then if we let $V_L = e^T P_L e$, we see after a simple calculation that $-V'_L = e^T Q_L e$ as desired.

This requires that P_L is invertible. However, not every choice of $P_L(t_0)$ and M will produce a positive definite P_L for the chosen Q_L. Articles [80, 113, 114] do not include a description of how $P_L(t_0)$, M, and Q_L should be defined. To examine this issue, we proceed as follows. A manipulation of (4.7) using the state transition matrix $\Phi(t, t_0)$ produces

$$\left(\Phi(t, t_0)^T P_L \Phi(t, t_0) \right)' = -\Phi(t, t_0)^T Q_L \Phi(t, t_0) + \Phi(t, t_0)^T C^T M C \Phi(t, t_0)$$

so that

$$\Phi(t, t_0)^T P_L \Phi(t, t_0) = P_L(t_0) - \int_{t_0}^{t} \Phi(s, t_0)^T Q_L(s) \Phi(s, t_0) \, ds$$

$$+ \int_{t_0}^{t} \Phi(s, t_0)^T C(s)^T M C(s) \Phi(s, t_0) \, ds.$$

The rightmost integral is related to observability Gramian (4.3), which is symmetric and positive definite. Thus, for some $P_L(t_0)$ and Q_L, P_L will remain positive definite if M is chosen large enough. We take M constant in this paper but it can be taken time-varying if that is desirable.

The positive definiteness of P_L may numerically come into question when one of the eigenvalues (or singular values) of P_L becomes very large. Even though the smallest eigenvalue of P_L may not be close to zero, it may appear to be, causing an inverse computation of P_L for L to be ill-conditioned. The solution is to use P_L^{-1} rather than P_L to construct a Riccati equation. If $S_L = P_L^{-1}$, then

$$S_L' = -P_L^{-1} P_L' P_L^{-1}$$
$$= \tilde{A} S_L + S_L \tilde{A}^T + S_L \left(Q_L - C^T M C \right) S_L.$$

If $Q_L = \overline{Q} + C^T M C$ for some real symmetric, positive definite matrix $\overline{Q}$, then S_L will remain real symmetric, positive definite if $S_L (t_0)$ is real symmetric, positive definite. The gain matrix becomes $L = \frac{1}{2} S_L C^T M$ for the Riccati equation with S_L.

The only requirements on $\tilde{A}$, $\tilde{B}$, and C mentioned in [80, 113, 114] were that they be continuous and bounded and that $\left(\tilde{A}, C \right)$ be uniformly completely observable. However, articles [3] (Theorem 5) and [4] also require that $\tilde{A}$ be exponentially asymptotically stable if a matrix P similar to P_L in (4.7) is to exist. This stability requirement is revisited in Sect. 7, including a look at its effect on the positive definiteness of S_L. Furthermore, the results in Sect. 7 reveal the full-order observer also asymptotically estimates the state if the condition on $\left(\tilde{A}, C \right)$ is relaxed to detectable. In addition to the detectable definition based on the standard form for the unobservable system presented in Sect. 4.1, a linear time-varying ODE is detectable if there exists a bounded gain matrix L such that (4.6) is uniformly asymptotically stable [134, 135] or exponentially stable [98]. Thus, if $\left(\tilde{A}, C \right)$ is not observable but an observer is still able to estimate the state, then the pair is detectable.

4.3   The Reduced-Order Observer (RO)

The construction of the reduced-order observer, proposed in [101, 102] and designed in [73], takes into consideration information about the state vector provided through the output equation. If a linear ODE with output equation, system (4.1) for example, has an $n \times 1$ state vector $\tilde{x}$, $m \times 1$ output vector $\tilde{y}$, and a full row rank output matrix C, then there is a coordinate system in which m state variables are known through the output equation. These m components of the state vector are described as measurable while the remaining $n - m$ components are referred to as unmeasurable [83, 101, 132]. The description reduced-order comes from constructing an observer to estimate only the unmeasurable state variables rather than the full state vector. The linear time-varying case presents computational challenges not present in the linear time-invariant case so we discuss the two cases separately.

4.3.1 RO: Linear Time-Invariant Case

The development of the reduced-order observer in this section is based on derivations described in [9, 83, 132]. The output should clearly identify m measurable
components. The desired structure for $\tilde{y}$ can be obtained by applying an appropriate
transformation to the system to be observed. If R is a nonsingular transformation
matrix such that $CR^{-1} = \begin{bmatrix} I & 0 \end{bmatrix}$, where I is an $m \times m$ identity matrix, then for
$\tilde{x} = R^{-1}q$,

$$
\tilde{y} = C\tilde{x} = CR^{-1}q = \begin{bmatrix} I & 0 \end{bmatrix}
\begin{bmatrix} q_1 \\ \vdots \\ q_m \\ q_{m+1} \\ \vdots \\ q_n \end{bmatrix}
= \begin{bmatrix} q_1 \\ \vdots \\ q_m \end{bmatrix} = q_{\mathrm{m}}.
$$

R can be computed from an SVD of C. After substituting into (4.1a) for $\tilde{x}$ and
$\tilde{x}' = R^{-1}q'$, we get

$$
q' = A_R q + B_R v, \tag{4.8}
$$

where $A_R = R\tilde{A}R^{-1}$ and $B_R = R\tilde{B}$. Output equation $\tilde{y} = q_{\mathrm{m}}$ equals the
measurable components of transformed system (4.8). The method chosen for
constructing $R = \begin{bmatrix} C \\ C_R \end{bmatrix}$ comes from [9] and defines C_R as the transpose of the
basis vectors for $N(C)$ found from calculating the singular value decomposition
of C.

Transformed system (4.8) can be rewritten so it appears in terms of its measurable
and unmeasurable components:

$$
q'_{\mathrm{m}} = A_{R_{11}} q_{\mathrm{m}} + A_{R_{12}} q_{\mathrm{u}} + B_{R_1} v \tag{4.9a}
$$

$$
q'_{\mathrm{u}} = A_{R_{21}} q_{\mathrm{m}} + A_{R_{22}} q_{\mathrm{u}} + B_{R_2} v. \tag{4.9b}
$$

In (4.9b) q_{m} and v are known. In (4.9a) everything is known except for $A_{R_{12}} q_{\mathrm{u}}$.
Thus (4.9a) can be viewed as the output of an ODE given by (4.9b). Equation (4.9b)
is then used with a correction term to construct a reduced-order observer. Unlike
full-order observer (4.5), the reduced-order observer is unable to incorporate
$L(\tilde{y} - \hat{y})$ as its correction term since $\tilde{y} - \hat{y} = \tilde{y} - \begin{bmatrix} I & 0 \end{bmatrix} \begin{bmatrix} q_{\mathrm{m}} \\ \hat{q}_{\mathrm{u}} \end{bmatrix} = \tilde{y} - q_{\mathrm{m}} = 0$.
Instead, the correction term comes from

$$
\tilde{y}' = q'_{\mathrm{m}} - A_{R_{11}} q_{\mathrm{m}} + A_{R_{12}} q_{\mathrm{u}} + B_{R_1} v.
$$

This equation for the first derivative of the output can be rearranged by grouping measurable and known terms on one side and unmeasurable terms on the other:

$$\tilde{y}' - A_{R_{11}} q_m - B_{R_1} v = A_{R_{12}} q_u.$$

Then for the correction term, $\tilde{y}' - A_{R_{11}} q_m - B_{R_1} v$ represents the plant output and $A_{R_{12}} \hat{q}_u$ represents the observer output, resulting in

$$\hat{q}'_u = A_{R_{22}} \hat{q}_u + (A_{R_{21}} q_m + B_{R_2} v) + L_R \left(\tilde{y}' - A_{R_{11}} q_m - B_{R_1} v - A_{R_{12}} \hat{q}_u \right). \tag{4.10}$$

A change of variable is required to keep $\tilde{y}'$ from appearing explicitly in (4.10):

$$\begin{aligned}
\hat{q}'_u - L_R \tilde{y}' &= A_{R_{22}} \hat{q}_u + (A_{R_{21}} q_m + B_{R_2} v) - L_R (A_{R_{11}} q_m + B_{R_1} v + A_{R_{12}} \hat{q}_u) \\
&= (A_{R_{22}} - L_R A_{R_{12}}) (\hat{q}_u - L_R \tilde{y}) \\
&\quad + ((A_{R_{22}} - L_R A_{R_{12}}) L_R + (A_{R_{21}} - L_R A_{R_{11}})) \tilde{y} \\
&\quad + (B_{R_2} - L_R B_{R_1}) v. \tag{4.11}
\end{aligned}$$

Letting $z = \hat{q}_u - L_R \tilde{y}$ and substituting z and $z' = \hat{q}'_u - L_R \tilde{y}'$ into (4.11) gives

$$\begin{aligned}
z' &= (A_{R_{22}} - L_R A_{R_{12}}) z \\
&\quad + ((A_{R_{22}} - L_R A_{R_{12}}) L_R + (A_{R_{21}} - L_R A_{R_{11}})) \tilde{y} + (B_{R_2} - L_R B_{R_1}) v.
\end{aligned}$$

After defining $D_R = A_{R_{22}} - L_R A_{R_{12}}$, $F_R = D_R L_R + (A_{R_{21}} - L_R A_{R_{11}})$, and $G_R = B_{R_2} - L_R B_{R_1}$, the reduced-order observer for the transformed system is

$$z' = D_R z + F_R \tilde{y} + G_R v \tag{4.12a}$$

$$\hat{q}_u = z + L_R \tilde{y}. \tag{4.12b}$$

The transformation $\hat{x} = R^{-1} \hat{q}$ is required to study the reduced-order observer results in terms of the original state vector.

From the error equation $e_R = q_u - \hat{q}_u$,

$$e'_R = q'_u - \hat{q}'_u = (A_{R_{22}} - L_R A_{R_{12}}) e_R$$

reveals the gain matrix L_R should be constructed so the difference between the unmeasurable components of the transformed system and the state variables being estimated by the reduced-order observer goes to zero as $t \longrightarrow \infty$. The eigenvalues of $A_{R_{22}} - L_R A_{R_{12}}$ can be placed so that $\lim_{t \to \infty} e_R = 0$ if $(A_{R_{22}}, A_{R_{12}})$ is observable or detectable.

4.3.2   RO: Linear Time-Varying Case

The following process for constructing a reduced-order observer when the linear ODE is time-varying comes from [113,114,149]. Similar to the linear time-invariant case, the system to be observed is transformed when necessary to affect the structure of the output equation. However, now the change of variables given by R may be time dependent. Also we now need to deal with the stability of time-varying subsystems. Again, letting transformation matrix R equal $\begin{bmatrix} C \\ C_R \end{bmatrix}$, where C_R is defined so R is nonsingular, results in $CR^{-1} = \begin{bmatrix} I & 0 \end{bmatrix}$. The derivative of $\tilde{x} = R^{-1}q$ is

$$\tilde{x}' = \left(R^{-1}\right)' q + R^{-1}q' = \left(-R^{-1}R'R^{-1}\right)q + R^{-1}q'.$$

Substituting into (4.1a) for $\tilde{x}$ and $\tilde{x}'$ produces

$$q' = \left(R\tilde{A}R^{-1} + R'R^{-1}\right)q + R\tilde{B}v, \tag{4.13a}$$

and with $\tilde{y} = q_{\mathrm{m}}$ from

$$\tilde{y} = C\tilde{x} = CR^{-1}q = \begin{bmatrix} I & 0 \end{bmatrix}\begin{bmatrix} q_{\mathrm{m}} \\ q_{\mathrm{u}} \end{bmatrix},$$

the transformed system is

$$q' = A_R q + B_R v \tag{4.14a}$$

$$\tilde{y} = q_{\mathrm{m}} \tag{4.14b}$$

for $A_R = R\tilde{A}R^{-1} + R'R^{-1}$ and $B_R = R\tilde{B}$.

If the output matrix R is time-invariant, the transformation matrix is also constant. However, for a time-varying output matrix C, the first derivative of transformation matrix R is needed to calculate the coefficient matrix of state vector q in (4.14a). Output matrix C is assumed smooth, but C_R needs to be smooth as well. In the time-invariant case, we chose to define C_R as the transpose of the basis vectors for $N(C)$ from a singular value decomposition of output matrix C. The smooth decomposition described in [91, 122] can be implemented in the time-varying case to find smooth basis vectors for $N(C)$. The process outlined in [91] also calculates the first derivatives of the basis vectors found during the smooth decomposition. An alternative method for determining a smooth C_R and its first derivative is to utilize either MATLAB's or Maple's commands for finding a nullspace and taking a derivative with t defined symbolically.

Unlike the linear time-invariant reduced-order observer in Sect. 4.3.1, the derivation of this section's reduced-order observer for transformed system (4.14) begins by defining

$$z = T\hat{q} = \begin{bmatrix} T_1 & T_2 \end{bmatrix} \begin{bmatrix} q_{\mathrm{m}} \\ \hat{q}_{\mathrm{u}} \end{bmatrix}. \tag{4.15}$$

Equation (4.15) represents an output for the observer-estimated unmeasurable components. When (4.15) is considered together with output Eq. (4.14b), we have

$$\begin{bmatrix} \tilde{y} \\ z \end{bmatrix} = \begin{bmatrix} q_{\mathrm{m}} \\ T\hat{q} \end{bmatrix} = \begin{bmatrix} I & 0 \\ T_1 & T_2 \end{bmatrix} \begin{bmatrix} q_{\mathrm{m}} \\ \hat{q}_{\mathrm{u}} \end{bmatrix}. \tag{4.16}$$

System (4.16) reveals T should be chosen so $\hat{q}$ can be solved for explicitly. If

$$\begin{bmatrix} I & 0 \\ T_1 & T_2 \end{bmatrix}^{-1} = \begin{bmatrix} V_1 & P_1 \\ V_2 & P_2 \end{bmatrix},$$

where V_1 is $m \times m$ and P_2 is $(n-m) \times (n-m)$, then

$$\begin{bmatrix} I & 0 \\ T_1 & T_2 \end{bmatrix} \begin{bmatrix} V_1 & P_1 \\ V_2 & P_2 \end{bmatrix} = \begin{bmatrix} I & 0 \\ 0 & I \end{bmatrix}$$

implies V_1 is an identity matrix, P_1 is a zero matrix, $T_1 + T_2 V_2 = 0$, and P_2 needs to be nonsingular since $T_2 P_2 = I$. Choosing $T = \begin{bmatrix} -P_2^{-1} V_2 & P_2^{-1} \end{bmatrix}$ appropriately is now dependent on the choices of matrices V_2 and P_2.

After substituting into (4.15) for T,

$$z = T\hat{q} = -P_2^{-1} V_2 q_{\mathrm{m}} + P_2^{-1} \hat{q}_{\mathrm{u}}.$$

The derivative of z is

$$\begin{aligned}
z' &= -P_2'^{-1} V_2 q_{\mathrm{m}} - P_2^{-1} V_2' q_{\mathrm{m}} - P_2^{-1} V_2 q_{\mathrm{m}}' + P_2'^{-1} \hat{q}_{\mathrm{u}} + P_2^{-1} \hat{q}_{\mathrm{u}}' \\
&= P_2^{-1} \left(A_{R_{22}} - V_2 A_{R_{12}} - P_2' P_2^{-1} \right) \hat{q}_{\mathrm{u}} \\
&\quad + P_2^{-1} \left(A_{R_{21}} - V_2 A_{R_{11}} + P_2' P_2^{-1} V_2 - V_2' \right) q_{\mathrm{m}} \\
&\quad + P_2^{-1} \left(-V_2 B_{R_1} + B_{R_2} \right) v,
\end{aligned}$$

where q_{m}' and $\hat{q}_{\mathrm{u}}'$ come from (4.14a) written in terms of its measurable and unmeasurable state variables (the linear time-varying version of (4.9)). The explicit solution of $\hat{q}$ provides $q_{\mathrm{m}} = \tilde{y}$ and $\hat{q}_{\mathrm{u}} = V_2 \tilde{y} + P_2 z$ so z' becomes

$$\begin{aligned}
z' &= \left(P_2^{-1} \left(A_{R_{22}} - V_2 A_{R_{12}} \right) P_2 - P_2^{-1} P_2' \right) z \\
&\quad + P_2^{-1} \left(A_{R_{21}} - V_2 A_{R_{11}} + A_{R_{22}} V_2 - V_2 A_{R_{12}} V_2 - V_2' \right) \tilde{y} \\
&\quad + P_2^{-1} \left(-V_2 B_{R_1} + B_{R_2} \right) v.
\end{aligned}$$

After defining

$$D_R = P_2^{-1} \left(A_{R_{22}} - V_2 A_{R_{12}} \right) P_2 - P_2^{-1} P_2',$$
$$F_R = P_2^{-1} \left(A_{R_{21}} - V_2 A_{R_{11}} + A_{R_{22}} V_2 - V_2 A_{R_{12}} V_2 - V_2' \right),$$
$$G_R = P_2^{-1} \left(-V_2 B_{R_1} + B_{R_2} \right),$$

the reduced-order observer for the transformed system is

$$z'(t) = D_R(t)z(t) + F_R(t)\tilde{y}(t) + G_R(t)v(t) \qquad (4.17a)$$

$$\hat{q}_u(t) = V_2(t)\tilde{y}(t) + P_2(t)z(t). \qquad (4.17b)$$

The transformation $\hat{x}_R = R^{-1}\hat{q}$ returns the reduced-order observer results in terms of the original state vector.

For reduced-order observer (4.17), the convergence of $e = q - \hat{q}$ to zero as time goes to infinity is dependent on the choice of V_2. The second error equation $\epsilon = Tq - z$, a measure of how well the reduced-order observer is estimating the unmeasurable components, is required to show this dependence. The first derivative of ϵ is

$$\epsilon' = T'q + Tq' - z'$$
$$= D_R\epsilon - \left(D_R T - TA_R - T' + F_R CR^{-1} \right) q + (TB_R - G_R) v. \qquad (4.18)$$

The coefficient of q equals zero,

$$D_R T - TA_R$$

$$= D_R P_2^{-1} \left[-V_2 \; I \right] + P_2^{-1} \left[V_2 A_{R_{11}} - A_{R_{21}} \; V_2 A_{R_{12}} - A_{R_{22}} \right]$$
$$= \left[- \left(P_2^{-1} V_2 \right)' \; \left(P_2^{-1} \right)' \right] - F_R \left[I \; 0 \right]$$
$$= T' - F_R CR^{-1},$$

as does the coefficient of v, since

$$G_R = \left[-P_2^{-1} V_2 \; P_2^{-1} \right] \begin{bmatrix} B_{R_1} \\ B_{R_2} \end{bmatrix} = TB_R.$$

Thus, (4.18) simplifies to $\epsilon' = D_R\epsilon$. Matrix P_2 can be taken constant as long as it is nonsingular, resulting in $\epsilon' = P_2^{-1} \left(A_{R_{22}} - V_2 A_{R_{12}} \right) P_2\epsilon$. This equation shows the choice of matrix V_2 affects whether or not $\lim_{t \to \infty} \epsilon = 0$.

If e is expressed as $\begin{bmatrix} e_1 \\ e_2 \end{bmatrix}$, where the row dimensions of e_1 and e_2 equal the number of measurable and unmeasurable components, respectively, then

$$e = q - \hat{q} = \begin{bmatrix} 0 \\ -V_2 q_{\mathrm{m}} + q_{\mathrm{u}} - P_2 z \end{bmatrix}$$

$$= \begin{bmatrix} 0 \\ P_2 \end{bmatrix} \left(\begin{bmatrix} -P_2^{-1} V_2 & P_2^{-1} \end{bmatrix} \begin{bmatrix} q_{\mathrm{m}} \\ q_{\mathrm{u}} \end{bmatrix} - z \right)$$

$$= \begin{bmatrix} 0 \\ P_2 \end{bmatrix} \epsilon.$$

Thus, $e_1 = 0$ and $e_2 = P_2 \epsilon$, and the derivative

$$e' = \begin{bmatrix} 0 \\ P_2 \epsilon' \end{bmatrix} = \begin{bmatrix} 0 & 0 \\ 0 & A_{R_{22}} - V_2 A_{R_{12}} \end{bmatrix} \begin{bmatrix} e_1 \\ e_2 \end{bmatrix}$$

demonstrates that for an appropriate V_2 (and a constant, nonsingular P_2), $\lim_{t \to \infty} e = 0$ and reduced-order observer (4.17) converges to the unmeasurable state variables.

The construction of gain matrix V_2 follows the process outlined for the construction of gain matrix L in Sect. 4.2 with $A_{R_{22}}$ and $A_{R_{12}}$ inserted in for $\tilde{A}$ and C, respectively. The resulting Riccati equation is

$$S'_L = A_{R_{22}} S_L + S_L A^T_{R_{22}} + S_L \left(Q_L - A^T_{R_{12}} M A_{R_{12}} \right) S_L$$

and is used to compute $V_2 = \frac{1}{2} S_L A^T_{R_{12}} M$. Additionally, the first derivative of V_2,

$$V'_2 = \frac{1}{2} \left(S'_L A^T_{R_{12}} M + S_L \left(A^T_{R_{12}} \right)' M + S_L A^T_{R_{12}} M' \right),$$

is required to determine F_R, the coefficient matrix of $\tilde{y}$ in (4.17a). The calculations of S'_L and M' are straightforward. The Riccati equation provides the first derivative of S_L. The user determines M and therefore has control over constructing a matrix for which M' exists and is known without differentiating any numerically computed quantities.

Recall, matrix $A_{R_{12}}$ is a submatrix of $A_R = R \tilde{A} R^{-1} + R' R^{-1}$, so knowing the first derivative of $\tilde{A}$ and the second derivative of R when the output matrix is time-varying is necessary for calculating $\left(A^T_{R_{12}} \right)'$. The process from [91] mentioned earlier in Sect. 4.3.2 for finding C_R and its first derivative has yet to be extended to finding higher order derivatives of matrices computed using the smooth decomposition method. Instead, the symbolic results from finding C_R and both its first and second derivatives using either MATLAB or Maple are implemented in our numerical algorithms when defining R and its derivatives.

To avoid introducing numerical error by differentiating the computed value of $\tilde{A}$ at each time t, $\tilde{A}'$ can be defined in terms of matrices with known derivatives for all three completions. For both the LSC and the SLSC, $\tilde{A}$ is taken as the first n rows of $-\mathcal{E}^\dagger \mathcal{F}$. Thus, $\tilde{A}'$ is taken as the first n rows of $-\left(\mathcal{E}^\dagger\right)' \mathcal{F} - \mathcal{E}^\dagger \mathcal{F}'$, where [41]

$$\left(\mathcal{E}^\dagger\right)' = -\mathcal{E}^\dagger \mathcal{E}' \mathcal{E}^\dagger + \left(I - \mathcal{E}^\dagger \mathcal{E}\right) \left(\mathcal{E}^T\right)' \left(\mathcal{E}^\dagger\right)^T \mathcal{E}^\dagger + \mathcal{E}^\dagger \left(\mathcal{E}^\dagger\right)^T \left(\mathcal{E}^T\right)' \left(I - \mathcal{E}\mathcal{E}^\dagger\right).$$

Recall, for the ASC,

$$\tilde{A} = -\begin{bmatrix} Z_{1,0}^T E \\ Z_2^T \mathcal{F} \end{bmatrix}^{-1} \begin{bmatrix} Z_{1,0}^T F \\ \left(Z_2^T \mathcal{F}\right)' + \lambda Z_2^T \mathcal{F} \end{bmatrix}.$$

The first derivative of $\tilde{A}$ requires the second derivative of matrix Z_2^T. However, if Z_2^T is calculated from a smooth decomposition as suggested in Sect. 2.3.3, then only the first derivative of the computed matrix is known. Alternatively, either MATLAB's or Maple's commands for determining range and nullspace bases can be used to symbolically define Z_2^T as well as T_2 (used to find $Z_{1,0}^T$) and $Z_{1,0}^T$. Thus, the ASC's coefficient matrix $\tilde{A}$ can be computed symbolically, allowing MATLAB or Maple to return the first derivative of $\tilde{A}$ as an expression in terms of t.

Once $\tilde{A}'$, R, and the necessary derivatives of R are defined, submatrix $A'_{R_{12}}$ is taken from

$$A'_R = R' \tilde{A} R^{-1} + R \tilde{A}' R^{-1} + R \tilde{A} \left(R^{-1}\right)' + R'' R^{-1} + R' \left(R^{-1}\right)',$$

where $\left(R^{-1}\right)' = -R^{-1} R' R^{-1}$. If the output matrix is time-invariant, the derivative of A_R simplifies to $A'_R = R \tilde{A}' R^{-1}$. Transposing $A'_{R_{12}}$ produces the desired matrix $\left(A_{R_{12}}^T\right)'$ for calculating V_2' without differentiating a numerically computed quantity.

4.4 The Maximally Reduced Observer (MR)

One goal of designing a reduced-order observer is to reduce the number of state variables the observer must estimate by utilizing the information the output provides about the physical system. A new observer introduced in [20] uses the constraints defining the solution manifold of a DAE to provide additional information about previously unmeasurable components so the order of the observer is reduced even more. We refer to these observers as maximally reduced observers.

Suppose the derivative array equations $\mathcal{E}w + \mathcal{F}x = \mathcal{B}v$ were found without using stabilized differentiation. Thus, a system of algebraic equations characterizing the solution manifold is given by

$$\mathcal{G}(t) \left(\mathcal{F}(t)x(t) - \mathcal{B}(t)v(t)\right) = 0, \tag{4.19}$$

where matrix $\mathcal{G}$ has maximal rank and $\mathcal{GE} = 0$. In order to simplify the algorithm in both the linear time-invariant and time-varying cases, the chosen method defines $\mathcal{G}$ from the results of a singular value decomposition of $\mathcal{E}$. Since $\mathcal{GE} = 0$ is equivalent to $\mathcal{E}^T \mathcal{G}^T = 0^T$, the columns of $\mathcal{G}^T$ are in the nullspace of $\mathcal{E}^T$.

Knowing matrices $\mathcal{G}$ and $\mathcal{B}$ and vector v, product $\mathcal{GB}v$ behaves as another output, permitting the formation of

$$y_{\mathcal{E}} = \begin{bmatrix} y \\ \mathcal{GB}v \end{bmatrix}_{\mathcal{E}} = \begin{bmatrix} C \\ \mathcal{GF} \end{bmatrix}_{\mathcal{E}} x = C_{\mathcal{E}}x, \tag{4.20}$$

where $C_{\mathcal{E}}$ is full row rank. We refer to (4.20) as the extended output equation and to $C_{\mathcal{E}}$ as the extended output matrix. The structure of the extended output equation is not unique. For example, the user may decide to save all of output matrix C and then complete the extension with the rows of $\mathcal{GF}$ linearly independent from the rows of C. Alternatively, all of $\mathcal{GF}$ could be saved and then extended with the appropriate rows of output matrix C. However the extension occurs, it is necessary to keep track of the rows of C and $\mathcal{GF}$ used to form extended output matrix $C_{\mathcal{E}}$ so the correct rows of y and $\mathcal{GB}v$ are selected to define $y_{\mathcal{E}}$.

It is important to note that while we now have more outputs than before that they are different from the previous outputs. For one thing, $\mathcal{G}, \mathcal{B}, \mathcal{F}$ are quantities that must be computed for each time of interest. This is in contrast to C which is often known explicitly and is sometimes constant. Whether the additional computational effort of the MR observer is justified will depend on the problem and its sensitivity.

Once $C_{\overline{R}}$ is determined so transformation matrix $\overline{R} = \begin{bmatrix} C_{\mathcal{E}} \\ C_{\overline{R}} \end{bmatrix}$ is nonsingular, the construction of the maximally reduced observer follows the process outlined in Sect. 4.3 for the construction of the reduced-order observer. Based on the observer development in Sect. 4.3.2, the first and second derivatives of $\overline{R} = \begin{bmatrix} C_{\mathcal{E}} \\ C_{\overline{R}} \end{bmatrix}$ are required for the construction of the maximally reduced observer. Thus, for our numerical algorithms, we again select either MATLAB's or Maple's symbolic commands to define $\mathcal{GF}$, $C_{\overline{R}}$, and their necessary derivatives.

4.5 Completion Stability Revisited

The discussion on completion stability in Sect. 2.3.4 described the effects a constant real scalar λ has on a linear time-invariant completion's additional dynamics. These effects influence the observability of the completion. If the only unobservable eigenvalues of a completion come from its additional dynamics eigenvalues, which equal $-\lambda$, the completion is detectable when $\lambda > 0$ and unobservable when $\lambda \leq 0$ (recall, $\lambda = 0$ for the LSC).

Although the additional dynamics also influence the observability of a linear time-varying completion, the consequences from choosing a particular λ are not as apparent. For a constant scalar λ in the linear time-invariant case, the Jordan canonical form of a completion's coefficient matrix $\tilde{A}$ reveals the potential for a repeated additional dynamics eigenvalue to have linearly independent eigenvectors. Repeated eigenvectors are known to make observability impossible unless there are sufficient outputs. One proposal to reduce the number of linearly independent eigenvectors and thereby increase the probability of having an observable system is to stabilize using a stabilization parameter matrix Λ that is a constant diagonal matrix with distinct entries. We suspect this modification to stabilized differentiation $\left(\frac{d}{dt} + \Lambda\right)$ also improves the chances for observability in the linear time-varying case. The results from implementing $\frac{d}{dt} + \Lambda$ for a linear time-varying example are analyzed in Sect. 7.

Ideally we would want to choose Λ so that we could have some idea of what the additional dynamics eigenvalues are. The recent study [36] shows this is possible with linear time-invariant index one and index two Hessenberg DAEs, but not for general DAEs.

5   Completions and Observability

This section examines several theoretical issues. Following the presentation of some new results, the subsections consider linear time-invariant and time-varying theory that connects these theoretical results with the construction of full-order and reduced-order observers.

5.1   Some Linear Time-Invariant Theory

An important question is how having different possible completions can affect such properties as observability and controllability. The introduction of Sect. 2.3 mentioned all completions of a linear DAE are related by a general form. Two completions of a homogeneous linear time-invariant DAE

$$Ex' + Fx = 0 \tag{5.1}$$

can be written as $\tilde{x}' = \tilde{A}\tilde{x}$ and $\check{x}' = \check{A}\check{x} = \left(\tilde{A} + \Delta M\right)\check{x}$ [34] where $Mx = 0$ for full row rank matrix M describes the solution manifold of (5.1) and Δ is a matrix multiplier. Recall from Sect. 4.4, the solution manifold of a linear DAE can be characterized from the derivative array equations found with differential polynomial $\frac{d}{dt}$ (notationally, $\mathcal{E}w + \mathcal{F}x = 0$ for (5.1)). For a maximal rank (preferably full row rank) matrix $\mathcal{G}$ such that $\mathcal{G}\mathcal{E} = 0$, the nullspace of matrix M, $N(M)$ such that $M = \mathcal{G}\mathcal{F}$, characterizes the solution manifold. Matrix Δ describes the relationship between the completions' coefficient matrices $\breve{A}$ and $\tilde{A}$; that is, $\breve{A}$ equals $\tilde{A}$ plus a multiple Δ of M.

In fact, Δ is arbitrary and parameterizes all possible completions obtainable from the derivative array. In the linear time-varying case the need to keep Δ smooth is one major factor in our choice of completions.

Theorem 5.1 *Suppose that* $Mx = 0$ *describes the solution manifold of* (5.1) *and* M *is full row rank. For some linear time-invariant output matrix* C, *let* $\mathcal{O}_{\Delta}$ *and* $\mathcal{O}_{\tilde{A}}$ *designate the observability matrices for the respective pairs* $(\tilde{A} + \Delta M, C)$ *and* $(\tilde{A}, C)$ *of two completions of* (5.1). *Also, let* $N(Y)$ *denote the nullspace of a general matrix* Y. *Then*

$$N(M) \cap N(\mathcal{O}_{\Delta}) = N(M) \cap N(\mathcal{O}_{\tilde{A}}). \tag{5.2}$$

Proof It is known that $N(M)$ is $\tilde{A}$-invariant [23]. Suppose $\phi \in N(M)$. The observability matrices for the pairs $(\tilde{A} + \Delta M, C)$ and $(\tilde{A}, C)$ are

$$\mathcal{O}_{\Delta} = \begin{bmatrix} C \\ C(\tilde{A} + \Delta M) \\ C(\tilde{A} + \Delta M)^2 \\ \vdots \\ C(\tilde{A} + \Delta M)^{n-1} \end{bmatrix}, \quad \mathcal{O}_{\tilde{A}} = \begin{bmatrix} C \\ C\tilde{A} \\ C\tilde{A}^2 \\ \vdots \\ C\tilde{A}^{n-1} \end{bmatrix}.$$

Clearly, the first rows of $\mathcal{O}_{\Delta}\phi$ and $\mathcal{O}_{\tilde{A}}\phi$ equal $C\phi$. The second row of $\mathcal{O}_{\Delta}\phi$,

$$C(\tilde{A} + \Delta M)\phi = C\tilde{A}\phi + C\Delta M\phi = C\tilde{A}\phi,$$

equals the second row of $\mathcal{O}_{\tilde{A}}\phi$ since $M\phi = 0$ from $\phi \in N(M)$. Similarly, the third row of $\mathcal{O}_{\Delta}\phi$,

$$C(\tilde{A} + \Delta M)^2 \phi = C(\tilde{A} + \Delta M)(\tilde{A} + \Delta M)\phi = C\tilde{A}^2\phi,$$

equals the third row of $\mathcal{O}_{\tilde{A}}\phi$ since $M\tilde{A}\phi = 0$. Continuing this row-by-row comparison shows $\mathcal{O}_{\Delta}\phi = \mathcal{O}_{\tilde{A}}\phi$ if $\phi \in N(M)$. Equation (5.2) follows from the $\tilde{A}$ invariance of $N(M)$.

Theorem 5.1 indicates changing the completion does not alter which matrix pencil eigenvalues are observable. Thus, given a completion of a linear time-invariant DAE and an output equation, if at least one matrix pencil eigenvalue is unstable and unobservable, our observer construction approach will never be able to observe the DAE with that particular output equation.

The following remark's theoretical development shows completions of a linear time-invariant DAE may have observability matrices with different ranks. Keeping the results from Theorem 5.1 in mind, we know any rank variation is dependent on the additional dynamics.

Remark 5.1 If $N(M) + N\left(\mathcal{O}_{\tilde{A}}\right) = \Re^n$, then $N\left(\mathcal{O}_{\tilde{A}}\right)$ may not equal $N\left(\mathcal{O}_{A}\right)$. In particular, observability properties can vary between completions of the same DAE.

Rather than given a specific example of Remark 5.1 we shall show analytically exactly how the variability of observability properties is possible. It is known that $N\left(\mathcal{O}_{\tilde{A}}\right)$ is $\tilde{A}$-invariant. Given $N(M) + N\left(\mathcal{O}_{\tilde{A}}\right) = \Re^n$, the Kalman form simplifies as follows. We have $\nu_1 \oplus \nu_2 \oplus \nu_3 = \Re^n$, where $\nu_3 = N(M) \cap N\left(\mathcal{O}_{\tilde{A}}\right)$, $\nu_1 \oplus \nu_3 = N\left(\mathcal{O}_{\tilde{A}}\right)$, and $\nu_2 \oplus \nu_3 = N(M)$. The new $\tilde{A}$ matrix is

$$
\tilde{A}_s = \begin{bmatrix} \tilde{A}_{11} & 0 & 0 \\ 0 & \tilde{A}_{22} & 0 \\ \tilde{A}_{31} & \tilde{A}_{32} & \tilde{A}_{33} \end{bmatrix},
$$

and the new C matrix is $C_s = \begin{bmatrix} 0 & C_2 & 0 \end{bmatrix}$. Similarly, $\mu_1 \in \nu_2 \oplus \nu_3$ implies

$$
M\mu_1 = \begin{bmatrix} M_1 & M_2 & M_3 \end{bmatrix} \begin{bmatrix} 0 \\ * \\ * \end{bmatrix} = 0,
$$

where $*$ denotes an arbitrary matrix. Thus, $M_2* + M_3* = 0$ is true for any μ_1 when $M_2 = 0$ and $M_3 = 0$. The simplified structure of matrix M is $M_s = \begin{bmatrix} M_1 & 0 & 0 \end{bmatrix}$.

Again consider $\mu_2 \in N\left(\mathcal{O}_{\tilde{A}}\right)$. Although $C_s\tilde{A}_s\mu_2 = 0$, since

$$
C_s\tilde{A}_s\mu_2 = \begin{bmatrix} 0 & C_2 & 0 \end{bmatrix} \begin{bmatrix} \tilde{A}_{11} & 0 & 0 \\ 0 & \tilde{A}_{22} & 0 \\ \tilde{A}_{31} & \tilde{A}_{32} & \tilde{A}_{33} \end{bmatrix} \begin{bmatrix} * \\ 0 \\ * \end{bmatrix} = \begin{bmatrix} 0 & C_2\tilde{A}_{22} & 0 \end{bmatrix} \begin{bmatrix} * \\ 0 \\ * \end{bmatrix} = 0,
$$

$C_s\left(\tilde{A}_s + \Delta M_s\right)\mu_2$ may not equal zero. Using

$$
\tilde{A}_s + \Delta M_s = \begin{bmatrix} \tilde{A}_{11} & 0 & 0 \\ 0 & \tilde{A}_{22} & 0 \\ \tilde{A}_{31} & \tilde{A}_{32} & \tilde{A}_{33} \end{bmatrix} + \begin{bmatrix} \Delta_1 \\ \Delta_2 \\ \Delta_3 \end{bmatrix} \begin{bmatrix} M_1 & 0 & 0 \end{bmatrix}
$$

$$
= \begin{bmatrix} \tilde{A}_{11} + \Delta_1 M_1 & 0 & 0 \\ \Delta_2 M_1 & \tilde{A}_{22} & 0 \\ \tilde{A}_{31} + \Delta_3 M_1 & \tilde{A}_{32} & \tilde{A}_{33} \end{bmatrix},
$$

the product

$$C_s \left(\tilde{A}_s + \Delta M_s \right) \mu_2 = \begin{bmatrix} 0 & C_2 & 0 \end{bmatrix} \begin{bmatrix} \tilde{A}_{11} + \Delta_1 M_1 & 0 & 0 \\ \Delta_2 M_1 & \tilde{A}_{22} & 0 \\ \tilde{A}_{31} + \Delta_3 M_1 & \tilde{A}_{32} & \tilde{A}_{33} \end{bmatrix} \begin{bmatrix} * \\ 0 \\ * \end{bmatrix}$$

$$= \begin{bmatrix} C_2 \Delta_2 M_1 & C_2 \tilde{A}_{22} & 0 \end{bmatrix} \begin{bmatrix} * \\ 0 \\ * \end{bmatrix}$$

$$= C_2 \Delta_2 M_1 *$$

does not have to equal zero. Therefore, μ_2 may not be an element of $N \left(\mathcal{O}_A \right)$, which means $N \left(\mathcal{O}_{\tilde{A}} \right)$ may not equal $N \left(\mathcal{O}_A \right)$. Additionally, in the lower triangular matrix $\tilde{A}_s + \Delta M_s$, the choice of Δ, particularly Δ_1, has the potential to make the rank of $\mathcal{O}_A$ different from the rank of $\mathcal{O}_{\tilde{A}}$. Thus, observability properties can vary between completions.

5.2 Full-Order and Reduced-Order Linear Time-Invariant Observers

Proofs of long established theory are sometimes difficult to locate. As a result, a proof of theory important to example analysis is included in this section. An original version of Proof Part 1, Theorem 5.2 can be found in [73], and a different but related proof on detectable pairs can be found in [133].

Consider a linear time-invariant ODE with output equation

$$x' = Ax + Bu$$

$$y = Cx,$$

where matrix A is $n \times n$ and $m \times n$ output matrix C is assumed full row rank. This system does not need to be a completion. If the full row rank condition is not met, that is rank$(C) = r < m$, then without losing any information provided by the output equation, r linearly independent rows of C can be taken to define a full row rank output matrix. An additional assumption for the following proofs is that output matrix C has structure $\begin{bmatrix} I & 0 \end{bmatrix}$. If the output matrix does not have this block form, then the transformation discussed in Sect. 4.3.1 for constructing the reduced-order observer obtains the desired structure. Matrix A is considered in terms of the block form $\begin{bmatrix} A_{11} & A_{12} \\ A_{21} & A_{22} \end{bmatrix}$, where submatrix A_{11} is $m \times m$ and submatrix A_{22} is $(n - m) \times (n - m)$.

Theorem 5.2 reveals that for a linear time-invariant ODE, the unobservable eigenvalues from the full-order (pair (A, C)) and the reduced-order (pair (A_{22}, A_{12})) systems are equivalent. That is, the full-order observer can be constructed to estimate the state if and only if the reduced-order observer can be constructed to estimate the state.

Theorem 5.2 *Pair (A, C) is observable if and only if pair (A_{22}, A_{12}) is observable.*

Proof The observability matrices for pair (A, C) and pair (A_{22}, A_{12}) are

$$
\mathcal{O}_F = \begin{bmatrix} C \\ CA \\ CA^2 \\ \vdots \\ CA^{n-1} \end{bmatrix}, \quad \mathcal{O}_R = \begin{bmatrix} A_{12} \\ A_{12}A_{22} \\ A_{12}A_{22}^2 \\ \vdots \\ A_{12}A_{22}^{(n-m)-1} \end{bmatrix},
$$

respectively.

Then the following set of equivalences hold:

$$(A, C) \text{ is observable} \Leftrightarrow y = 0 \Rightarrow x = 0 \tag{5.3}$$

$$\Leftrightarrow x_1 = 0 \Rightarrow x_2 = 0 \tag{5.4}$$

$$\Leftrightarrow \{x_1 = 0 \text{ and } A_{12}x_2 = 0\} \Rightarrow x_2 = 0 \tag{5.5}$$

$$\Leftrightarrow (A_{22}, A_{12}) \text{ is observable.} \tag{5.6}$$

The first and last equivalences are just the definitions of observability for (A, C) and (A_{22}, A_{12}), respectively.

As a consequence of this theorem, $N(\mathcal{O}_F) = N(\mathcal{O}_R)$ so the unobservable eigenvalues of matrices A and A_{22} are equivalent. Thus, (A, C) is detectable if and only if (A_{22}, A_{12}) is detectable.

5.3 Comments on Linear Time-Varying Theory

Additional research on the completions of a linear time-varying DAE and their observability is required before drawing theoretical conclusions similar to those in Sect. 5.

The theory if pair (A, C) is observable, then (A_{22}, A_{12}) is observable is proven in [113] by first considering the pairs at a specific value of time t and then continuing with a rank-dependent linear time-invariant proof. Our suggested time-varying explanation for this relationship and its reverse implication focuses on two descriptions of observable:

1. a linear ODE is observable at t_0 if its state $x(t_0) = 0$ is the only unobservable state at t_0;
2. if the output $y = C\Phi(t, t_0) x(t_0) = 0$ for every $t \geq t_0$ when $u = 0$, then $x(t_0)$ is unobservable at t_0. ($\Phi(t, t_0)$ is the state transition matrix of the linear ODE) [5].

Thus, a linear ODE can be shown to be observable at t_0 if $x(t_0) = 0$ is implied when $u = 0$ and $y = C\Phi(t, t_0) x(t_0) = 0$ for $t \geq t_0$.

We assume that C is full row rank for all t. Then after a linear time-varying coordinate change we have $C = \begin{bmatrix} I & 0 \end{bmatrix}$ with an $m \times m$ identity matrix, and the full-order system $x' = Ax + Bu$ with output equation $y = Cx$ can be written as

$$x_1' = A_{11}x_1 + A_{12}x_2 + B_1 u \tag{5.7a}$$

$$x_2' = A_{21}x_1 + A_{22}x_2 + B_2 u \tag{5.7b}$$

$$y = x_1. \tag{5.7c}$$

We continue our development based on (5.7). System (5.7)'s associated reduced-order system is

$$x_2' = A_{22}x_2 + \bar{u} \tag{5.8a}$$

$$\bar{y} = A_{12}x_2, \tag{5.8b}$$

where available information on x_1 from output (5.7c) is used to define new output $\bar{y} = x_1' - A_{11}x_1 - B_1 u$ and new input $\bar{u} = A_{21}x_1 + B_2 u$.

Theorem 5.2 holds also in the linear time-varying case and the proof is very similar to the linear time-invariant proof except that we need things to hold for all t greater than t_0. A more detailed development of this result goes as follows.

Given (5.7) is observable, let $\bar{u} = 0$ and $\bar{y} = A_{12}\Phi_R(t, t_0) x_2(t_0) = 0$ for $t \geq t_0$, where $x_2(t_0)$ is arbitrary and $\Phi_R(t, t_0)$ is the state transition matrix of (5.8). The reduced-order system simplifies to

$$x_2' = A_{22}x_2 \tag{5.9a}$$

$$0 = A_{12}x_2. \tag{5.9b}$$

However, for some $\bar{x} = 0$, system (5.9) also has structure

$$\bar{x}' = A_{11}\bar{x} + A_{12}x_2 \tag{5.10a}$$

$$x_2' = A_{21}\bar{x} + A_{22}x_2 \tag{5.10b}$$

$$y = \bar{x} = 0, \tag{5.10c}$$

which is equivalent to (5.7) when $u = 0$ and $y = x_1 = 0$. Thus, $x_2(t_0) = 0$ from $x(t_0) = \begin{bmatrix} x_1(t_0) \\ x_2(t_0) \end{bmatrix} = 0$ since the full-order system is observable. Therefore,

$\bar{u} = 0$ and $\bar{y} = A_{12}\Phi_R(t, t_0) x_2(t_0) = 0$ for $t \geq t_0$ implies $x_2(t_0) = 0$, so the reduced-order system is observable at t_0. This observability check holds for all $t \geq t_0$, resulting in pair (A_{22}, A_{12}) being observable.

Conversely, given (5.8) is observable, let $u = 0$ and $y = C\Phi(t, t_0) x(t_0) = 0$ for $t \geq t_0$, where $x_2(t_0)$ from $x(t_0) = \begin{bmatrix} x_1(t_0) \\ x_2(t_0) \end{bmatrix}$ is arbitrary and $\Phi(t, t_0)$ is the state transition matrix of (5.7). The initial condition for $x_1(t_0)$ is not arbitrary since the assumption $y = 0$ implies $x_1 = 0$, so $x_1(t_0) = 0$. But when $y = x_1 = 0$, the full-order system simplifies to (5.9), which is equivalent to (5.8) when $\bar{u} = 0$ and $\bar{y} = 0$. Thus, $x_2(t_0) = 0$ since the reduced-order system is observable. Therefore, $u = 0$ and $y = C\Phi(t, t_0) x(t_0) = 0$ for $t \geq t_0$ implies $x(t_0) = 0$, so the full-order system is observable at t_0. With the observability check holding for all $t \geq t_0$, pair (A, C) is observable.

Although there is no reason to suspect the detectability implication between the full-order and reduced-order observers does not also hold in the time-varying case, the linear time-invariant proof uses eigenvectors and is not applicable. Instead, a different idea from control theory can be used to reason why the detectability implication holds. The zero dynamics of an ODE are its solutions that produce a zero output [10, 79], making these states indistinguishable from the zero solution. Thus, the zero dynamics are unobservable. From this theory, the basis vectors for the unobservable subspace have structure $b_f = \begin{bmatrix} 0_{(n-m) \times 1} \\ b_{f2} \end{bmatrix}$ for the full-order system and $b_r = \alpha b_{f2}$, α some multiple, for the reduced-order system. Due to these basis vectors' structural relationship, a detectable full-order or reduced-order system implies its counterpart is also detectable.

6   Linear Time-Invariant Example

We illustrate some of the preceding ideas about completion-based observers on a linear time-invariant example. All code and its documentation as well as many more computational studies can be found in [17].

The example system

$$x_1' = x_2 \tag{6.1a}$$

$$x_2' = Kx_1 + Sx_2 + H^T x_3 + Gu_1 \tag{6.1b}$$

$$0 = Hx_1 + u_2 \tag{6.1c}$$

is representative of a constrained mechanical system [19, 20]. State variables x_1 and x_2 are position and velocity, respectively, while (6.1c) is a physical constraint that produces the force $H^T x_3$. The input u applies a force through Gu_1 in (6.1b) and affects constraint (6.1c) through u_2. In some physical problems $u_2 = 0$ and we just have u_1. Additionally, matrix H is assumed full row rank. There is also an output equation

$$y = Cx. \tag{6.1d}$$

A linear or nonlinear DAE in Hessenberg form of size k is

$$f_1' = F_1(f_1, f_2, \ldots, f_k, t) \tag{6.2a}$$

$$f_2' = F_2(f_1, f_2, \ldots, f_{k-1}, t) \tag{6.2b}$$

$$\vdots \tag{6.2c}$$

$$f_i' = F_i(f_{i-1}, f_i, \ldots, f_{k-1}, t), \quad 3 \leq i \leq k-1, \tag{6.2d}$$

$$\vdots \tag{6.2e}$$

$$0 = F_k(f_{k-1}, t) \tag{6.2f}$$

and has index k if $(\partial F_k / \partial f_{k-1})(\partial F_{k-1}/\partial f_{k-2}) \cdots (\partial F_2/\partial f_1)(\partial F_1/\partial f_k)$ is nonsingular [23]. Here the index is the differentiation index or equivalently the minimal k such that assumptions A1–A3 hold. System (6.1) is in Hessenberg form of size 3 with index 3.

In a physical system y is an output and we do not need to simulate the DAE (6.1). But for our example here, we will need to integrate the DAE to produce the output y. Due to this DAE's Hessenberg form and the row rank assumption on H, the solution of (6.1) can be determined from an index 1 DAE with the same solutions. Let $H^\dagger = H^T (HH^T)^{-1}$, the Moore–Penrose pseudoinverse of H [41], and $P_H = (I - H^\dagger H)$. Differentiation and some algebra give

$$x_1' = x_2 \tag{6.3a}$$

$$x_2' = P_H K x_1 + P_H S x_2 + P_H G u_1 - H^\dagger u_2'' \tag{6.3b}$$

$$x_3 = -(HH^T)^{-1}(HKx_1 + HSx_2 + HGu_1 + u_2'') \tag{6.3c}$$

$$0 = Hx_1 + u_2 \tag{6.3d}$$

$$0 = Hx_2 + u_2'. \tag{6.3e}$$

Equations (6.3a)–(6.3c) are an index one DAE in x_1, x_2, x_3. Equations (6.3c)–(6.3e) describe the solution manifold. We can use (6.3c)–(6.3e) to find a consistent initial condition and then use (6.3a)–(6.3c) to simulate the state and get y.

For purposes of illustration we specify K, S, H, and G in (6.1) as

$$x_1' = x_2 \tag{6.4a}$$

$$x_2' = \begin{bmatrix} -2 & 1 \\ 1 & -2 \end{bmatrix} x_1 + \frac{1}{4} \begin{bmatrix} 1 & 0 \\ 0 & 1 \end{bmatrix} x_2 + \begin{bmatrix} 1 \\ -1 \end{bmatrix} x_3 + \begin{bmatrix} 1 \\ 1 \end{bmatrix} u_1 \tag{6.4b}$$

$$0 = \begin{bmatrix} 1 & -1 \end{bmatrix} x_1 + u_2 \tag{6.4c}$$

with input $u = \begin{bmatrix} u_1 \\ u_2 \end{bmatrix} = \begin{bmatrix} \sin \\ \sin \end{bmatrix}$. System (6.4) is in the form $Ex' + Fx = Bu$, where

$$E = \begin{bmatrix} 1 & 0 & 0 & 0 & 0 \\ 0 & 1 & 0 & 0 & 0 \\ 0 & 0 & 1 & 0 & 0 \\ 0 & 0 & 0 & 1 & 0 \\ 0 & 0 & 0 & 0 & 0 \end{bmatrix}, \quad F = \begin{bmatrix} 0 & 0 & -1 & 0 & 0 \\ 0 & 0 & 0 & -1 & 0 \\ 2 & -1 & -\frac{1}{4} & 0 & -1 \\ -1 & 2 & 0 & -\frac{1}{4} & 1 \\ -1 & 1 & 0 & 0 & 0 \end{bmatrix}, \quad B = \begin{bmatrix} 0 & 0 \\ 0 & 0 \\ 1 & 0 \\ 1 & 0 \\ 0 & 1 \end{bmatrix}.$$

We take the output matrix as $C = \begin{bmatrix} 0 & 0 & 1 & 0 & 0 \\ 0 & 0 & 0 & 1 & 0 \end{bmatrix}$ for which (6.4) is observable [42]. This output matrix is designated as C_a and is considered in Case 1 of Chap. 5 in [17].

The initial conditions for solving (6.4), finding the completions, and constructing the observers are defined at time $t_0 = 0$. From constraints (6.3d) and (6.3e) and equation (6.3c), a consistent initial condition is $x(0) = \tilde{x}(0) = \begin{bmatrix} 0 & 0 & -0.5 & 0.5 & 0.125 \end{bmatrix}^T$. Vector $\hat{x}(0) = \begin{bmatrix} 6 & 7 & 8 & 9 & 10 \end{bmatrix}^T$ is chosen arbitrarily (except that $\hat{x}(0) \neq x(0)$) as the initial condition for the full-order observer and is transformed accordingly for the other two observers discussed in Sect. 4. Note that $\hat{x}(0)$ need not be consistent with the DAE which can be useful.

System (6.4) is a regular linear time-invariant DAE. The matrix pencil eigenvalues are $0.125000 \pm 0.992157i$ and, hence, unstable. Whenever these eigenvalues are unobservable, any of the observers fail to converge to the true state. Similarly, for the LSC, the additional dynamics eigenvalues are not stable since they equal 0. The SLSC and the ASC are stabilized with $\lambda = 2$, so their additional dynamics eigenvalues equal -2 and are stable.

For all linear time-invariant observers, the observer gain is found in the examples by transforming the system to observer form in order to extract the observable subsystem. Then the transpose of this subsystem is put into controller form and then the feedback is computed from the controller form as described in [5].

The following matrices are the completions' coefficient matrices (notationally, $\tilde{A}$, $\tilde{B}$ from $\tilde{x}' = \tilde{A}\tilde{x} + \tilde{B}v$, the linear time-invariant version of (2.13)) and are included to illustrate how the coefficients can differ between completions.

Stabilized Least Squares Completion: $\tilde{A}$, $\tilde{B}$

$$
\begin{bmatrix}
-0.52 & 0.52 & 0.41 & 0.59 & -0.17 \\
0.52 & -0.52 & 0.59 & 0.41 & 0.17 \\
-1.78 & 0.78 & -0.13 & 0.38 & 0.74 \\
0.78 & -1.78 & 0.38 & -0.13 & -0.74 \\
4.54 & -4.54 & -2.32 & 2.32 & -4.44
\end{bmatrix},
\begin{bmatrix}
0.0 & -0.78\ 0 & -0.57\ 0 & -0.09\ 0 & 0.0 \\
0.0 & 0.78\ 0 & 0.57\ 0 & 0.09\ 0 & 0.0 \\
1.0 & -0.17\ 0 & -0.35\ 0 & -0.13\ 0 & 0.0 \\
1.0 & 0.17\ 0 & 0.35\ 0 & 0.13\ 0 & 0.0 \\
0.0 & -2.13\ 0 & -3.26\ 0 & -2.10\ 0 & -0.5
\end{bmatrix}.
$$

Alternative Stabilized Completion: $\tilde{A}$, $\tilde{B}$

$$
\begin{bmatrix}
-1.0 & 1.0 & 0.50 & 0.50 & 0.0 \\
1.0 & -1.0 & 0.50 & 0.50 & 0.0 \\
-0.5 & -0.5 & -0.88 & 1.13 & 0.0 \\
-0.5 & -0.5 & 1.13 & -0.88 & 0.0 \\
0.0 & 0.0 & 0.00 & 0.00 & -2.0
\end{bmatrix},
\begin{bmatrix}
0.0 & -1.0\ 0 & -0.50\ 0 & 0.00\ 0 & 0.0 \\
0.0 & 1.0\ 0 & 0.50\ 0 & 0.00\ 0 & 0.0 \\
1.0 & 0.0\ 0 & -1.00\ 0 & -0.50\ 0 & 0.0 \\
1.0 & 0.0\ 0 & 1.00\ 0 & 0.50\ 0 & 0.0 \\
0.0 & -3.0\ 0 & -1.25\ 0 & -0.88\ 0 & -0.5
\end{bmatrix}.
$$

Least Squares Completion: $\tilde{A}$, $\tilde{B}$

$$
\begin{bmatrix}
0.00 & 0.00 & 0.67 & 0.33 & 0.00 \\
0.00 & 0.00 & 0.33 & 0.67 & 0.00 \\
-1.40 & 0.40 & 0.20 & 0.05 & 0.60 \\
0.40 & -1.40 & 0.05 & 0.20 & -0.60 \\
0.23 & -0.23 & 0.48 & -0.48 & -0.15
\end{bmatrix},
\begin{bmatrix}
0.0\ 0\ 0 & -0.33\ 0 & 0.00\ 0 & 0.0 \\
0.0\ 0\ 0 & 0.33\ 0 & 0.00\ 0 & 0.0 \\
1.0\ 0\ 0 & 0.00\ 0 & -0.20\ 0 & 0.0 \\
1.0\ 0\ 0 & 0.00\ 0 & 0.20\ 0 & 0.0 \\
0.0\ 0\ 0 & -1.00\ 0 & 0.05\ 0 & -0.5
\end{bmatrix}.
$$

In this section the ODEs are solved by MATLAB's ode45 solver with a relative tolerance error set at 1e–6. For all figures in this section, each color corresponds with a particular state variable: blue for the first component of x_1, green for the second component of x_1, red for the first component of x_2, gray for the second component of x_2, and magenta for x_3.

Figure 1a displays x, the solution of (6.4) the state observers are attempting to estimate on the $t \in [0, 15]$ interval. The final time $t_f = 15$ is arbitrary. For the consistent initial condition identified earlier, all three completions produce the same dynamics except for a slight error growth in the LSC since it has unstable additional dynamics. Parameter ρ is the desired eigenvalue for the eigenvalue placement of observable eigenvalues by the gain matrix in the observer and is taken as -1.

Note, "full-order system" and "completion with output equation" are interchangeable. Also, reduced-order system describes the system associated with pair $(A_{R_{22}}, A_{R_{12}})$. In order to specify a pair for a particular completion, subscripts are included: $(\cdot, \cdot)_{\mathrm{SL}}$ for the SLSC, $(\cdot, \cdot)_{\mathrm{A}}$ for the ASC, and $(\cdot, \cdot)_{\mathrm{L}}$ for the LSC.

The rank of observability matrix $\mathcal{O}$ for pair $(\tilde{A}, C)_{\mathrm{SL}}$ is 5, so all eigenvalues of $\tilde{A}$ are observable. Remark 5.1 showed observability matrices for different completions of the same linear time-invariant DAE may not have the same rank. Although rank $(\mathcal{O}) = 3$ for pair $(\tilde{A}, C)_{\mathrm{A}}$ and rank $(\mathcal{O}) = 4$ for pair $(\tilde{A}, C)_{\mathrm{L}}$, the matrix

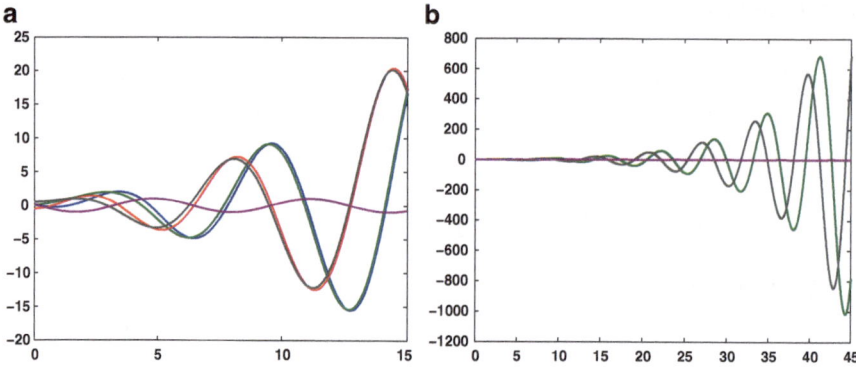

Fig. 1 The solution x of (6.4) (**a**) $t \in [0, 15]$ (**b**) $t \in [0, 45]$

pencil eigenvalues are observable for both completions due to the observability of the SLSC with output matrix C and Theorem 5.1.

Since C is full row rank, measurements on two state variables are available through the output. Thus, the reduced-order observer must estimate only three unmeasurable components. Transformation matrix $R = \begin{bmatrix} C \\ C_R \end{bmatrix}$, where

$C_R = \begin{bmatrix} -1 & 0 & 0 & 0 & 0 \\ 0 & -1 & 0 & 0 & 0 \\ 0 & 0 & 0 & 0 & 1 \end{bmatrix}$, provides the desired $CR^{-1} = \begin{bmatrix} I & 0 \end{bmatrix}$ structure. The pair $(A_{R_{22}}, A_{R_{12}})_{\mathrm{SL}}$ is observable since the full-order system is observable if and only if the reduced-order system is observable. The pair $(A_{R_{22}}, A_{R_{12}})_{\mathrm{A}}$ is detectable since the full-order system is detectable if and only if the reduced-order system is detectable. Therefore, the reduced-order system also has two unobservable eigenvalues equaling -2, so the observability matrix has rank 1. Finally, the pair $(A_{R_{22}}, A_{R_{12}})_{\mathrm{L}}$ is unobservable since the full-order and reduced-order systems have the same unobservable eigenvalues. The one unobservable eigenvalue remains unobservable and since it equals zero, the LSC does not produce a reduced-order observer that asymptotically estimates the state. This illustrates the advantage of having different completions available.

In the linear time-invariant case, the dimension of the solution manifold equals the number of matrix pencil eigenvalues. Therefore, matrix $\mathcal{G}$ from the characterization of the solution manifold $\mathcal{G}(\mathcal{F}x - \mathcal{B}v) = 0$ is expected to be 3×5 and have rank 3. For this example, the extended output matrix provides information on four state variables so the order of the maximally reduced observer is one. The observability matrix is 4×1 with rank 1 for all three completions, indicating the only eigenvalue of their $A_{\overline{R}_{22}}$ submatrices is observable. Therefore, each of the three completions can be used to construct a maximally reduced observer that estimates the state. This result shows it may be possible to construct the maximally reduced observer even if the full-order and reduced-order systems are unobservable.

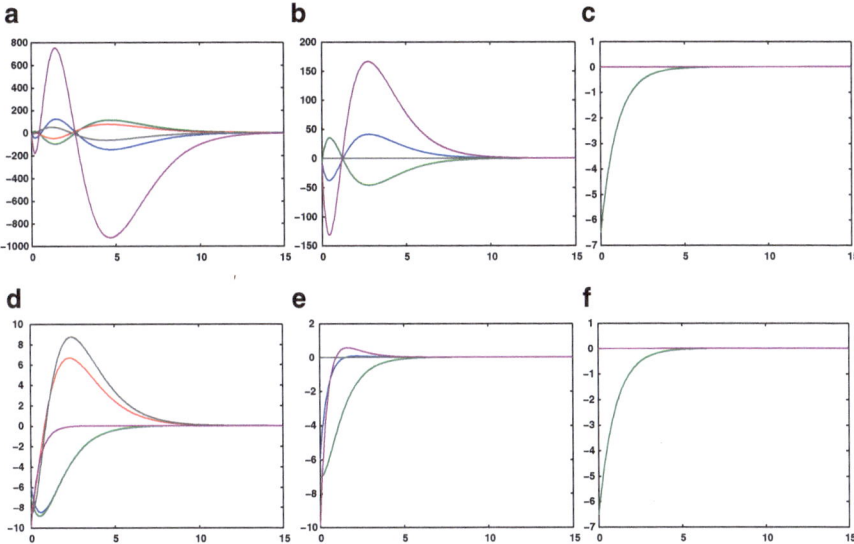

Fig. 2 $x - \hat{x}$ comparison. $(C, \lambda = 2, \rho = -1)$ (**a**) SLSC/FO (**b**) SLSC/RO (**c**) SLSC/MR (**d**) ASC/FO (**e**) ASC/RO (**f**) ASC/MR

Based on the magnitude of the estimation error $x - \hat{x}$ and on the time it takes $x - \hat{x}$ to converge to zero, Fig. 2a,d reinforce the observation that with output matrix C, $\lambda = 2$, and $\rho = -1$, the full-order observer constructed using the alternative stabilized completion (ASC/FO observer) provides a better estimate of x than the full-order observer constructed using the stabilized least squares completion (SLSC/FO observer). For both completions, reducing the observer's order to three produces a more accurate estimate of the state than when observing the full state vector. Figure 2b,e show the reduced-order observer constructed using the alternative stabilized completion (ASC/RO observer) is closer to x than the reduced-order observer constructed using the stabilized least squares completion (SLSC/RO observer).

Figure 2c,f illustrate that the maximally reduced observer is independent of the completion used in its construction (the maximally reduced observers constructed using the stabilized least squares completion and the alternative stabilized completion are abbreviated SLSC/MR and ASC/MR, on on respectively). Recall, λ comes from the stabilization in the completion. All additional dynamics no matter the completion get removed from the maximally reduced observer's error dynamics. Any observable eigenvalues from the original dynamics get placed at ρ by the feedback in the observer. The only effect the choice of completion has is on a simulation's generation of the DAE response. There is no effect on the maximally reduced observer's behavior.

Figure 3 shows the plot of both the state and the estimated state for the ASC/FO observer illustrating the convergence. The estimates are dashed lines and the state

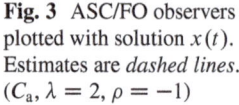

Fig. 3 ASC/FO observers plotted with solution $x(t)$. Estimates are *dashed lines*. $(C_a, \lambda = 2, \rho = -1)$

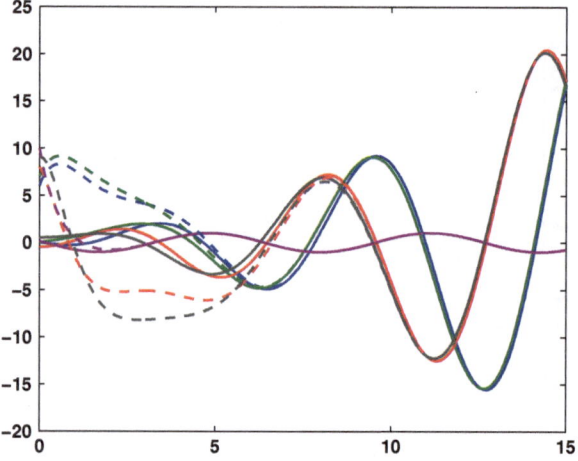

is given by solid lines. The SLSC/FO observer also converges but the error goes larger before smaller as noted in Fig. 2a. This illustrates how the ASC completion can sometimes be advantageous when designing observers.

7 Linear Time-Varying Example

Observer construction is more difficult in the linear time-varying case. Also, the literature often states a quantity particular to the topic being discussed should be taken large enough, or something similar, without really addressing the analytical and numerical issues involved. Accordingly, we give more detail and discussion for the linear time-varying example than we gave for the linear time-invariant example.

To illustrate the linear time-varying case we chose a system based on an example circuit in [105, 123]. The circuit in Fig. 4 can be described by the index 2 DAE

$$(C_1 e_1)' - i_{r_1} + i_{r_2} - i_v = 0 \tag{7.1a}$$

$$(C_2 e_2)' + i_l + i_{r_1} = 0 \tag{7.1b}$$

$$(L i_l)' - e_2 = 0 \tag{7.1c}$$

$$-e_1 + e_2 - R_1 i_{r_1} = 0 \tag{7.1d}$$

$$e_1 - R_2 i_{r_2} = 0 \tag{7.1e}$$

$$e_1 = -V. \tag{7.1f}$$

Ref represents ground (or the reference point); L is an inductor; C_1, C_2 are capacitors; R_1, R_2 are resistors; V represents a voltage source; e_1, e_2 are nodes at which to measure voltage; and i_l, i_{r_1}, i_{r_2}, i_v are currents through L, R_1, R_2, and

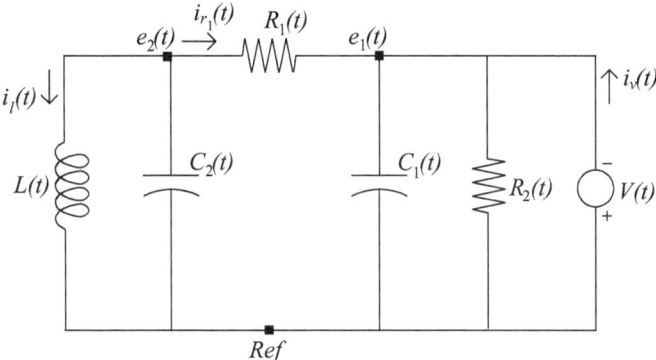

Fig. 4 Circuit example

V, respectively. C_1, C_2, L, R_1, and R_2 are assumed nonsingular. System (7.1) is in the form $Ex' + Fx = Bu$, where $u = V$ and

$$
x = \begin{bmatrix} e_1 \\ e_2 \\ i_l \\ i_{r_1} \\ i_{r_2} \\ i_v \end{bmatrix}, \quad
E = \begin{bmatrix} C_1 & 0 & 0 & 0 & 0 & 0 \\ 0 & C_2 & 0 & 0 & 0 & 0 \\ 0 & 0 & L & 0 & 0 & 0 \\ 0 & 0 & 0 & 0 & 0 & 0 \\ 0 & 0 & 0 & 0 & 0 & 0 \\ 0 & 0 & 0 & 0 & 0 & 0 \end{bmatrix},
$$

$$
F = \begin{bmatrix} C_1' & 0 & 0 & -I & I & -I \\ 0 & C_2' & I & I & 0 & 0 \\ 0 & -I & L' & 0 & 0 & 0 \\ -I & I & 0 & -R_1 & 0 & 0 \\ I & 0 & 0 & 0 & -R_2 & 0 \\ I & 0 & 0 & 0 & 0 & 0 \end{bmatrix}, \quad
B = \begin{bmatrix} 0 \\ 0 \\ 0 \\ 0 \\ 0 \\ -I \end{bmatrix}.
$$

The entries $\{e_1, e_2, i_l, i_{r_1}, i_{r_2}, i_v\}$ of state vector x are referred to as state variables or as components. The original circuit in [105, 123] does not have a voltage source, but we add one to the example system so that we can get a state trajectory that does not go to zero. This provides a better challenge for the observers.

In order to illustrate the observer construction approach on this circuit, we define the capacitors, inductor, and voltage source as $C_1(t) = 3 + \cos(t/3)$, $C_2(t) = 2 - \cos(2t)$, $L(t) = 2 - \exp(-t)$, and $V(t) = 4\cos(2t)\sin(t/5)$, respectively. Two sets of resistors are considered. The system investigated in Sect. 7.1 has positive resistors $R_1(t) = 4 + 2\sin(t)$ and $R_2(t) = 2 + \sin(t)$, while the system examined in Sect. 7.2 has their negative counterparts, $R_1(t) = -(4 + 2\sin(t))$ and $R_2(t) = -(2 + \sin(t))$. A negative resistor is a device that puts energy into a system instead of removing it and can therefore affect stability.

Although a system of equations is said to observe a physical system if the difference between the true state and the estimated state goes to zero as time goes to infinity, an effective observer has sufficiently small error by a specific time. There has been work on observers that converge in finite time. With discrete time they are called dead beat observers. The continuous time case is considered in [106]. We do not consider finite time convergence in this paper. We are interested in convergence within a specified tolerance on a specified interval. In our examples, the goal is for the estimation error to be within the interval $(-10^{-3}, 10^{-3})$ by $t_f = 45$. Notation $e \approx 0$ is used to indicate when this condition is met. At times, results are considered on the extended time interval $t \in [0, 135]$ to provide a better understanding of the variables' behavior.

The observer construction code for this linear time-varying example DAE was primarily programmed and executed using MATLAB version 7.0.1.15 (R14). For our numerical algorithms, MATLAB's ode45 solver is called to solve the ODEs. The relative error tolerance (RelTol) is set at 1e–9 instead of its default 1e–3.

7.1 Example with Positive Resistors

The linear time-varying DAE with the positive resistors is

$$\frac{d}{dt}\left((3 + \cos(t/3)) e_1(t)\right) - i_{r_1}(t) + i_{r_2}(t) - i_v(t) = 0 \tag{7.2a}$$

$$\frac{d}{dt}\left((2 - \cos(2t)) e_2(t)\right) + i_l(t) + i_{r_1}(t) = 0 \tag{7.2b}$$

$$\frac{d}{dt}\left((2 - \exp(-t)) i_l(t)\right) - e_2(t) = 0 \tag{7.2c}$$

$$-e_1(t) + e_2(t) - (4 + 2\sin(t)) i_{r_1}(t) = 0 \tag{7.2d}$$

$$e_1(t) - (2 + \sin(t)) i_{r_2}(t) = 0 \tag{7.2e}$$

$$e_1(t) = -4\cos(2t)\sin(t/5). \tag{7.2f}$$

The output matrices considered are

$$C_a = \begin{bmatrix} 0 & 1 & 0 & 0 & 0 & 0 \\ 0 & 0 & 1 & 0 & 0 & 0 \\ 0 & 0 & 0 & 1 & 0 & 0 \\ 0 & 0 & 0 & 0 & 0 & 1 \end{bmatrix}, \quad C_b = \begin{bmatrix} 0 & 0 & 1 & 0 & 0 & 1 \end{bmatrix}. \tag{7.3}$$

Output matrix C_b corresponds with output matrix C_c in [17]. $\tilde{x}(0) = \begin{bmatrix} 0 & 0 & 0 & 0 & 0 & 16/5 \end{bmatrix}^T$ is selected as the consistent initial condition when solving the completions. Recall, the completion's $\tilde{x}$ is the resulting trajectory to be estimated.

Except for the desire that $\hat{x}(0) \neq \tilde{x}(0)$, vector $\hat{x}(0) = \begin{bmatrix} 1 & 2 & 3 & 4 & 5 & 6 \end{bmatrix}^T$ is an arbitrary initial condition for the full-order observer and is transformed accordingly for the reduced-order and maximally reduced observers. Subscript notation identifies a particular completion's pair: $(\cdot, \cdot)_{\text{SL}}$ for the SLSC, $(\cdot, \cdot)_A$ for the ASC, and $(\cdot, \cdot)_L$ for the LSC.

The solution manifold is 2-dimensional for circuit example (7.1), so $\mathcal{G}$ from $\mathcal{G}(\mathcal{F}\tilde{x} - \mathcal{B}v) = 0$ is expected to have rank 4. The extended output matrices for constructing the maximally reduced observers in this section have structure $C_{\bar{z}} = \begin{bmatrix} \mathcal{G}\mathcal{F} \\ C_x \end{bmatrix}$. MATLAB's null command returns a $\mathcal{G}$ for a symbolically defined t such that

$$(\mathcal{G}\mathcal{F})(t) = \begin{bmatrix} -1 & 1 & 0 & -(4 + 2\sin(t)) & 0 & 0 \\ 1 & 0 & 0 & 0 & -(2 + \sin(t)) & 0 \\ 1 & 0 & 0 & 0 & 0 & 0 \\ -(1/3)\sin(t) & (t/3) & 0 & 0 & -1 & 1 & -1 \end{bmatrix}.$$

(7.4)

The same command in Maple gives a matrix $\mathcal{G}$ with singularities. The full row rank $\mathcal{G}\mathcal{F}$ in (7.4) is used in the construction of each extended output matrix. If an output matrix C with rank ≥ 2 has at least two rows linearly independent from the rows of (7.4) for all time, then enough information is known to determine the state variables without constructing an observer. That is, the state vector can be solved for explicitly using (4.20) when the extended output matrix $C_{\bar{z}}$ is nonsingular. Otherwise, the extended output equation is used to construct the maximally reduced observer.

When selecting a stabilization parameter matrix for either the SLSC (Λ_{SL}) or the ASC (Λ_A), there are a number of things to consider. One is whether the completion and its additional dynamics are asymptotically stable. For this stability analysis, we checked if $\lim_{t\to\infty} \|\Phi(t, 0)\| = 0$ [5], where $\Phi(t, 0)$ is the completion's state transition matrix. Article [72] offers an alternative characterization of asymptotic stability not considered here. Another point to consider is if $\tilde{A}$ and $\tilde{B}$ are bounded. This boundedness was used in the proofs of convergence for observer construction. Finally, rank (W_o) for $(\tilde{A}, C)$ is checked to determine if the completion is observable. Since the observability Gramian is found by integrating a system of differential equations with a zero initial condition, it may take a while before the numerical rank is correct. For this example it took up to 1.5 units of time for some choices of Λ. After some experimentation we took $\Lambda_{\text{SL}} = \text{diag}\{1.2, 1.4, 1.6, 1.8, 2.0, 2.2\}$ and $\Lambda_A = \text{diag}\{4.0, 3.5, 3.0, 2.5\}$.

The literature differs on whether or not an exponentially asymptotically stable $\tilde{A}$ is required for estimating the state. Coefficient matrix $\tilde{A}$ from the LSC of (7.2) is not exponentially asymptotically stable as shown in Fig. 5. However, the coefficient matrices of the LSC are bounded.

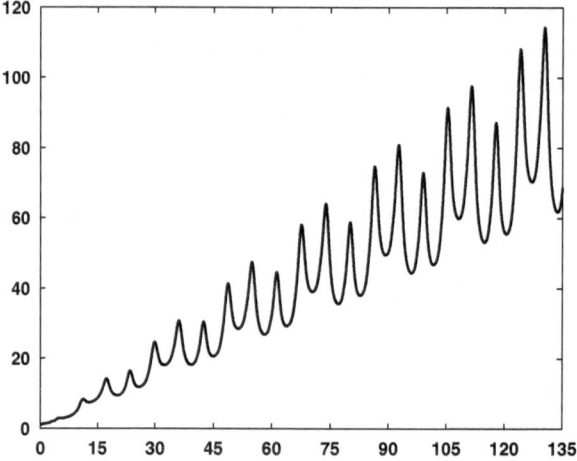

Fig. 5 $\|\Phi(t, 0)\|$ fails to converge to zero for the LSC of (7.2)

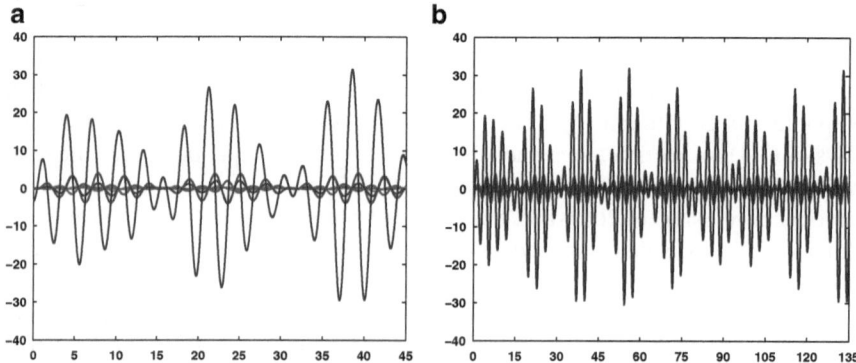

Fig. 6 The observers discussed in Sect. 7.1 estimate the solution of (7.2), seen here on two intervals of time (**a**) $t \in [0, 45]$ (**b**) $t \in [0, 135]$

For the remaining figures in Sect. 7.1, each color corresponds with a particular state variable: blue for e_1, green for e_2, red for i_l, gray for i_{r_1}, magenta for i_{r_2}, and brown for i_v. Figure 6 displays the solution of the system with positive resistors that the observers are attempting to estimate.

The computation of the gain matrix requires the user to define three matrices: $S_L(t_0)$, M, and $\overline{Q}$. Initial condition $S_L(t_0)$ is defined as δI for some scalar δ and an $n \times n$ identity matrix. We decided to make M constant and defined it as ηI for some scalar η and an $m \times m$ identity matrix. Recall, $Q_L = \overline{Q} + C^T M C$ for the full-order observer, $Q_L = \overline{Q} + A_{R_{12}}^T M A_{R_{12}}$ for the reduced-order observer, and $Q_L = \overline{Q} + A_{\overline{R}_{12}}^T M A_{\overline{R}_{12}}$ for the maximally reduced observer. $\overline{Q}$ is taken to be ξI, with ξ a scalar. Although Q_L is constant for the full-order observer since our choices for output matrix are time-invariant, Q_L may not be constant for the other observers if $A_{R_{12}}$ or $A_{\overline{R}_{12}}$ varies with time. The selection of δ, η, and ξ has been trial and error, but for the following results, $\delta = \xi = 0.1$ unless otherwise noted. The choices of η vary and are specified with each observer. The construction of the reduced-order and maximally reduced observers in the time-varying case also requires the selection of a nonsingular matrix P_2. Since this matrix may be constant, we selected $P_2 = I$.

Case 1 (output matrix C_a) begins by assessing the full-order observers constructed using the SLSC and the ASC. The consequences from $\|\Phi(t, 0)\|$ failing to go to zero are realized while attempting to construct the full-order observer using the LSC. This case also constructs reduced-order observers using the stabilized completions.

Case 2 (output matrix C_b) examines the full-order, reduced-order, and maximally reduced observers constructed using the SLSC.

7.1.1   Case 1: C_a with Positive Resistors

Pairs $(\tilde{A}, C_a)_{SL}$, $(\tilde{A}, C_a)_A$, and $(\tilde{A}, C_a)_L$ are observable. Figure 7 presents the estimation error for the full-order observer constructed using the stabilized least squares completion (SLSC/FO observer) when $\eta = 1$. The SLSC/FO observer visibly fails to estimate $\tilde{x}$ sufficiently accurately by $t_f = 45$. However, this SLSC/FO observer converges to $\tilde{x}$ with the required accuracy on the extended time interval $t \in [0, 135]$ in Fig. 7b.

Increasing η decreases the amount of time it takes the SLSC/FO observer to converge to $\tilde{x}$, confirmed visually by Fig. 8a and numerically with $e(45) \approx 0$. This larger η also improves the SLSC/FO observer's estimate of $\tilde{x}$, indicated by the smoother difference in Fig. 8a compared with the oscillating difference in Fig. 7. Figure 8b shows the estimation error on a smaller interval to more clearly show the

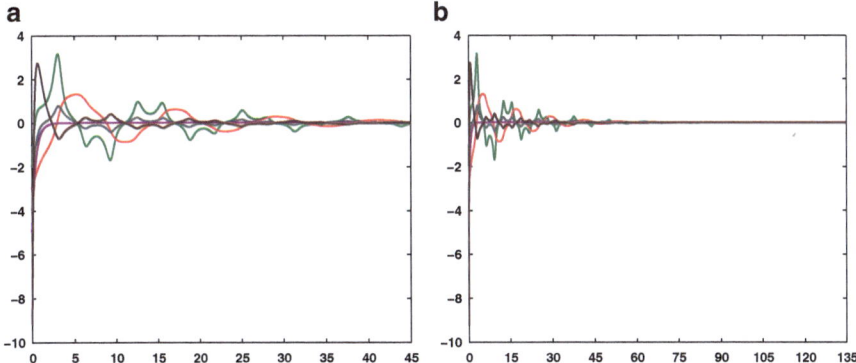

Fig. 7 SLSC/FO observer's $\tilde{x} - \hat{x}$ on two intervals of time. $e \approx 0$ by $t = 135$. (C_a, Λ_{SL}, $\delta = 0.1$, $\eta = 1$, $\xi = 0.1$) (**a**) $t \in [0, 45]$ (**b**) $t \in [0, 135]$

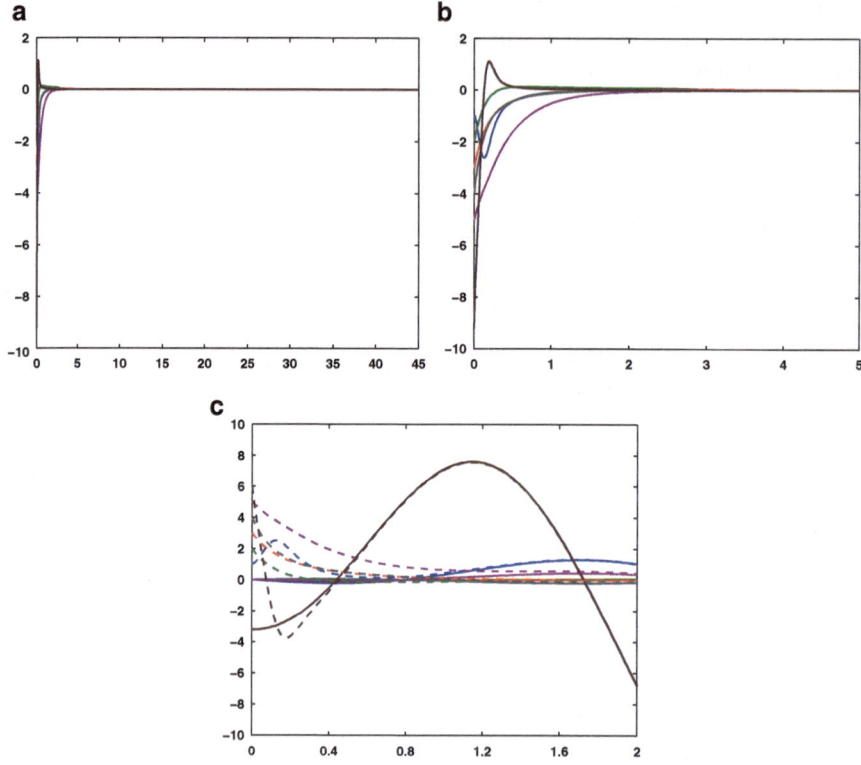

Fig. 8 SLSC/FO observer's $\tilde{x}(t) - \hat{x}(t)$ on two intervals of time; SLSC solution and FO observer plotted together. $e(t) \longrightarrow 0$ by $t_f = 45$. $(C_a, \Lambda_{4SL}, \delta = 0.1, \eta = 100, \xi = 0.1)$ **(a)** $\tilde{x}(t) - \hat{x}(t)$, $t \in [0, 45]$ **(b)** $\tilde{x}(t) - \hat{x}(t)$, $t \in [0, 5]$ **(c)** Sol (*solid*), Obs (*dashed*)

convergence of the different variables. Figure 8c plots both the estimates and the states being estimated.

The estimation error for the full-order observer constructed using the alternative stabilized completion (ASC/FO observer) when $\eta = 100$ appears in Fig. 9a. Note $\tilde{x} - \hat{x}$ now converges to zero monotonically for the ASC/FO observer. A monotonic estimation error is more common with ASCs than with SLSCs.

The stabilization parameter matrix can influence whether or not a particular combination of gain matrix parameters causes computational difficulties [17]. Different combinations of δ, η, and ξ result in observers with a variety of convergence properties and rates, as well as observers that cannot be constructed due to computational instabilities. Some trial and error is necessary when deciding on the values of δ, η, and ξ for producing an acceptable observer.

As mentioned earlier, matrix $\bar{A}$ for the LSC is not exponentially asymptotically stable, which affects the computation of S_L. Using gain matrix parameters $\delta = 0.1$, $\eta = 100$, and $\xi = 0.1$, for which both the SLSC/FO and the ASC/FO observers had

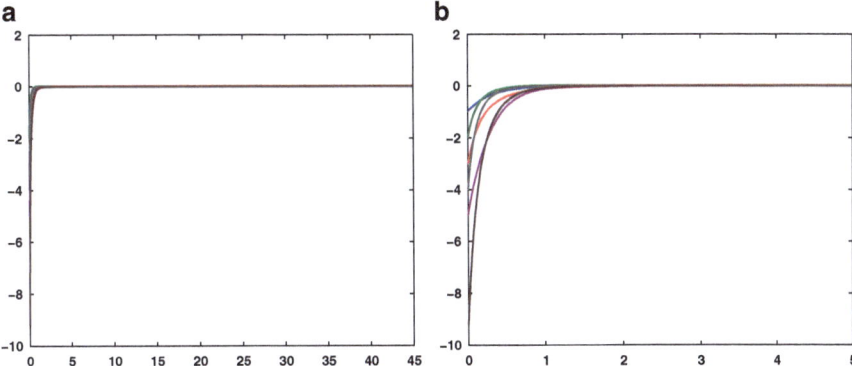

Fig. 9 ASC/FO observer's $\tilde{x} - \hat{x}$ on two intervals of time. $e \approx 0$ by $t_f = 45$. (C_a, Λ_A, $\delta = 0.1$, $\eta = 100$, $\xi = 0.1$) (**a**) $t \in [0, 45]$ (**b**) $t \in [0, 5]$

sufficiently small estimation errors by the desired final time, the code for the full-order observer constructed using the least squares completion (LSC/FO observer) ends prematurely at $t = 10.36$ due to what we refer to as the solver's integration tolerance error. At this time, the error tolerances of the numerical integrator can no longer be met as the largest eigenvalue of S_L has blown up to 5.32×10^{12}. The same error did occur while attempting to construct the SLSC/FO observer with $\delta = 1$, $\eta = 100$, and $\xi = 1$, but decreasing δ to 0.1 successfully controlled the growth of S_L. Alternatively, letting $\overline{Q} = \xi I$ for a smaller value of ξ reduces the effect of the rightmost term in the Riccati equation

$$S'_L = \tilde{A} S_L + S_L \tilde{A}^T + S_L \overline{Q} S_L, \tag{7.5}$$

which also slows the growth of S_L without minimizing the influence of η in $L = \frac{1}{2} S_L C_a^T M$. Thus, for Case 1, in addition to delaying the solver's integration tolerance error, decreasing ξ instead of δ produces observers with better convergence properties.

However, even with this understanding of how to manipulate the gain matrix parameters to obtain desired results, the construction of the LSC/FO observer is not as straightforward as the construction of the two full-order observers using completions with stabilized additional dynamics. The attempted gain matrix parameter combinations failed to produce an LSC/FO observer that converged to $\tilde{x}$ with the required accuracy by the desired final time. Even if such a combination exists, the program's run time becomes a concern. When constructing a full-order observer to estimate the state of the circuit example using output matrix C_a, the LSC/FO observer is not recommended.

Gain matrix parameter η can be taken as 1 for the reduced-order observer constructed using the stabilized least squares completion (SLSC/RO observer) and for the reduced-order observer constructed using the alternative stabilized

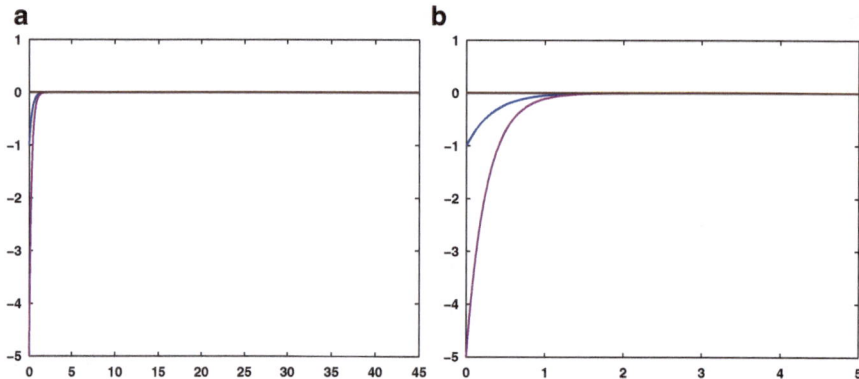

Fig. 10 ASC/RO observer's $\tilde{x} - \hat{x}$ on two intervals of time. The order of the observer is reduced from 6 to 2. $(C_a, \Lambda_A, \delta = 0.1, \eta = 1, \xi = 0.1)$ (**a**) $t \in [0, 45]$ (**b**) $t \in [0, 5]$

completion (ASC/RO observer). These observers provide similar estimates of the circuit example's state. The ASC/RO observer's estimation error is shown in Fig. 10.

7.1.2   Case 2: C_b with Positive Resistors

In Case 1, increasing η either improved or maintained the convergence rate of the SLSC/FO observer depending on the state variable being considered. However, this influence on convergence cannot be assumed for all cases. For observable pair $\left(\tilde{A}, C_b\right)_{SL}$, increasing η decreases the SLSC/FO observer's convergence rate in four of the state variables. Although the gain matrix M should be chosen large, there may be an upper bound on how large M can be if there is a time by which the observer needs to converge. Further investigation into a possible upper bound of M is required.

When considering output matrix C_b, increasing η does not affect the SLSC/RO observer the same way it does the SLSC/FO observer. Increasing η for the SLSC/RO observer with output matrix C_b decreases the amount of time it takes the estimation error to meet the desired estimation tolerance. However, the magnitude of the difference before convergence increases as shown in Fig. 11.

Figure 12 displays the estimation error for the maximally reduced observer constructed using the stabilized least squares completion (SLSC/MR observer) when $\eta = 1$. At first glance, the oscillations may make the estimate look unsatisfactory, but the vertical scale reveals this order 1 observer provides a close estimate of $\tilde{x}$.

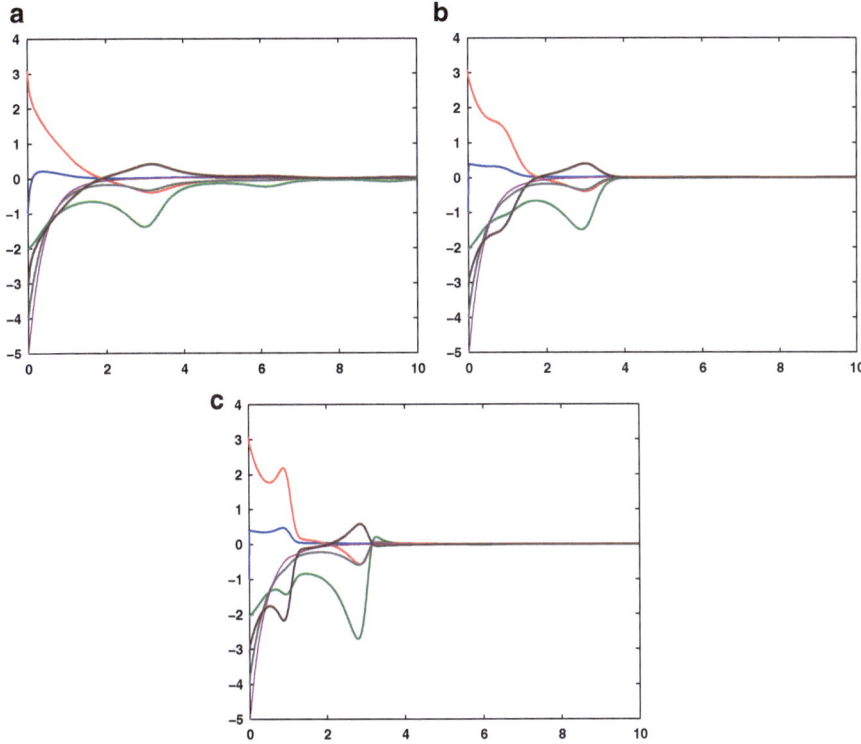

Fig. 11 SLSC/RO observers' $\tilde{x} - \hat{x}$ as η is varied. For output matrix C_b, increasing η exchanges an estimate's accuracy for faster convergence. (Λ_{SL}, $\delta = 0.1$, $\xi = 0.1$) (**a**) $\eta = 1$ (**b**) $\eta = 10$ (**c**) $\eta = 100$

7.2 Example with Negative Resistors

Switching the resistors from positive to negative adds energy into the system, affecting the computation of the gain matrix and, hence, the ability to construct the observers. The linear time-varying DAE for this section is the same as (7.2) except for Eqs. (7.2d) and (7.2e), which are now

$$-e_1 + e_2 + (4 + 2\sin(t)) i_{r_1} = 0$$

$$e_1 + (2 + \sin(t)) i_{r_2} = 0.$$

Output matrices C_a from (7.3) and $C_c = \begin{bmatrix} 0 & 0 & 0 & 1 & 0 & 0 \end{bmatrix}$ are selected to construct the output equations with this negative resistor system. Output matrix C_c corresponds with output matrix C_f in [17]. This modified linear time-varying DAE is observable for each of these output matrices [42].

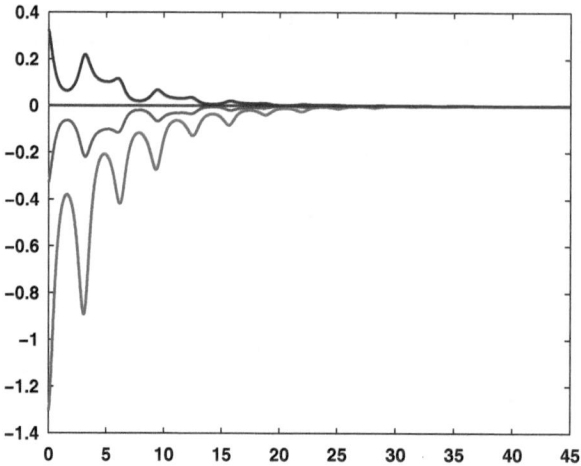

Fig. 12 SLSC/MR observer's $\tilde{x} - \hat{x}$. $(C_b, \Lambda_{SL}, \delta = 0.1, \eta = 1, \xi = 0.1)$

Changing the resistors does not alter the consistent initial conditions but does modify $\mathcal{G}\mathcal{F}$ from the constraints characterizing the solution manifold. $\tilde{x}(0)$ remains $\begin{bmatrix} 0 & 0 & 0 & 0 & 0 & -16/5 \end{bmatrix}^T$. The initial condition for the full-order observer is again taken as $\hat{x}(0) = \begin{bmatrix} 1 & 2 & 3 & 4 & 5 & 6 \end{bmatrix}^T$ and is transformed accordingly for the other two observers. With the resistors being negative instead of positive,

$$(\mathcal{G}\mathcal{F})(t) = \begin{bmatrix} -1 & 1 \, 0 \, (4 + 2\sin(t)) & 0 & 0 \\ 1 & 0 \, 0 & 0 & (2 + \sin(t)) & 0 \\ 1 & 0 \, 0 & 0 & 0 & 0 \\ -(1/3)\sin(t/3) \, 0 \, 0 & -1 & 1 & -1 \end{bmatrix}. \quad (7.6)$$

Note, $\mathcal{G}\mathcal{B}$ for this example with negative resistors is unchanged from the positive resistor case. In order to distinguish notation between the two time-varying examples, an underline is added to identify the pairs for this example with negative resistors: $(\cdot, \cdot)_{\underline{SL}}$ for the SLSC, $(\cdot, \cdot)_{\underline{A}}$ for the ASC, and $(\cdot, \cdot)_{\underline{L}}$ for the LSC. The selection of the stabilization parameter matrices for the SLSC $(\Lambda_{\underline{SL}})$ and for the ASC $(\Lambda_{\underline{A}})$ is especially delicate since energy is entering the system. Following an examination of different possibilities we took $\Lambda_{\underline{SL}} = \text{diag}\{3.5, 3.0, 2.5, 2.0, 1.5, 1.0\}$ and $\Lambda_{\underline{A}} = \text{diag}\{8.0, 7.0, 6.0, 5.0\}$.

In this section's figures, the color and state variable combinations remain the same as those listed in Sect. 7.1. Figure 13 displays the solution of the system with negative resistors that the observers are attempting to estimate.

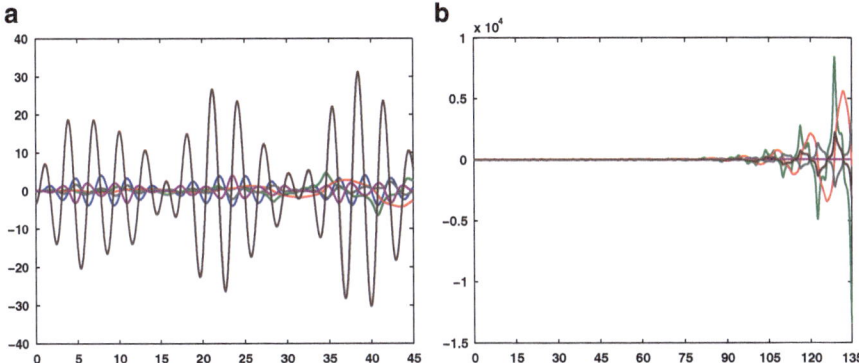

Fig. 13 The observers discussed in Sect. 7.2 estimate the solution of the system with negative resistors, seen here on two intervals of time (**a**) $t \in [0, 45]$ (**b**) $t \in [0, 135]$

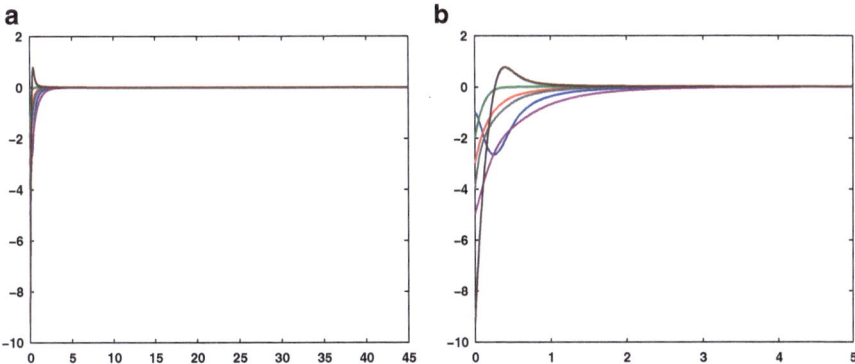

Fig. 14 SLSC/FO observer's $\tilde{x} - \hat{x}$ on two intervals of time. $e \approx 0$ by $t_f = 45$. (C_a, $\Lambda_{\underline{SL}}$, $\delta = 0.1$, $\eta = 100$, $\xi = 0.001$) (**a**) $t \in [0, 45]$ (**b**) $t \in [0, 5]$

7.2.1 Case 3: C_a with Negative Resistors

Figure 14 presents the estimation error on two intervals for the SLSC/FO observer when $\delta = 0.1$, $\eta = 100$, and $\xi = 0.001$. On the $t \in [0, 45]$ interval, the observer appears to converge to the completion by the desired final time, while the $t \in [0, 5]$ interval shows the estimation error monotonically converges to zero in four of the six state variables. Pulling from generalizations formed during plot analysis for the example with positive resistors, this observer's convergence properties are indicative of an η that is larger than necessary for $e(45) \in (-10^{-3}, 10^{-3})$ to occur. Figure 15b,c display the estimation errors for the SLSC/FO observers when $\eta = 10$ and $\eta = 1$, respectively. The results for the ASC/FO observer are similar.

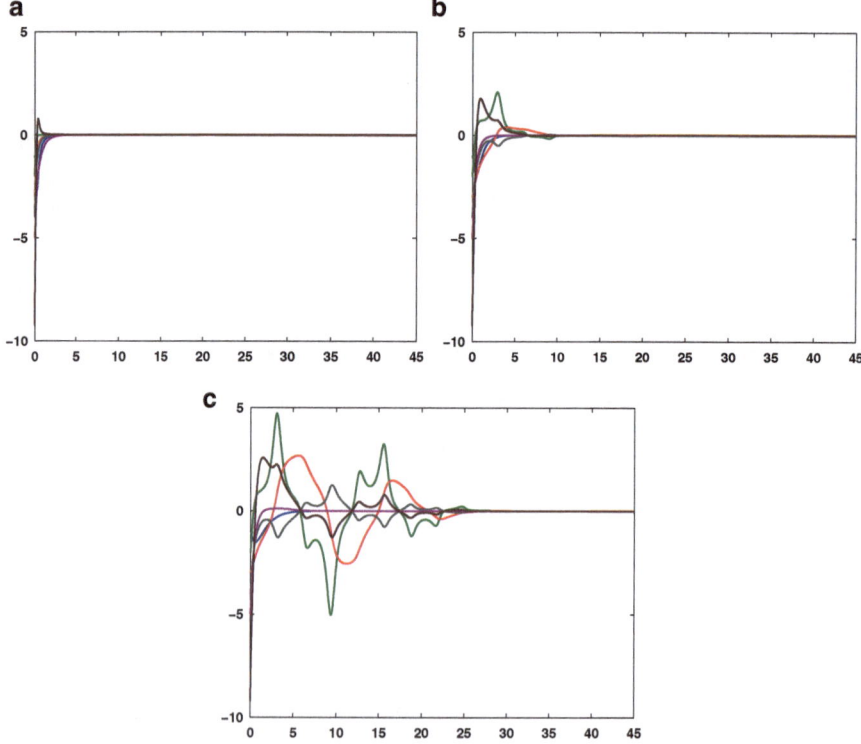

Fig. 15 SLSC/FO observers' $\tilde{x} - \hat{x}$ as η is varied. $e \approx 0$ by $t_f = 45$ even as η is decreased. (C_a, Λ_{SL}, $\delta = 0.1$, $\xi = 0.001$) (**a**) $\eta = 100$ (**b**) $\eta = 10$ (**c**) $\eta = 1$

The stabilized completions are transformed to construct the reduced-order observer using transformation matrix $R = \begin{bmatrix} C_a \\ C_R \end{bmatrix}$, where C_R equals $\begin{bmatrix} 0 & 0 & 0 & 0 & 1 & 0 \\ 1 & 0 & 0 & 0 & 0 & 0 \end{bmatrix}$ for output matrix C_a. The reduced-order systems of both stabilized completions with output matrix C_a are observable on $t \in (0, 135]$; that is, the observability Gramians of pairs $(A_{R_{22}}, A_{R_{12}})_{\text{SL}}$ and $(A_{R_{22}}, A_{R_{12}})_{\text{A}}$ numerically have rank 2 by $t = 0.01$. When the gain matrix parameters are $\delta = 0.1$, $\eta = 1$, and $\xi = 0.001$, the SLSC/RO observer's estimation error is sufficiently small by the desired final time. Due to the structure of output matrix C_a, only two of the state variables have a nonzero estimation error for the interval of time under consideration. Figure 16a shows how well this reduced-order observer estimates state variables e_1 and i_{r_2}. The code constructing this observer terminates at $t = 48.81$ due to the solver's integration tolerance error. The reduced-order system is not exponentially asymptotically stable on the extended time interval, demonstrated by the plot in Fig. 16b of $\|\Phi_R(t,0)\|$, where $\Phi_R(t,0)$ is the state transition matrix of the reduced-order system. This growth translates into an eigenvalue of S_L equaling 3.70×10^{12} by the time the error occurs.

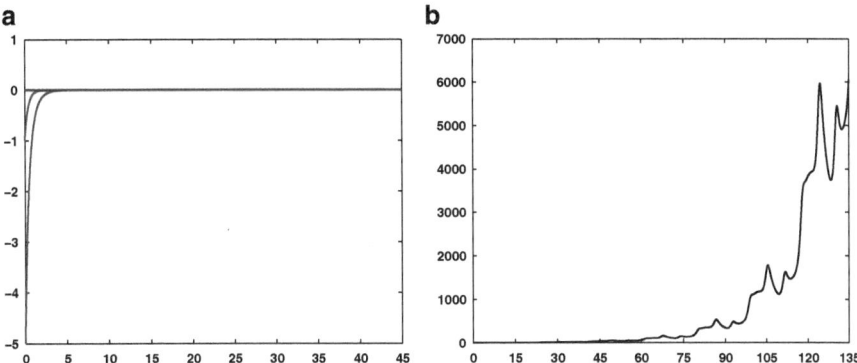

Fig. 16 SLSC/RO observer's $\tilde{x} - \hat{x}$; $e \approx 0$ by $t_f = 45$ but $\|\Phi_R(t,0)\|$ does not go to zero by $t = 135$. (C_a, $\Lambda_{\underline{SL}}$, $\delta = 0.1$, $\eta = 1$, $\xi = 0.001$) (**a**) $\tilde{x} - \hat{x}$ (**b**) $\|\Phi_R(t,0)\|$

The ASC/RO observer with output matrix C_a also converges to the completion by $t_f = 45$ when using the same gain matrix parameters as the SLSC/RO observer. Unlike the SLSC/RO observer, this ASC/RO observer can be constructed to estimate the state on the entire extended time interval. The singular values of S_L for this completion's reduced-order observer do not grow like they do for the SLSC/RO observer. Based on computational experience, the relaxation of assumptions on S_L to real symmetric, positive semi-definite seems appropriate and has already been suggested in [2, 14] for detectable systems.

7.2.2 Case 4: C_c with Negative Resistors

Neither the SLSC, the ASC, nor the LSC is observable when paired with output matrix C_c. The convergence of $\|\Phi_3(t,0)\|$ to zero, where $\Phi_3(t,0)$ represents an unobservable system's state transition matrix, indicates the stabilized completions are detectable. Therefore, with a suitable combination of gain matrix parameters, the SLSC/FO and ASC/FO observers can be designed to estimate the state. The LSC with output matrix C_c is unobservable, so neither the LSC/FO observer nor the reduced-order observer constructed using the LSC can be designed to estimate $\tilde{x}$.

The graphs of the SLSC/FO observer's estimation errors when $\eta = 50$, $\eta = 100$, and $\eta = 150$ are included in Fig. 17. Increasing η from 50 to 100 reduces the estimation error's oscillatory behavior and improves the observer's convergence rate. However, the convergence rate decreases when η is increased from 100 to 150. Although the estimation errors appear to converge to zero by $t_f = 45$, the numerical measurement $e(45)$ is within $\left(-10^{-3}, 10^{-3}\right)$ for only $\eta = 50$.

For gain matrix parameters $\delta = 0.1$, $\eta = 100$, and $\xi = 0.1$, the SLSC/RO observer appears to converge to the completion by $t_f = 45$ in Fig. 18a, and $e(45) \approx 0$ upholds the visual check for convergence. For the same gain matrix parameters, Fig. 18b reveals the ASC/RO observer is unable to estimate components e_2 and i_l

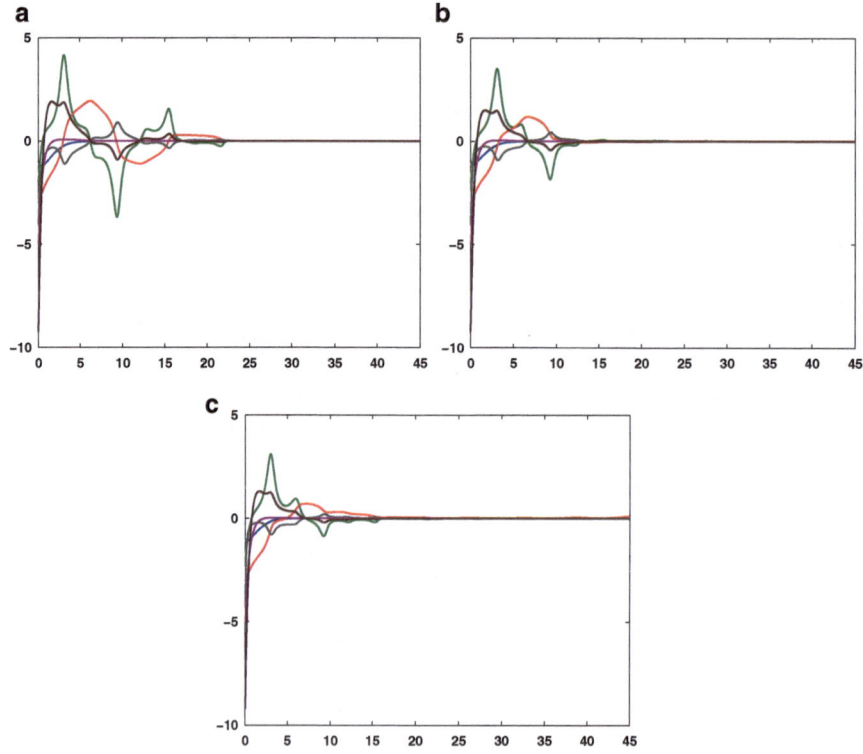

Fig. 17 SLSC/FO observers' $\tilde{x} - \hat{x}$ as η is varied. When $\eta = 50$, $e(45) \in (-10^{-3}, 10^{-3})$. ($C_c$, $\Lambda_{\underline{SL}}$, $\delta = 0.1$, $\xi = 0.001$) (**a**) $\eta = 50$ (**b**) $\eta = 100$ (**c**) $\eta = 150$

by $t_f = 45$. Instead, decreasing η from 100 to 10 when δ and ξ equal 0.1 reduces the amount of time it takes the ASC/RO observer to converge to $\tilde{x}$. An upper bound seems to exist for η if convergence is to occur by a specific time.

While the full-order observer has order 6 and the reduced-order observer has order 5, the maximally reduced observer has order 1. Output matrix C_c is linearly independent from the rows of (7.6), so with $C_x = C_c$, extended output equation $\tilde{y} = C_{\mathcal{E}}\tilde{x}$ provides information on five state variables. After transforming each completion using transformation matrix $\overline{R} = \begin{bmatrix} C_{\mathcal{E}} \\ C_{\overline{R}} \end{bmatrix}$ with $C_{\overline{R}} = [0\ 0\ 1\ 0\ 0\ 0]$, pairs $\left(A_{\overline{R}_{22}}, A_{\overline{R}_{12}}\right)_{\underline{SL}}$, $\left(A_{\overline{R}_{22}}, A_{\overline{R}_{12}}\right)_{\underline{A}}$, and $\left(A_{\overline{R}_{22}}, A_{\overline{R}_{12}}\right)_{\underline{L}}$ are observable. Each of the three observability Gramians has rank 1 for $t \in (0, 135]$.

The differences $\tilde{x} - \hat{x}$ in Fig. 19 are for the maximally reduced observers constructed with gain matrix parameters $\delta = 0.1$, $\eta = 1000$, and $\xi = 0.1$ (the maximally reduced observers constructed using the alternative stabilized completion and the least squares completion are abbreviated ASC/MR and LSC/MR, respectively). The maximally reduced observer's estimate is independent of the completion or

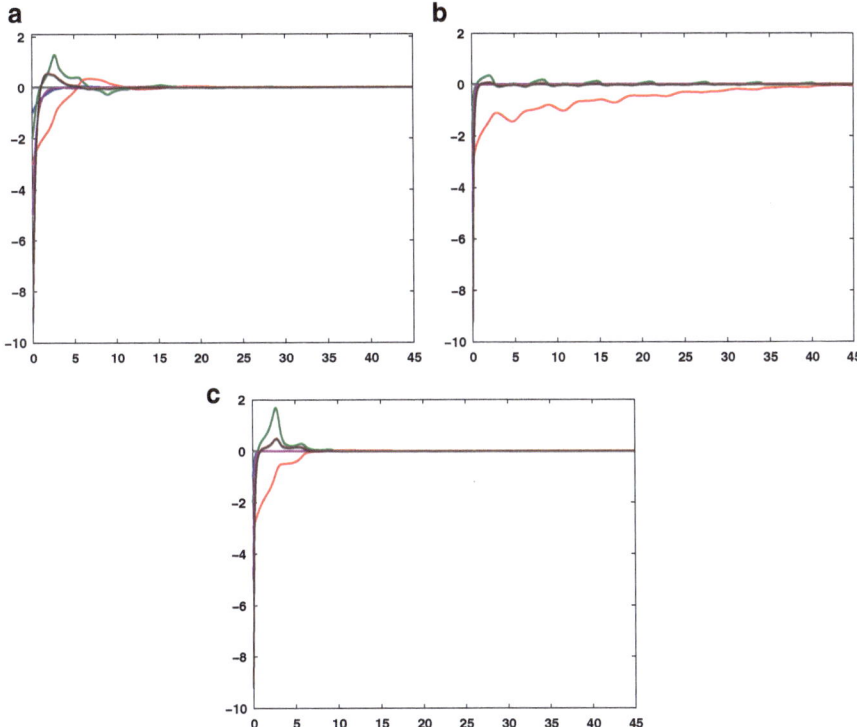

Fig. 18 $\tilde{x} - \hat{x}$ as η and the completions are varied. $e \approx 0$ for the SLSC/RO observer when $\eta = 100$ and for the ASC/RO observer when $\eta = 10$. $(C_c, \Lambda_{\underline{SL}}, \Lambda_{\underline{A}}, \delta = 0.1, \xi = 0.1)$ **(a)** SLSC/RO, $\eta = 100$ **(b)** ASC/RO, $\eta = 100$ **(c)** ASC/RO, $\eta = 10$

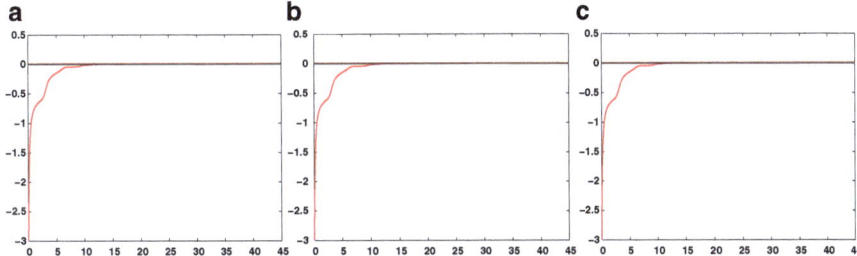

Fig. 19 $\tilde{x} - \hat{x}$ for SLSC/MR, ASC/MR, and LSC/MR observers. $(C_c, \delta = 0.1, \eta = 1000, \xi = 0.1)$ **(a)** SLSC $(\Lambda_{\underline{SL}})$ **(b)** ASC $(\Lambda_{\underline{A}})$ **(c)** LSC

stabilization parameter matrix. Furthermore, this case provides a linear time-varying example where the LSC/MR observer can be constructed even though the full-order and reduced-order systems are unobservable.

8 Extending the Results to Nonlinear Systems

In the first part of this paper we surveyed some results on DAEs and observers. Then we discussed constructing Luenberger-type observers for DAEs based on the construction of completions and stabilized completions. This work all focused on the linear case. An advantage of the completion-based observers is the potential for extension to nonlinear DAEs. We conclude this paper with some comments on this extension. Much of the nonlinear theory on completions necessary for this extension has already been developed and is listed in the cited literature.

In developing such a nonlinear extension, one should expect to have local regions of convergence for a nonlinear system's stabilization results although global convergence is sometimes possible. This local observer construction contrasts with the global convergence we saw for the linear case. In principle, the completion can be stabilized as before by doing a weighted least squares solution of the derivative array. However, instead of getting a formula involving a generalized inverse, we get a nonlinear equation that is the limit of a weighted generalized Newton iterative process.

Another difference with the linear time-invariant case, which is shared with the linear time-varying case, is $\lim_{t \to \infty} G(x, u, t) = 0$, where $G = 0$ are the constraints defining the solution manifold, instead of the stabilized completions' additional dynamics converging to the solution manifold. This limiting condition does not imply that the additional dynamics converge to solutions of the original DAE. A simple example is given by $(e^{-3t}x)' = -2(e^{-3t}x)$; the $\lim_{t \to \infty}(e^{-3t}x) = 0$ since $(e^{-3t}x) = e^{-2t}x(0)$. However, x does not go to zero as t goes to infinity. Thus, additional assumptions on the conditioning of subJacobians of the derivative array are needed to have the additional dynamics asymptotically approach the original dynamics.

Acknowledgements Research supported in part by NSF Grants DMS-0907832 and DMS-1209251. The authors would like to thank the reviewers for their many helpful comments and careful reading of each version of this paper. The paper was greatly improved thanks to their efforts.

References

1. Alaviani, S.S., Shafiee, M.: Exponential stability and stabilization of linear time-varying singular systems. In: Proceedings of International Multi Conference of Engineers and Computer Scientists, vol. II. Hong Kong (2009)
2. Anderson, B.D.O., Moore, J.B.: Detectability and stabilizability of time-varying discrete-time linear systems. SIAM J. Control Optim. **19**, 20–32 (1981)
3. Anderson, B.D.O., Moore, J.B.: New results in linear system stability. SIAM J. Control **7**, 398–414 (1969)
4. Anderson, B.D.O., Moore, J.B.: Time-varying version of the lemma of Lyapunov. Electron. Lett. **3**, 293–294 (1967)

5. Antsaklis, P.J., Michel, A.N.: Linear Systems. Birkhäuser, Boston (2006)
6. Aplevich, J.D.: Implicit linear systems. Lecture Notes in Control and Information Sciences, vol. 152. Springer, Berlin (1991)
7. Åslund, J., Frisk, E.: An observer for non-linear differential-algebraic systems. Automatica **42**, 959–965 (2006)
8. Baumgarte, J.: Stabilization of constraints and integrals of motion in dynamical systems. Comput. Meth. Appl. Mech. Eng. **1**, 1–16 (1972)
9. Bay, J.S.: Fundamentals of Linear State Space Systems. WCB/McGraw-Hill, New York (1999)
10. Berger, T., Ilchmann, A.: Zero dynamics of linear time-varying linear systems. Ilmenau University of Technology, Technical Report (2010)
11. Berger, T., Reis, T.: Controllability of linear differential-algebraic systems—a survey. In: Surveys in Differential-Algebraic Equations I, DAE Forum, pp. 1–62. Springer, Heidelberg (2013)
12. Biehn, N., Campbell, S.L., Delebecque, F., Nikoukhah, R.: Observer design for linear time varying descriptor systems: numerical algorithms. In: Proceedings of 37th IEEE Conference on Decision and Control, Tampa, pp. 3801–3806 (1998)
13. Biehn, N., Campbell, S.L., Nikoukhah, R., Delebecque, F.: Numerically constructible observers for linear time-varying descriptor systems. Automatica **37**, 445–452 (2001)
14. Bittanti, S., Bolzern, P., Colaneri, P.: The extended periodic Lyapunov lemma. Automatica **21**, 603–605 (1985)
15. Blajer, W.: Index of differential-algebraic equations governing the dynamics of constrained mechanical systems. Appl. Math. Model. **16**, 70–77 (1992)
16. Blanchini, F.: Eigenvalue assignment via state observer for descriptor systems. Kybernetika **27**, 384–392 (1991)
17. Bobinyec, K.: Observer construction for systems of differential algebraic equations using completions. Ph.D. Dissertation, Department of Mathematics, North Carolina State University, Raleigh (2013)
18. Bobinyec, K., Campbell, S.L., Kunkel, P.: Constructing observers for linear time varying DAEs. In: Proceedings of 51st IEEE Conference on Decision and Control, Maui, pp. 5749–5754 (2012)
19. Bobinyec, K., Campbell, S.L., Kunkel, P.: Full order observers for linear DAEs. In: Proceedings of 50th IEEE Conference on Decision and Control and European Control Conference, Orlando, pp. 4011–4016 (2011)
20. Bobinyec, K., Campbell, S.L., Kunkel, P.: Maximally reduced observers for linear time varying DAEs. In: Proceedings of 2011 IEEE Multi-Conference on Systems and Control, Denver, pp. 1373–1378 (2011)
21. Bobinyec, K., Campbell, S.L., Kunkel, P.: Stabilized completions of differential algebraic equations and the design of observers. In: Proceedings of Neural, Parallel, and Scientific Computations, Atlanta (2010)
22. Boutat, D., Zheng, G., Boutat-Baddas, L., Darouach, M.: Observers design for a class of nonlinear singular systems. In: Proceedings of 51st IEEE Conference on Decision and Control, Maui, pp. 7407–7412 (2012)
23. Brenan, K.E., Campbell, S.L., Petzold, L.R.: Numerical Solution of Initial-Value Problems in Differential-Algebraic Equations. SIAM, Philadelphia (1996)
24. Brown, P.N., Hindmarsh, A.C., Petzold, L.R.: Consistent initial condition calculation for differential-algebraic systems. SIAM J. Sci. Comput. **19**, 1495–1512 (1998)
25. Byrne, G.D., Ponzi, P.R.: Differential-algebraic systems, their applications and solutions. Comput. Chem. Eng. **12**, 377–382 (1988)
26. Campbell, S.L.: A computational method for general higher index nonlinear singular systems of differential equations. In: Proceedings of 12th IMACS World Congress on Scientific Computation, Paris, pp. 178–180 (1988)
27. Campbell, S.L.: A general form for solvable linear time varying singular systems of differential equations. SIAM J. Math. Anal. **18**, 1101–1115 (1987)

28. Campbell, S.L.: High-index differential algebraic equations. Mech. Struct. Mach. **23**, 199–222 (1995)
29. Campbell, S.L.: Index two linear time varying singular systems of differential equations. Circuits Syst. Sig. Process **5**, 97–107 (1986)
30. Campbell, S.L.: Least squares completions for nonlinear differential algebraic equations. Numer. Math. **65**, 77–94 (1993)
31. Campbell, S.L.: Least squares completions of nonlinear higher index Hessenberg DAEs. In: Computational and Applied Mathematics, II: Differential Equations: Selected and Revised Papers from the IMACS 13th World Congress, Dublin, Ireland, pp. 117–126. North-Holland, New York (1992)
32. Campbell, S.L.: Least squares completions of nonlinear index three Hessenberg DAEs. In: Proceedings of 13th IMACS World Congress on Computation and Applied Mathematics, Dublin, Ireland, pp. 1145–1148 (1991)
33. Campbell, S.L.: Local realizations of time varying descriptor systems. In: Proceedings of 26th IEEE Conference on Decision and Control, Los Angeles, pp. 1129–1130 (1987)
34. Campbell, S.L.: Uniqueness of completions for linear time varying differential algebraic equations. Lin. Alg. Appl. **161**, 55–67 (1992)
35. Campbell, S.L., Griepentrog, E.: Solvability of general differential algebraic equations. SIAM J. Sci. Comput. **16**, 257–270 (1995)
36. Campbell, S.L., Holte, L.E.: Eigenvalue placement in completions of DAEs. Electron. J. Linear Algebra **26**, 520–534 (2013)
37. Campbell, S.L., Hollenbeck, R.: Automatic differentiation and implicit differential equations. In: Computational Differentiation: Techniques, Applications, and Tools, pp. 215–227. SIAM, Philadelphia (1996)
38. Campbell, S.L., Kunkel, P.: Completions of nonlinear DAE flows based on index reduction techniques and their stabilization. J. Comput. Appl. Math. **233**, 1021–1034 (2009)
39. Campbell, S.L., Leimkuhler, B.: Differentiation of constraints in differential-algebraic equations. Mech. Struct. Mach. **19**, 19–39 (1991)
40. Campbell, S.L., März, R.: Direct transcription solution of high index optimal control problems and regular Euler–Lagrange equations. J. Comput. Appl. Math. **202**, 186–202 (2007)
41. Campbell, S.L., Meyer, C.D.: Generalized Inverses of Linear Transformations. SIAM, Philadelphia (2009)
42. Campbell, S.L., Terrell, W.J.: Observability of linear time-varying descriptor systems. SIAM J. Matrix Anal. Appl. **12**, 484–496 (1991)
43. Campbell, S.L., Delebecque, F., Nikoukhah, R.: Observer design for linear time varying descriptor systems. In: Proceedings of Control Industrial Systems (CIS97), Belfort, pp. 507–512 (1997)
44. Campbell, S.L., Kelley, C.T., Yeomans, K.D.: Consistent initial conditions for unstructured higher index DAEs: a computational study. In: Proceedings of Computational Engineering in Systems Applications, Lille, pp. 416–421 (1996)
45. Campbell, S.L., Kunkel, P., Mehrmann, V.: Regularization of linear and nonlinear descriptor systems. In: Control and Optimization with Differential-Algebraic Constraints, pp. 17–36. SIAM, Philadelphia (2012)
46. Campbell, S.L., Moore, E., Zhong, Y.: Utilization of automatic differentiation in control algorithms. IEEE Trans. Autom. Control **39**, 1047–1052 (1994)
47. Campbell, S.L., Nichols, N.K., Terrell, W.J.: Duality, observability, and controllability for linear time-varying descriptor systems. Circuits Syst. Sig. Process **10**, 455–470 (1991)
48. Campbell, S.L., Hollenbeck, R., Yeomans, K., Zhong, Y.: Mixed symbolic-numerical computations with general DAEs I: System properties. Numer. Algorithm. **19**, 73–83 (1998)
49. Chahlaoui, Y., Van Dooren, P.: Estimating gramians of large-scale time-varying systems. In: Proceedings of 15th IFAC World Congress, vol. 2440. Barcelona (2002)
50. Chai, W., Loh, N.K.: Design of minimal-order state observers for time-varying multivariable systems. Int. J. Syst. Sci. **23**, 581–592 (1992)

51. Chai, W., Loh, N.K., Hu, H.: Observer design for time-varying systems. Int. J. Syst. Sci. **22**, 1177–1196 (1991)
52. Clark, K.D.: A structural form for higher-index semistate equations I: theory and applications to circuit and control theory. Lin. Alg. Appl. **98**, 169–197 (1988)
53. Corradini, M.L., Cristofaro, A., Pettinari, S.: Robust FDI filters and fault sensitivity analysis in continuous-time descriptor systems. In: Proceedings of 51st IEEE Conference on Decision and Control, Maui, pp. 1220–1225 (2012)
54. Dafis, C.J., Nwankpa, C.O.: Addressing nonlinear observability issues in power systems. In: Proceedings of 14th Power Systems Computation Conference, Seville (2002)
55. Dai, L.: Singular control systems. Lecture Notes in Control and Information Science, vol. 118. Springer, Berlin (1989)
56. Darouach, M.: Functional observers for linear descriptor systems. In: Proceedings of 17th Mediterranean Conference on Control and Automation, Thessaloniki, pp. 1535–1539 (2009)
57. Darouach, M., Benzaouia, A.: Constrained observer based control for linear singular systems. In: Proceedings of 18th Mediterranean Conference on Control and Automation, Marrakech, pp. 29–33 (2010)
58. Darouach, M., Boutat-Baddas, L.: Observers for a class of nonlinear singular systems. IEEE Trans. Autom. Control **53**, 2627–2633 (2008)
59. Darouach, M., Boutayeb, M.: Design of observers for descriptor systems. IEEE Trans. Autom. Control **40**, 1323–1327 (1995)
60. Darouach, M., Zasadzinski, M., Hayar, M.: Reduced-order observer design for descriptor systems with unknown inputs. IEEE Trans. Autom. Control **41**, 1068–1072 (1996)
61. El-Tohami, M., Lovass-Nagy, V., Mukundan, R.: On the design of observers for generalized state space systems using singular value decomposition. Int. J. Control **38**, 673–683 (1983)
62. Engelsone, A., Campbell, S.L., Betts, J.T.: Direct transcription solution of higher-index optimal control problems and the virtual index. Appl. Numer. Math. **57**, 281–296 (2007)
63. Evard, J.-C.: On the existence of bases of class C^p of the kernel and the image of a matrix function. Lin. Alg. Appl. **135**, 33–67 (1990)
64. Fahmy, M.M., O'Reilly, J.: Observers for descriptor systems. Int. J. Control **49**, 2013–2028 (1989)
65. Fliess, M., Join, C., Sira-Ramírez, H.: Non-linear estimation is easy. Int. J. Model. Identif. Control **4**, 12–27 (2008)
66. Fliess, M., Lévine, J., Rouchon, P.: Index of an implicit time-varying linear differential equation: a noncommutative linear algebraic approach. Lin. Alg. Appl. **186**, 59–71 (1993)
67. Fliess, M., Lévine, J., Martin, P., Rouchon, P.: Implicit differential equations and Lie-Bäcklund mappings. In: Proceedings of 34th IEEE Conference on Decision and Control, New Orleans, pp. 2704–2709 (1995)
68. Friedland, B.: A nonlinear observer for estimating parameters in dynamic systems. Automatica **33**, 1525–1530 (1997)
69. Führer, C., Leimkuhler, B.J.: Numerical solution of differential-algebraic equations for constrained mechanical motion. Numer. Math. **59**, 55–69 (1991)
70. Gantmacher, F.R.: The Theory of Matrices, vol. 2. AMS Chelsea Publishing, Providence (2000)
71. Gao, Z., Wang, H.: Descriptor observer approaches for multivariable systems with measurement noises and application in fault detection and diagnosis. Syst. Control Lett. **55**, 304–313 (2006)
72. Garcia, G., Peres, P.L.D., Tarbouriech, S.: Necessary and sufficient numerical conditions for asymptotic stability of linear time-varying systems. In: Proceedings of 47th IEEE Conference on Decision and Control, Cancun, pp. 5146–5151 (2008)
73. Gopinath, B.: On the control of linear multiple input-output systems. Bell Syst. Tech. J. **50**, 1063–1081 (1971)
74. Hao, F.: Full-order observer design for descriptor systems with delayed state and unknown inputs. In: Proceedings of 25th Chinese Control Conference, Harbin, pp. 765–770 (2006)

75. Hill, D.J., Mareels, I.M.Y.: Stability theory for differential/algebraic systems with application to power systems. IEEE Trans. Circuits Syst. **37**, 1416–1423 (1990)
76. Hou, M., Müller, P.C.: Observer design for descriptor systems. IEEE Trans. Autom. Control **44**, 164–168 (1999)
77. Hou, M., Schmidt, T., Schüpphaus, R., Müller, P.C.: Normal form and Luenberger observer for linear mechanical descriptor systems. J. Dyn. Syst. Meas. Control **115**, 611–620 (1993)
78. Ilchmann, A., Mehrmann, V.: A behavioral approach to time-varying linear systems. Part 2: descriptor systems. SIAM J. Control Optim. **44**, 1748–1765 (2005)
79. Ilchmann, A., Mueller, M.: Time-varying linear systems: relative degree and normal form. IEEE Trans. Autom. Control **52**, 840–851 (2007)
80. Johnson, G.W.: A deterministic theory of estimation and control. IEEE Trans. Autom. Control **14**, 380–384 (1969)
81. Kalman, R.E., Bucy, R.S.: New results in linear filtering and prediction theory. J. Basic Eng. **83**, 95–108 (1961)
82. Kidane, N., Yamashita, Y., Nishitani, H.: Observer based I/O-linearizing control of high index DAE systems. In: Proceedings of American Control Conference, Denver, pp. 3537–3542 (2003)
83. Kisačanin, B., Agarwal, G.C.: Linear Control Systems: with Solved Problems and MATLAB Examples. Kluwer Academic/Plenum Publishers, New York (2001)
84. Koenig, D.: Unknown input proportional multiple-integral observer design for linear descriptor systems: application to state and fault estimation. IEEE Trans. Autom. Control **50**, 212–217 (2005)
85. Koenig, D., Mammar, S.: Design of proportional-integral observer for unknown input descriptor systems. IEEE Trans. Autom. Control **47**, 2057–2062 (2002)
86. Kumar, A., Daoutidis, P.: Control of Nonlinear Differential Algebraic Equation Systems. Chapman and Hall/CRC, New York (1999)
87. Kunkel, P., Mehrmann, V.: Canonical forms for linear differential-algebraic equations with variable coefficients. J. Comput. Appl. Math. **56**, 225–251 (1994)
88. Kunkel, P., Mehrmann, V.: Differential-Algebraic Equations: Analysis and Numerical Solution. European Mathematical Society, Zürich (2006)
89. Kunkel, P., Mehrmann, V.: A new class of discretization methods for the solution of linear differential-algebraic equations with variable coefficients. SIAM J. Numer. Anal. **33**, 1941–1961 (1996)
90. Kunkel, P., Mehrmann, V.: Regular solutions of nonlinear differential-algebraic equations and their numerical determination. Numer. Math. **79**, 581–600 (1998)
91. Kunkel, P., Mehrmann, V.: Smooth factorizations of matrix valued functions and their derivatives. Numer. Math. **60**, 115–131 (1991)
92. Kunkel, P., Mehrmann, V., Rath, W., Weickert, J.: A new software package for linear differential-algebraic equations. SIAM J. Sci. Comput. **18**, 115–138 (1997)
93. Lamour, R.: Index determination and calculation of consistent initial values for DAEs. Comput. Math. Appl. **50**, 1125–1140 (2005)
94. Lamour, R., März, R., Tischendorf, C.: Differential-Algebraic Equations: A Projector Based Analysis. Springer, Heidelberg (2013)
95. LeFevre, V.C.: The design and implementation of a time-varying observer and an EMA controller for linear systems. In: Proceedings of Twenty-Ninth Southeastern Symposium on System Theory, pp. 297–301. IEEE Conference Publications (1997)
96. Leimkuhler, B., Petzold, L.R., Gear, C.W.: Approximation methods for the consistent initialization of differential-algebraic equations. SIAM J. Numer. Anal. **28**, 205–226 (1991)
97. Lewis, F.L.: A survey of linear singular systems. Circuits Syst. Sig. Process **5**, 3–36 (1986)
98. Li, X., Zhou, K.: A time domain approach to robust fault detection of linear time-varying systems. Automatica **45**, 94–102 (2009)
99. Liu, C.-S., Peng, H.: Inverse-dynamics based state and disturbance observers for linear time-invariant systems. J. Dyn. Syst. Meas. Control **124**, 375–381 (2002)

100. Lovass-Nagy, V., Miller, R.J., Mukundan, R.: On the application of matrix generalized inverses to the design of observers for time-varying and time-invariant linear systems. IEEE Trans. Autom. Control **AC-25**, 1213–1218 (1980)
101. Luenberger, D.G.: Observers for multivariable systems. IEEE Trans. Autom. Control **AC-11**, 190–197 (1966)
102. Luenberger, D.G.: Observing the state of a linear system. IEEE Trans. Mil. Electron. **8**, 74–80 (1964)
103. Luenberger, D.G.: Time-invariant descriptor systems. Automatica **14**, 473–480 (1978)
104. Luo, J., Pongratananukul, N., Abu-Qahouq, J.A., Batarseh, I.: Time-varying current observer with parameter estimation for multiphase low-voltage high-current voltage regulator modules. In: Proceedings of Eighteenth Annual IEEE Applied Power Electronics Conference and Exposition, Miami Beach, pp. 444–450 (2003)
105. März, R., Riaza, R.: Linear differential-algebraic equations with properly stated leading term: Regular points. J. Math. Anal. Appl. **323**, 1279–1299 (2006)
106. Menold, P.H., Findeisen, R., Allgöwer, F.: Finite time convergent observers for linear time-varying systems. In: Proceedings of 11th IEEE Mediterranean Conference on Control and Automation, Rhodes (2003)
107. Minamide, N., Arii, N., Uetake, Y.: Design of observers for descriptor systems using a descriptor standard form. Int. J. Control **50**, 2141–2149 (1989)
108. Newcomb, R.W.: The semistate description of nonlinear time-variable circuits. IEEE Trans. Circ. Syst. **CAS-28**, 62–71 (1981)
109. Newcomb, R.W., Dziurla, B.: Some circuits and systems applications of semistate theory. Circuits Syst. Sig. Process **8**, 235–260 (1989)
110. Nguyen, C.C.: Design of reduced-order state estimators for linear time-varying multivariable systems. Int. J. Control **46**, 2113–2126 (1987)
111. Nguyen, C., Lee, T.N.: Design of a state estimator for a class of time-varying multivariable systems. IEEE Trans. Autom. Control **AC-30**, 179–182 (1985)
112. Nikoukhah, R., Campbell, S.L., Delebecque, F.: Observer design for general linear time-invariant systems. Automatica **34**, 575–583 (1998)
113. O'Reilly, J.: Observers for Linear Systems. Academic Press, London (1983)
114. O'Reilly, J., Newmann, M.M.: Minimal-order observer-estimators for continuous-time linear systems. Int. J. Control **22**, 573–590 (1975)
115. Okay, I.: The additional dynamics of the least squares completions of linear differential algebraic equations. Ph.D. Dissertation, Department of Mathematics, North Carolina State University, Raleigh (2008)
116. Okay, I., Campbell, S.L., Kunkel, P.: The additional dynamics of least squares completions for linear differential algebraic equations. Lin. Alg. Appl. **425**, 471–485 (2007)
117. Okay, I., Campbell, S.L., Kunkel, P.: Completions of implicitly defined linear time varying vector fields. Lin. Alg. Appl. **431**, 1422–1438 (2009)
118. Pearson, D.W., Chapman, M.J., Shields, D.N.: Partial singular-value assignment in the design of robust observers for discrete-time descriptor systems. IMA J. Math. Control Inf. **5**, 203–213 (1988)
119. Rabier, P.J., Rheinboldt, W.C.: Nonholonomic Motion of Rigid Mechanical Systems from a DAE Viewpoint. SIAM, Philadelphia (2000)
120. Reger, J., Jouffroy, J.: On algebraic time-derivative estimation and deadbeat state reconstruction. In: Proceedings of 48th IEEE Conference on Decision and Control and 28th Chinese Control Conference, Shanghai, pp. 1740–1745 (2009)
121. Reissig, G.: The index of the standard circuit equations of passive RLCTG-networks does not exceed 2. In: Proceedings of 1998 IEEE International Symposium on Circuits and Systems, IEEE Conference Publications, pp. 419–422 (1998)
122. Rheinboldt, W.C.: On the computation of multi-dimensional solution manifolds of parametrized equations. Numer. Math. **53**, 165–181 (1988)
123. Riaza, R.: Differential-algebraic systems. In: Analytical Aspects and Circuit Applications. Word Scientific, Hackensack (2008)

124. Scott, J.R., Campbell, S.L.: Observer based fault detection and identification in differential algebraic equations. In: Proceedings of ASME Dynamic Systems and Control Conference, Palo Alto, pp. V002T24A001; 10 pages (2013)
125. Scott, J.R., Campbell, S.L.: Observer based fault detection in differential algebraic equations. In: Proceedings of SIAM Control Conference, San Diego, pp. 176–183 (2013)
126. Seal, C.E., Stubberud, A.R.: Canonical forms for multiple-input time-variable systems. IEEE Trans. Autom. Control **14**, 704–707 (1969)
127. Shafai, B., Carroll, R.L.: Design of a minimal-order observer for singular systems. Int. J. Control **45**, 1075–1081 (1987)
128. Shafai, B., Carroll, R.L.: Design of proportional-integral observer for linear time-varying multivariable systems. In: Proceedings of 24th Conference on Decision and Control, Ft. Lauderdale, pp. 597–599 (1985)
129. Shafai, B., Carroll, R.L.: Minimal-order observer designs for linear time-varying multivariable systems. IEEE Trans. Autom. Control **AC-31**, 757–761 (1986)
130. Shin, K.-C., Kabamba, P.T.: Observation and estimation in linear descriptor systems with application to constrained dynamical systems. J. Dyn. Syst. Meas. Control **110**, 255–265 (1988)
131. Silverman, L.M., Meadows, H.E.: Controllability and observability in time-variable linear systems. J. SIAM Control **5**, 64–73 (1967)
132. Stefani, R.T., Shahian, B., Savant, C.J. Jr., Hostetter, G.H.: Design of Feedback Control Systems. Oxford University Press, New York (2002)
133. Sundarapandian, V.: Reduced order observer design for discrete-time nonlinear systems. Appl. Math. Lett. **19**, 1013–1018 (2006)
134. Tai, H.-M.: Equivalent characterizations of detectability and stabilizability for a class of linear time-varying systems. Syst. Control Lett. **8**, 425–428 (1987)
135. Tai, H.-M.: Linear time-varying systems: algebraic structure, system properties and control. Ph.D. Dissertation, Texas Tech University (1987)
136. Tanwani, A., Trenn, S.: On observability of switched differential-algebraic equations. In: Proceedings of 49th IEEE Conference in Decision and Control, pp. 5658–5661 (2010)
137. Terrell, W.J.: Observability of nonlinear differential algebraic systems. Circuits Syst. Sig. Process **16**, 271–285 (1997)
138. Terrell, W.J.: The output-nulling space, projected dynamics, and system decomposition for linear time-varying singular systems. SIAM J. Control Optim. **32**, 876–889 (1994)
139. Trumpf, J.: Observers for linear time-varying systems. Lin. Alg. Appl. **425**, 303–312 (2007)
140. Van Dooren, P.: Reduced order observers: a new algorithm and proof. Syst. Control Lett. **4**, 243–251 (1984)
141. Vemuri, A.T., Polycarpou, M.M., Ciric, A.R.: Fault diagnosis of differential-algebraic systems. IEEE Trans. Syst. Man Cybern. Part A Syst. Humans **31**, 143–152 (2001)
142. Verhaegen, M.H., Van Dooren, P.: A reduced order observer for descriptor systems. Syst. Control Lett. **8**, 29–37 (1986)
143. von Wissel, D., Nikoukhah, R., Campbell, S.L., Delebecque, F.: Nonlinear observer design using implicit system descriptions. In: Proceedings of Computational Engineering in Systems Applications, Lille, pp. 404–409 (1996)
144. Wang, C.-J.: Controllability and observability of linear time-varying singular systems. IEEE Trans. Autom. Control **44**, 1901–1905 (1999)
145. Wu, A.-G., Duan, G.-R.: Design of generalized PI observers for descriptor linear systems. IEEE Trans. Circuits Syst. Regul. Pap. **53**, 2828–2837 (2006)
146. Wu, M.Y.: A note on stability of linear time-varying systems. IEEE Trans. Autom. Control **19**, 162 (1974)
147. Xu, A., Zhang, Q.: Fault detection and isolation based on adaptive observers for linear time varying systems. In: Proceedings of 15th IFAC World Congress, Barcelona (2002)
148. Yang, C., Zhang, Q., Chai, T.: Observer design for a class of nonlinear descriptor systems. In: Proceedings of Joint 48th IEEE Conference on Decision and Control and 28th Chinese Control Conference, Shanghai, pp. 8232–8237 (2009)

149. Yüksel, Y.Ö., Bongiorno, J.J. Jr.: Observers for linear multivariable systems with applications. IEEE Trans. Autom. Control **AC-16**, 603–613 (1971)
150. Zimmer, G., Meier, J.: On observing nonlinear descriptor systems. Syst. Control Lett. **32**, 43–48 (1997)

DAEs in Model Reduction of Chemical Processes: An Overview

Prodromos Daoutidis

Abstract Differential algebraic equation (DAE) systems of semi-explicit type arise naturally in the modeling of chemical engineering processes. The differential equations typically arise from dynamic conservation equations, while the algebraic constraints from constitutive equations, rate expressions, equilibrium relations, stoichiometric constraints, etc. Of particular interest are DAE systems of high index, i.e., those for which the algebraic constraints are singular and cannot be eliminated through appropriate substitutions. In this paper we provide an overview of generic classes of fast-rate chemical process models, which in the limit of infinitely fast rates, generate equilibrium-based models that are high-index DAE systems. These slow approximations of multi-time-scale systems can be obtained rigorously via singular perturbations. Two classes of nonstandard singularly perturbed systems leading to high-index DAEs are identified and analyzed. The first class arises in processes with fast rates of reaction or transport. We focus in particular on chemical reaction systems which often exhibit dynamics in multiple time-scales due to reaction rate constants that vary over widely different orders of magnitude. For such systems, we describe the sequential application of singular perturbations arguments for deriving nonlinear DAE models of the dynamics in the different time-scales. The second class arises in the modeling of tightly integrated process networks, i.e., those with large rates of recovery and recycle of material or energy. For such systems we describe a similar model reduction method for deriving DAE models of the slow network dynamics and discuss control-relevant considerations.

Keywords Chemical processes • Chemical reaction systems • Model reduction • Multiple-time-scale systems • Nonstandard singularly perturbed form • Process networks • Singular perturbations

Mathematics Subject Classification (2010) 34C20, 34C40, 34D15, 34E13, 34E15, 93A15, 93C10, 93C30, 93C70, 93C83

P. Daoutidis (✉)
Department of Chemical Engineering and Materials Science, University of Minnesota, Minneapolis, MN 55455, USA
e-mail: daout001@umn.edu

© Springer International Publishing Switzerland 2015   69
A. Ilchmann, T. Reis (eds.), *Surveys in Differential-Algebraic Equations II*,
Differential-Algebraic Equations Forum, DOI 10.1007/978-3-319-11050-9_2

1   Introduction

Differential algebraic equation (DAE) systems of semi-explicit type arise naturally in the modeling of chemical engineering processes. The differential equations typically arise from dynamic balances of mass, energy and momentum, while the algebraic constraints from constitutive equations, rate expressions, equilibrium relations, stoichiometric constraints, etc. There exists a wide variety of chemical process models for which the algebraic constraints are singular [48]. Such DAE systems of high index are known to pose significant challenges in numerical simulation [13, 43, 56, 68] (e.g., consistent choice of initial conditions, numerical inaccuracies in enforcing the constraints etc.) and control [15, 21, 28, 49]. The occurrence of high-index DAE systems in chemical process models has been long documented [14, 27, 54, 59, 61], although often attributed to improper modeling assumptions. There has also been significant recent activity on the analysis, optimization, and control of DAEs in chemical engineering, not necessarily of high index (see, e.g., [10, 16, 22, 23, 25, 36, 65]). A recent broad exposition on DAE optimization and control topics can be found in [11].

Our previous research has identified broad classes of models of processes involving fast rates of mass transfer, heat transfer, reaction, or material/energy flows which, in the limit as these rates become infinite and the rate expressions are replaced by equilibrium relations, give rise to high-index DAE systems. The incentive for considering the limiting, equilibrium behavior of these rate-based models is twofold:

- the rate-based models are modeled by stiff ordinary differential equation (ODEs), which are costly to simulate, whereas the resulting DAE systems result in equivalent ODEs that are non-stiff.
- model-based controllers designed on the basis of stiff rate-based models are highly ill-conditioned and may even lead to instability, whereas those designed on the basis of the non-stiff equilibrium-based models are generally well-conditioned.

Singular perturbations provide a natural mathematical framework to analyze the properties of these fast-rate models. The fast rates are usually associated with large mass/heat transfer coefficients, reaction rate constants, or mass/energy flows. The resulting models exhibit dynamics in two or even multiple time-scales. There exists an extensive literature on the application of singular perturbation theory for the model reduction, analysis, and control of systems with multiple time-scales (see, e.g., [41, 42, 60]). Most of it has focused on two-time-scale systems modeled in the so-called "standard" singularly perturbed form; in this form, the fast and slow variables are explicitly separated due to the presence of a small parameter ϵ (the singular perturbation parameter) that multiplies the time derivative of the vector of fast state variables (see, e.g., [41, 42]). However, modeling a two-time-scale process in the standard singularly perturbed form is a nontrivial task. In some processes, there is a priori knowledge of the variables with slow and fast dynamics (usually these are associated with large and small holdups or heat capacities). This allows modeling such processes directly in the standard singularly perturbed form, by defining ϵ appropriately. However, for chemical processes where the two-time-scale behavior is due to a fast rate of heat/mass transfer, reaction etc., although a singular perturbation parameter can be naturally defined (e.g., the inverse of large heat/mass transfer coefficients, reaction rate coefficients etc.), the corresponding dynamic models are not in the standard singularly perturbed form; equivalently the fast and slow dynamics cannot be associated with distinct state variables.

In what follows, we provide an overview of results on model reduction of two-time-scale and multi-time-scale systems using singular perturbations, emphasizing the connection between high-index DAEs and nonstandard singularly perturbed systems of chemical process models. Two classes of nonstandard singularly perturbed systems leading to high-index DAEs are identified and analyzed. The first class arises in processes with fast rates of reaction or transport. We focus in particular on chemical reaction systems which often exhibit dynamics in multiple time-scales due to reaction rate constants that vary over widely different orders of magnitude. For such systems, we describe the sequential application of singular perturbations for deriving DAE models of the dynamics in the different time-scales. The second class arises in the modeling of tightly integrated process networks, i.e., those with large rates of recovery and recycle of material or energy. For such systems we describe a similar model reduction procedure for deriving DAE models of the slow network dynamics and discuss control-relevant considerations in the context of plant-wide control.

2 Two-Time-Scale Processes

2.1 Standard Singularly Perturbed Form

Consider a two-time-scale system of the following form:

$$\dot{\zeta} = F(\zeta, \eta, u, \epsilon)$$
$$\epsilon \dot{\eta} = G(\zeta, \eta, u, \epsilon) \tag{2.1}$$

where $\zeta \in \mathbb{R}^n$ and $\eta \in \mathbb{R}^m$ are the state variables, $u \in \mathbb{R}^q$ is the vector of manipulated inputs, ϵ is a small positive parameter known as the singular perturbation parameter, and F, G are smooth vector fields of dimensions n and m, respectively. In the limiting case when $\epsilon \to 0$, the m differential equations for η are reduced to a set of algebraic equations $G(\zeta, \eta, u, 0) = 0$, i.e., the dynamic order of the system in Eq. 2.1 degenerates from $n + m$ to n. Early research on two-time-scale systems focused on deriving conditions for regular degeneration in terms of the properties of the Jacobian $(\partial G(\zeta, \eta, u, 0)/\partial \eta)$. Specifically, the singularly perturbed system of Eq. 2.1 degenerates regularly if $0 = G(\zeta, \eta, u, 0)$ has $k \geq 1$

isolated real roots $\eta_i = \sigma_i(\zeta, u), i = 1, \ldots, k$, i.e., the Jacobian $(\partial G(\zeta, \eta, u, 0)/\partial \eta)$ is nonsingular along the solution of the system obtained by setting $\epsilon = 0$ (for details see [30,66]). In this case, the system in Eq. 2.1 is said to be in the standard singularly perturbed form.

A system in the form of Eq. 2.1 which degenerates regularly is characterized by an explicit time-scale separation, with the states η being the fast ones and the states ζ being the slow ones. A time-scale decomposition of the singularly perturbed system of Eq. 2.1 yields reduced-order representations for the slow and fast subsystems. More specifically in the limit $\epsilon \to 0$, the fast dynamics become instantaneous in the slow time-scale t, the corresponding differential equations reduce to algebraic equations, and a model of the slow dynamics can be obtained as:

$$\dot{\zeta} = F(\zeta, \eta, u, 0)$$
$$0 = G(\zeta, \eta, u, 0) \tag{2.2}$$

The DAE system in Eq. 2.2 is of index one and can be readily reduced to an ODE system. The algebraic equations $0 = G(\zeta, \eta, u, 0)$ can be solved for a quasi-steady-state solution $\eta = \sigma(\zeta, u)$, and a representation of the slow subsystem can be obtained as:

$$\dot{\zeta} = F(\zeta, \sigma(\zeta, u), u, 0) \tag{2.3}$$

The slow subsystem of Eq. 2.3 is also referred to as a *reduced* or *quasi-steady-state* subsystem.

A model of the fast dynamics is obtained in a "stretched" fast time-scale $\tau = t/\epsilon$, where the system of Eq. 2.1 takes the following form:

$$\frac{d\zeta}{d\tau} = \epsilon F(\zeta, \eta, u, 0)$$
$$\frac{d\eta}{d\tau} = G(\zeta, \eta, u, 0) \tag{2.4}$$

In the limit $\epsilon \to 0$, the dynamics of the slow variables ζ become negligible and a reduced-order model for the fast dynamics is obtained as:

$$\frac{d\eta}{d\tau} = G(\zeta, \eta, u, 0) \tag{2.5}$$

where the slow variables ζ are "frozen" at their initial condition $\zeta(0)$ and treated as constant parameters. This fast subsystem is also referred to as the *boundary-layer* subsystem. It is often implicit in this characterization that this fast subsystem is stable.

The asymptotic properties of the system in Eq. 2.1 with a small, nonzero ϵ, can be inferred from the analysis of the asymptotic properties of the reduced-order slow (Eq. 2.3) and fast (Eq. 2.5) subsystems [42]. A fundamental result to this end is the one of Tikhonov [66], which states that if the fast subsystem (Eq. 2.5) is locally exponentially stable and the slow subsystem has a unique solution on the time interval $t \in [0, t_1]$, then for a sufficiently small $\epsilon < \epsilon^*$, the solutions of the slow (Eq. 2.3) and fast (Eq. 2.5) subsystems yield $O(\epsilon)$ approximation of the true solution of the system in Eq. 2.1, uniformly for $t \in [0, t_1]$. The result holds on the infinite time interval if the slow subsystem is also locally exponentially stable (for details, see, e.g., [39]). Tikhonov's theorem enables the use of the reduced-order representations of the slow and fast subsystems for analysis purposes, i.e., considering only the behavior of the system in the limit $\epsilon \rightarrow 0$ in the separate time-scales, thereby avoiding the singularity with respect to $\epsilon = 0$.

Within the framework of singular perturbations, controller design is addressed through a combination of separate fast and slow controllers designed to stabilize the fast dynamics, if they are unstable, and achieve desired closed-loop performance objectives on the basis of the slow subsystem which essentially governs the input/output behavior of the system (see, e.g., [17, 40]). Such a controller design approach, involving combination of slow and fast controllers acting in the respective time-scales is often referred to as composite control [41].

2.2 Nonstandard Singularly Perturbed Form

For most chemical processes with fast heat/mass transfer and fast reactions, the fast and slow dynamics cannot be associated with distinct state variables, and the corresponding rate-based dynamic models are not in the standard singularly perturbed form. Early research on nonlinear two-time-scale systems with a small parameter ϵ, not necessarily in the standard singularly perturbed form, focused on studying geometric properties to obtain a coordinate-free characterization of time-scale multiplicity [24]. This characterization was subsequently used to derive necessary and sufficient geometric conditions for the existence of an ϵ-independent change of coordinates that yields a standard singularly perturbed form [58].

A class of nonlinear singularly perturbed systems in nonstandard form that arise as rate-based models of fast-rate chemical processes was identified in [44] and has the following form:

$$\dot{x} = f(x) + g(x)u + \frac{1}{\epsilon}b(x)k(x) \tag{2.6}$$

where $x \in \mathbb{R}^n$ is the vector of state variables, $f(x), k(x)$ are smooth vector fields of dimensions n and p ($p < n$), $b(x), g(x)$ are matrices of dimension $n \times p$ and $n \times q$, respectively, $b(x)$ and $(\partial k(x)/\partial x)$ have full column rank and full row rank, respectively and $k(x)$ is $\mathcal{O}(\epsilon)$. The $1/\epsilon$ term in Eq. 2.6 represents a large parameter in the dynamic model corresponding to a large heat/mass transfer coefficient, large reaction rate constant, etc. The assumption on the rank of $b(x)$ is not restrictive. If its rank is less than p, say $\bar{p}$, then there always exists a nonsingular $p \times p$ matrix $E(x)$ such that:

$$b(x)k(x) = b(x)E(x)E(x)^{-1}k(x) = \left[\bar{b}(x) \ \ 0\right]\bar{k}(x) \tag{2.7}$$

where $\bar{b}(x)$ has full column rank $\bar{p}$. Discarding the $p - \bar{p}$ zero columns and the corresponding components of $\bar{k}(x)$, we obtain a modified system where the new matrix $\bar{b}(x)$ has full column rank. The assumption on the column rank of the Jacobian of $k(x)$ assures that the constraints we obtain in the limit as $\epsilon \to 0$ are linearly independent and that a finite index for the resulting DAE system exists. If it is not satisfied, elementary operations as above can be performed leading to redefined $b(x), k(x)$ terms for which the assumption holds.

In the subsequent discussion we will use $L_b k(x)$ to denote the matrix whose $(i, j)^{th}$ component is the standard Lie (directional) derivative $L_{b_j} k_i(x) = (\partial k_i(x)/\partial x) b_j(x)$, where $b_j(x)$ is the jth column of $b(x)$ and k_i is the ith component of $k(x)$, and $L_f k(x)$ to denote the row vector whose ith element is the Lie derivative $L_f k_i(x)$.

Separate representations of the slow and fast subsystems of the system in Eq. 2.6 can be formally obtained in the limit $\epsilon \to 0$. In the slow time-scale, multiplying Eq. 2.6 by ϵ and considering the limit $\epsilon \to 0$, the following quasi-steady-state constraints are obtained:

$$k_i(x) = 0, i = 1, \dots, p \tag{2.8}$$

which must be satisfied in the slow subsystem. Defining $\lim_{\epsilon \to 0} \dfrac{k_i(x)}{\epsilon} = z_i$ and taking the limit $\epsilon \to 0$ in the system of Eq. 2.6, the following system which describes the slow dynamics of Eq. 2.6 is obtained:

$$\dot{x} = f(x) + g(x)\,u + b(x)\,z$$
$$0 = k(x) \tag{2.9}$$

The system of Eq. 2.9 is a DAE system of high index, as we do not have algebraic equations to evaluate the algebraic variables z. A solution for z can be obtained through index reduction, i.e., by differentiating the constraints $k(x) = 0$ a sufficient number of times until the resulting algebraic equations are solvable in z. For most chemical processes the z variables can be obtained after just one differentiation of the algebraic constraints, i.e., the index is two. In such a case, the $(p \times p)$ matrix $L_b k(x)$ is nonsingular and the solution for the variables z is:

$$z = -(L_b k(x))^{-1} \left\{L_f k(x) + L_g k(x)\,u\right\} \tag{2.10}$$

An ODE representation of the DAE system of Eq. 2.9 can be readily obtained as:

$$\dot{x} = f(x) + g(x)u - b(x)(L_b k(x))^{-1}\{L_f k(x) + L_g k(x)u\}$$
$$0 = k(x) \tag{2.11}$$

where the algebraic constraints specify the region in state space (a low-dimensional manifold) where the state variables evolve. A minimal-order realization of this ODE system can be obtained by employing a nonlinear change of variables which incorporates $k(x)$. Starting from Eq. 2.6 and defining the fast time-scale $\tau = t/\epsilon$, a model of the fast dynamics is obtained in the limit $\epsilon \to 0$ as:

$$\frac{dx}{d\tau} = b(x)\,k(x) \tag{2.12}$$

Note that though Eqs. 2.9 and 2.12 represent approximations of the fast and slow dynamics of Eq. 2.6, the fast and slow variables are still not explicitly separated. For the system in Eq. 2.6, conditions for the existence and explicit forms of nonlinear coordinate changes have been derived that transform the system into a standard singularly perturbed form [44].

Example 2.1 We consider a reaction system comprising of the following reactions in an isothermal constant volume continuously stirred tank reactor(CSTR):

$$A \xrightarrow{k_1} B$$
$$B \xrightarrow{k_{-1}} A$$
$$B \xrightarrow{k_2} C \tag{2.13}$$

where the reaction rates as functions of concentrations are given by $r_1 = k_1 C_A$, $r_{-1} = k_{-1} C_B$, and $r_2 = k_2 C_B$. We assume that the reversible reaction from A to B is much faster than the irreversible reaction to C. This implies the following relation for the reaction rate constants: $k_1, k_{-1} \gg k_2$. We also let $K_{eq} = \dfrac{k_1}{k_{-1}}$ denote the equilibrium constant for the reversible reaction. The dynamic model of this system, comprising of material balances for A and B has the form:

$$\dot{C}_A = \frac{F}{V}(C_{A0} - C_A) - k_1\left(C_A - \frac{C_B}{K_{eq}}\right)$$
$$\dot{C}_B = \frac{F}{V}(-C_B) + k_1\left(C_A - \frac{C_B}{K_{eq}}\right) - k_2 C_B \tag{2.14}$$

where F, V are the inlet/outlet flowrate and reactor volume, respectively, and C_{A0} is the inlet concentration of A. This system can be put in the form of Eq. 2.6 by defining the singular perturbation parameter as the inverse of the large reaction rate constant k_1, $\epsilon = \dfrac{1}{k_1}$. In this case, the DAE system describing the slow dynamics of the system has the form:

$$\dot{C}_A = \frac{F}{V}(C_{A0} - C_A) - z$$

$$\dot{C}_B = \frac{F}{V}(-C_B) + z - k_2 C_B$$

$$C_A - \frac{C_B}{K_{eq}} = 0 \qquad\qquad (2.15)$$

where the algebraic variable z is the net forward reaction rate of the reversible reaction, $k_1(C_A - \dfrac{C_B}{K_{eq}})$ in the limit as $k_1 \to \infty$. In this description, the algebraic constraint corresponds to the reaction equilibrium condition for the fast reversible reaction. After a single differentiation of the constraint we get:

$$\frac{F}{V}(C_{A0} - C_A) - z = \frac{F}{K_{eq}V}(-C_B) + \frac{z - k_2 C_B}{K_{eq}} \qquad\qquad (2.16)$$

from which an expression for z can be readily obtained.

In a series of publications [20, 44–47, 49, 50, 71] we have illustrated the advantages of using the equilibrium-based DAE models as the basis for deriving nonlinear model-based controllers for the corresponding fast-rate chemical processes. Specifically, the original two-time-scale stiff models are not appropriate for the synthesis of model-based controllers. For example, standard inversion-type or optimization-type controllers designed on the basis of such stiff models are inherently ill-conditioned: they contain the large parameters of the model, which act as high gains and amplify the effect of even small modeling or measurement errors, with detrimental consequences on stability and performance (see, e.g., [17, 41, 49]). Furthermore, the interaction of slow and fast dynamics in such systems very often leads to non-minimum-phase characteristics (see, e.g., [49]), that pose well-known limitations in the controller design. In contrast, using the slow DAE models for controller design, within a composite control framework, allows deriving well-conditioned controllers with well-characterized stability and performance properties. The controller synthesis problem in the above papers was addressed within the classical framework of smooth solutions (for consistent initial conditions), on the basis of the minimal-order realizations of the equivalent ODE systems.

3 Multiple-Time-Scale Processes

Many real processes are modeled by dynamic models containing *several* small/large parameters that arise due to the presence of more than one large reaction rate constants, heat/mass transfer coefficients, time constants, and other physical constants. If the large parameters are of different order of magnitudes, then the system may exhibit multiple-time-scale dynamics [38]. The modeling, analysis, and control of multiple-time-scale systems has received rather little attention; almost all of it has focused on systems in standard singularly perturbed form.

3.1 Standard Singularly Perturbed Form

A standard singularly perturbed form of multiple-time-scale systems can be expressed as follows:

$$\dot{\zeta} = F(\zeta, \eta_1, \ldots, \eta_M, u, \epsilon)$$

$$\epsilon_j \dot{\eta}_j = G_j(\zeta, \eta_1, \ldots, \eta_M, u, \epsilon), \qquad j = 1, \ldots, M \qquad (3.1)$$

where $\zeta \in \mathbb{R}^n$ and $\eta_j \in \mathbb{R}^{m_j}$ are the state variables, $u \in \mathbb{R}^q$ is the vector of manipulated inputs, $F, G_j \in \mathbb{R}^{m_j}$ are smooth vector fields of dimensions n and m_j respectively, and $\epsilon = [\epsilon_1, \ldots, \epsilon_M]^T$ is a vector of small positive parameters which satisfy:

$$\frac{\epsilon_{j+1}}{\epsilon_j} \to 0 \text{ as } \epsilon_j \to 0, \qquad j = 1, \ldots, M \qquad (3.2)$$

In analogy with two-time-scale systems, conditions for regular degeneration for multiple-time-scale systems are expressed in terms of the properties of the Jacobian matrices in the individual time-scales. Specifically, they require that the matrix $(\partial G_j(\zeta, \eta_1, \ldots, \eta_M, u, 0)/\partial \eta_j)$ for $j = 1, \ldots, M$ is nonsingular, and additionally that this condition is satisfied with ϵ replaced by ϵ_j for $j = 1, \ldots, M$. Under these conditions, the system exhibits M distinct fast time-scales and 1 slow time-scale [30], with the variable η_{j+1} being faster than the variable η_j, for $j = 1, \ldots, M - 1$. Such a hierarchy of fast subsystems is a characteristic feature that distinguishes multi-time-scale systems from two-time-scale ones.

In the limit as $\epsilon_1 \to 0$ the dynamic order of the system of Eq. 3.1 degenerates from $(n + \sum_j m_j)$ to n, and the slow subsystem is obtained as:

$$\dot{\zeta} = F(\zeta, \eta_1, \ldots, \eta_M, u, 0)$$

$$0 = G_j(\zeta, \eta_1, \ldots, \eta_M, u, 0), \qquad j = 1, \ldots, M \qquad (3.3)$$

Quasi-steady-state solutions $\eta_j = \sigma_j(\zeta, u)$ for $j = 1, \ldots, M$ can be obtained from the algebraic equations $0 = G_j(\zeta, \eta_1, \ldots, \eta_M, u, 0)$ and upon substitution, the slow subsystem takes the form:

$$\dot{\zeta} = F(\zeta, \sigma_1(\zeta, u), \ldots, \sigma_M(\zeta, u), u, 0) \tag{3.4}$$

Models of the fast and slow dynamics in each time-scale can be obtained through a nested application of singular perturbation arguments, starting from the fastest time-scale (corresponding to ϵ_M).

In general, the introduction of the "stretched" lth fast time-scale, where $1 \leq l \leq M$, $\tau_l = t/\epsilon_l$ results in the following description of the system in Eq. 3.1:

$$\frac{d\zeta}{d\tau_l} = \epsilon_l F(\zeta, \eta_1, \ldots, \eta_M, u, \epsilon)$$

$$\frac{d\eta_j}{d\tau_l} = \frac{\epsilon_l}{\epsilon_j} G_j(\zeta, \eta_1, \ldots, \eta_M, u, \epsilon), \qquad j = 1, \ldots, l-1$$

$$\frac{d\eta_l}{d\tau_l} = G_l(\zeta, \eta_1, \ldots, \eta_M, u, \epsilon)$$

$$\frac{\epsilon_j}{\epsilon_l} \frac{d\eta_j}{d\tau_l} = G_j(\zeta, \eta_1, \ldots, \eta_M, u, \epsilon), \qquad j = l+1, \ldots, M \tag{3.5}$$

In the limit $\epsilon_l \to 0$, the dynamics of the slow variables ζ become negligible, and since $\frac{\epsilon_j}{\epsilon_l} \to 0$ for $j = l+1, \ldots, M$, and $\frac{\epsilon_l}{\epsilon_j} \to 0$ for $j = 1, \ldots, l-1$, the differential equations for η_j for $j = l+1, \ldots, M$ are replaced by a set of algebraic equations $0 = G_j(\zeta, \eta_1, \ldots, \eta_M, u, 0)$, $j = l+1, \ldots, M$. The representation of the lth boundary-layer subsystem corresponding to the fast variables η_l is then obtained as:

$$\frac{d\eta_l}{d\tau_l} = G_l(\zeta, \eta_1, \ldots, \eta_M, u, 0)$$

$$0 = G_j(\zeta, \eta_1, \ldots, \eta_M, u, 0), \qquad j = l+1, \ldots, M \tag{3.6}$$

where the slow variables ζ and η_j for $j = 1, \ldots, l-1$ are "frozen" at their initial conditions $\zeta(0), \eta_j(0)$ and treated as constant parameters, and the variables η_j, $j = l+1, \ldots, M$ are obtained as quasi-steady-state solutions of the algebraic equations $G_j(\zeta, \eta_1, \ldots, \eta_M, u, 0) = 0$, $j = l+1, \ldots, M$. More details can be found, e.g., in [30, 60].

3.2 Nonstandard Singularly Perturbed Form

In what follows, we consider a nonstandard singularly perturbed form of multiple-time-scale systems which generalizes the one in Eq. 2.6 for two-time-scale systems. For this class of systems, we derive DAE representations of the subsystems describing the dynamics in the individual time-scales [69, 70].

Specifically, we consider systems with the following general form:

$$\dot{x} = f(x) + g(x)u + \sum_{j=1}^{M} \frac{1}{\epsilon_j} b_j(x) k_j(x) \qquad (3.7)$$

where $x \in \mathbb{R}^n$ is the vector of state variables, $f(x)$ is smooth vector field of dimension n, $g(x)$ represents a matrix of dimension $n \times q$, $k_j(x)$ denote smooth vector fields of dimensions p_j, $b_j(x)$ denote matrices of dimensions $n \times p_j$, and $\sum_j p_j < n$. We assume that the matrices $b_j(x)$ and the Jacobian matrices $(\partial k_j(x)/\partial x)$ have full column rank and full row rank, respectively. As discussed previously, these assumptions are not restrictive. In what follows, we will assume that the small parameters ϵ_j satisfy the relationship of Eq. 3.2, and thus the system of Eq. 3.7 is a multiple-time-scale one.

Let us proceed with the derivation of models of the system dynamics in the different time scales following a sequential application of formal singular perturbation arguments. We begin by defining the fastest time-scale $\tau_M = t/\epsilon_M$, and considering the limit $\epsilon_M \to 0$. Observing that $\lim\limits_{\epsilon_M \to 0} \frac{\epsilon_M}{\epsilon_j} = 0$ for $j < M$, we obtain the following fast model in this time-scale:

$$\frac{dx}{d\tau_M} = b_M(x) k_M(x) \qquad (3.8)$$

In the slow time-scale, t, multiplying Eq. 3.7 by ϵ_M and considering the limit $\epsilon_M \to 0$, the following constraints are obtained:

$$k_{M_i}(x) = 0, \quad i = 1, \ldots, p_M \qquad (3.9)$$

where $k_{M_i}(x)$ denotes the ith component of $k_M(x)$. Note that these constraints are independent (their co-vector fields are linearly independent), as $[\frac{\partial k_M(x)}{\partial x}]$ has full rank. Defining $\lim\limits_{\epsilon_M \to 0} \frac{k_{M_i}(x)}{\epsilon_M} = z_{M_i}$ and taking the limit $\epsilon_M \to 0$ in the system of Eq. 3.7, the following system is obtained:

$$\dot{x} = f(x) + g(x)u + \sum_{j=1}^{M-1} \frac{1}{\epsilon_j} b_j(x) k_j(x) + b_M(x) z_M$$

$$0 = k_M(x) \qquad (3.10)$$

which describes the slow dynamics in this fastest time-scale, where z_M denotes the p_M-dimensional vector comprising of the variables z_{M_i}. The system of Eq. 3.10 is a DAE system of high index, similarly to the two-time-scale case. Assuming that the matrix $(L_{b_M} k_M(x))$ is nonsingular (typically the case in practical problems), the variables z_M can be obtained after one differentiation of the constraints $k(x)$, as:

$$z_M = -(L_{b_M} k_M(x))^{-1} \left\{ L_f k_M(x) + L_g k_M(x) u + \sum_{j=1}^{M-1} \frac{1}{\epsilon_j} (L_{b_j} k_M(x)) k_j(x) \right\}$$

$$(3.11)$$

A state-space realization of the DAE system of Eq. 3.10 can be readily obtained as:

$$\dot{x} = f(x) + g(x)u + \sum_{j=1}^{M-1} \frac{1}{\epsilon_j} b_j(x) k_j(x)$$

$$- b_M(x) (L_{b_M} k_M(x))^{-1} \left\{ L_f k_M(x) + L_g k_M(x) u + \sum_{j=1}^{M-1} \frac{1}{\epsilon_j} (L_{b_j} k_M(x)) k_j(x) \right\}$$

$$0 = k_M(x) \qquad\qquad (3.12)$$

We can now proceed to obtain a model of the dynamics in the $(M-1)$th fast time-scale. To this end, we initially rearrange the system in Eq. 3.12 by collecting together terms containing the parameter ϵ_{M-1} as:

$$\dot{x} = \left(f(x) - b_M(x) (L_{b_M} k_M(x))^{-1} L_f k_M(x) \right)$$

$$+ \left(g(x) - b_M(x) (L_{b_M} k_M(x))^{-1} L_g k_M(x) \right) u$$

$$+ \left\{ \sum_{j=1}^{M-2} \frac{1}{\epsilon_j} b_j(x) k_j(x) - b_M(x) (L_{b_M} k_M(x))^{-1} \sum_{j=1}^{M-2} \frac{1}{\epsilon_j} (L_{b_j} k_M(x)) k_j(x) \right\}$$

$$+ \frac{1}{\epsilon_{M-1}} \left\{ b_{M-1}(x) k_{M-1}(x) - b_M(x) (L_{b_M} k_M(x))^{-1} (L_{b_{M-1}} k_M(x)) k_{M-1}(x) \right\}$$

$$0 = k_M(x) \qquad\qquad (3.13)$$

Furthermore, introducing the $(M - 1)$th fast time-scale $\tau_{M-1} = \dfrac{t}{\epsilon_{M-1}}$ and considering the limit $\epsilon_{M-1} \to 0$, we obtain the following description of the $(M - 1)$th fast dynamics of the system in Eq. 3.7:

$$\frac{dx}{d\tau_{M-1}} = \begin{bmatrix} b_{M-1}(x) \mid b_M(x) \end{bmatrix} \begin{bmatrix} k_{M-1}(x) \\ -(L_{b_M} k_M(x))^{-1} (L_{b_{M-1}} k_M(x)) k_{M-1}(x) \end{bmatrix}$$

$$0 = k_M(x) \tag{3.14}$$

Assuming that the matrix $\begin{bmatrix} b_{M-1}(x) \mid b_M(x) \end{bmatrix}$ has full column rank, the constraints obtained, in addition to $k_M(x) = 0$, are $k_{M-1}(x) = 0$.

Moreover, considering the limit $\epsilon_{M-1} \to 0$ in Eq. 3.13 results in the following description of the slow dynamics in this time-scale:

$$\dot{x} = \left(f(x) - b_M(x) (L_{b_M} k_M(x))^{-1} L_f k_M(x) \right)$$

$$+ \left(g(x) - b_M(x) (L_{b_M} k_M(x))^{-1} L_g k_M(x) \right) u$$

$$+ \left\{ \sum_{j=1}^{M-2} \frac{1}{\epsilon_j} b_j(x) k_j(x) - b_M(x) (L_{b_M} k_M(x))^{-1} \sum_{j=1}^{M-2} \frac{1}{\epsilon_j} (L_{b_j} k_M(x)) k_j(x) \right\}$$

$$+ \left[b_{M-1}(x) - b_M(x) (L_{b_M} k_M(x))^{-1} (L_{b_{M-1}} k_M(x)) \right] z_{M-1}$$

$$0 = k_{M-1}(x)$$
$$0 = k_M(x) \tag{3.15}$$

where z_{M-1} denotes the p_{M-1}-dimensional vector comprising of the variables z_{M-1_i} defined as, $z_{M-1_i} = \lim\limits_{\epsilon_{M-1} \to 0} \dfrac{k_{M-1_i}(x)}{\epsilon_{M-1}}, i = 1, \ldots, p_{M-1}$.

Note that the additional constraints $k_{M-1}(x) = 0$ obtained after the $(M - 1)$th boundary layer are the same as the ones that would be obtained in the limit $\epsilon_{M-1} \to 0$ from Eq. 3.10. This implies that the term $\dfrac{1}{\epsilon_{M-1}} (L_{b_{M-1}} k_M(x)) k_{M-1}(x)$ in Eq. 3.11 does not introduce additional constraints in the subsequent slow time scales. This indeed is the case as in the limit as $\epsilon_{M-1} \to 0$, we obtain $(L_{b_{M-1}} k_M(x)) k_{M-1}(x) = 0$ from Eq. 3.11, which is automatically satisfied for $k_{M-1}(x) = 0$.

Proceeding in a similar fashion as above, the slow system after the lth fast time-scale can be obtained. Specifically, assuming that the $(n \times \sum_{j=l}^{M} p_j)$ matrix $[\, b_l(x) \mid \cdots \mid b_M(x)\,]$ has full column rank the following system is obtained:

$$\dot{x} = f(x) + g(x)u + \sum_{j=1}^{l-1} \frac{1}{\epsilon_j} b_j(x)\, k_j(x) + \sum_{j=l}^{M} b_j(x)\, z_j$$

$$0 = k_j(x) \quad j = l, \ldots, M \tag{3.16}$$

which describes the slow dynamics (after the lth boundary layer) of Eq. 3.7 and where for simplicity we have not substituted the solutions for the algebraic variables z_j, $j = l + 1, \cdots, M$. Equation 3.16 is again a DAE system of nontrivial index, as the algebraic variables z_j, $j = l, \ldots, M$ are indeterminate. We assume that the $\sum_{j=l}^{M} p_j \times \sum_{j=l}^{M} p_j$ matrix $(L_b k(x))_l$ defined as:

$$(L_b k(x))_l := \begin{bmatrix} L_{b_M} k_M & \cdots & L_{b_M} k_l \\ \vdots & & \vdots \\ L_{b_l} k_M & \cdots & L_{b_l} k_l \end{bmatrix} \tag{3.17}$$

is nonsingular. The nonsingularity of the matrix $(L_b k(x))_l$ implies that all principal minors of $(L_b k(x))_l$ are nonzero, which ensures the solution for all the variables z_l, after just one differentiation of the corresponding constraints. It can indeed be shown [69] that the resulting model is identical to the one that we would obtain had we explicitly substituted the solutions for z_j, $j = l + 1, \cdots, M$ in the preceding reduction steps, and solved only for z_l in the current step.

Proceeding in an analogous manner, a description of the slow dynamics after the time-scale corresponding to $l = 1$ can be obtained. This will have the form:

$$\dot{x} = f(x) + g(x)u + \sum_{j=1}^{M} b_j(x)\, z_j$$

$$0 = k_j(x) \quad j = 1, \ldots, M \tag{3.18}$$

with the variables z_j suitably defined.

Note that the above approach to derive representations of dynamics in individual time-scales does not identify slow and fast variables associated with the individual time-scales. The derivation of nonlinear changes of coordinates which allow the transformation of system in Eq. 3.7 into a standard singularly perturbed form has been addressed in [70].

4 Model Reduction of Chemical Reaction Systems

The literature on model reduction approaches for reaction systems is quite extensive. The approaches can be broadly classified as (i) lumping, (ii) sensitivity analysis, and (iii) time-scale analysis. In lumping, the original composition vector is lumped into a low-dimensional composition vector such that the kinetic model is simplified (see, e.g., [1, 2, 55, 73]). However, the difficulty in selecting/finding an appropriate lumping approach increases manifold for large nonlinear reaction systems, whereas there is also an inherent loss of information about individual species and reactions. Sensitivity analysis seeks to determine and retain only significant reactions and species (see, e.g., [63, 64, 67]). However, the solution of the full-order model (or extensive data) is required to evaluate the sensitivity matrices with precision in order to ensure the accuracy of the reduced model, whereas also the achieved model order reduction decreases as we desire to retain more information. Time-scale analysis appears to be a natural framework for model order reduction of reaction systems with fast and slow reactions. The basic idea in this approach is to assume the fast dynamics to be in quasi-steady-state and obtain the slow dynamics subject to the corresponding quasi-steady-state constraints. In fact, the classical pseudo-steady state hypothesis (PSSH) and partial equilibrium approximations (see, e.g., [62, 74]) in essence are based on time-scale analysis arguments [12, 26]; however the applicability of these approximations has been typically restricted to simple reaction systems and for specific regions of initial and/or operating conditions [74]. For complex reaction systems, the intuitive identification of independent constraints corresponding to quasi-steady-state approximations for the fast dynamics and its subsequent enforcement in the reduced-order model describing the slow dynamics are nontrivial. Numerical approaches to model reduction based on time-scale analysis have also been proposed as an alternative and applied to large reaction systems. We note the method of algebraically approximating the equilibrium manifold in [57]. We also note the popular approach of computational singular perturbations (CSP) [53], which involves linearizing the dynamic model in each time step and applying essentially linear modal decomposition to separate the slow and fast modes. Note, however, that the inherent nonlinearity of chemical reaction systems requires a continuous update of the linearized model, and a corresponding update of the basis vector thats separates the slow and fast modes of the model, hence making this approach computationally intensive. Clearly, there is a need for a general approach that retains the insight and appeal of the analytical approaches and is (practically) applicable to reaction systems of high dimension. In the next subsection we describe how a general non-isothermal reaction system with fast and slow reactions can be modeled in a form similar to Eq. 3.7, which makes it amenable to the model reduction method described in the previous section. This discussion generalizes the analysis for two-time-scale reaction systems in [29, 72].

4.1 Modeling and Model Reduction of Multi-Time-Scale Reaction Systems

We consider spatially homogeneous non-isothermal, continuous chemical reactors, where the following R reactions involving S species take place: $\sum_{j=1}^{S} v_{ij} A_j = 0, i = 1, \ldots, R$, where A_j denotes the species j and v_{ij} denotes the stoichiometric coefficient of the species j in the reaction i. Let r_i denote the reaction rate of reaction i (in moles per unit time per unit volume) and ΔH_i denote the heat of reaction with the usual convention of $\Delta H_i < 0$ for an exothermic reaction. Assuming (for simplicity) constant volume of the reacting mixture, the material and energy balances describing the evolution of the species compositions and temperature take the form:

$$
\begin{bmatrix} \dot{C}_1 \\ \vdots \\ \dot{C}_j \\ \vdots \\ \dot{C}_S \\ \dot{T} \end{bmatrix} = \begin{bmatrix} v_{11} & \cdots & v_{i1} & \cdots & v_{R1} \\ \vdots & \ddots & \vdots & \ddots & \vdots \\ v_{1j} & \cdots & v_{ij} & \cdots & v_{Rj} \\ \vdots & \ddots & \vdots & \ddots & \vdots \\ v_{1S} & \cdots & v_{iS} & \cdots & v_{RS} \\ \dfrac{-\Delta H_1}{\rho C_p} & \cdots & \dfrac{-\Delta H_i}{\rho C_p} & \cdots & \dfrac{-\Delta H_R}{\rho C_p} \end{bmatrix} \begin{bmatrix} r_1 \\ \vdots \\ r_i \\ \vdots \\ r_R \end{bmatrix}
$$

$$
+ \left[\begin{pmatrix} C_1^{in} - C_1 \\ \vdots \\ C_j^{in} - C_j \\ \vdots \\ C_S^{in} - C_S \\ \dfrac{C_p^{in}}{C_p}(T^{in} - T) \end{pmatrix} \dfrac{F}{\rho V} + \begin{pmatrix} 0 \\ \vdots \\ 0 \\ \vdots \\ 0 \\ (T^h - T) \end{pmatrix} \dfrac{UA}{\rho V} \right] \quad (4.1)
$$

The model of Eq. 4.1 can be written in the following general form:

$$
\dot{x} = \mathcal{V}(x)\, r(x) + h(x) \quad (4.2)
$$

where $x = (C_1, C_2, \ldots, C_S, T)^T$ is the vector of state variables of dimension $n = S + 1$, $\mathcal{V}(x)$ is the $(n \times R)$ "generalized" stoichiometric matrix, $r(x)$ is the R-dimensional reaction rate vector, and $h(x)$ denotes the n-dimensional vector comprising of terms associated with mass flow and heat transfer. The reaction rates r_i are expressed as a product of a reaction rate constant, $k_i(T)$ (typically an exponential function of temperature), and a nonlinear (often polynomial) function of concentrations, $\bar{r}_i(C)$, as $r_i(x) = k_i(T)\bar{r}_i(C)$. In adopting such a rate expression for each reaction i, it is implicitly assumed (without loss of generality) that in the case of reversible reactions, the forward and backward reactions are treated separately.

We assume that the reaction rate constants of the reactions can be classified according to their magnitude (at some reference temperature T_0) in M distinct groups. Assuming that the residence time of the reactor and the characteristic time constant of heat transfer are of the same order of magnitude as the slow reaction rate constants, the dynamic model of Eq. 4.1 can be expressed as:

$$\dot{x} = f(x) + \sum_{l=1}^{M} \frac{1}{\epsilon_l} \mathcal{V}_{fl}(x) \, \bar{r}_{fl}(x) \tag{4.3}$$

where $\epsilon_l := \dfrac{1}{k_{fl}^*}$, for $l = 1, \ldots, M$ denotes a small parameter such that $\epsilon_{l+1} \ll \epsilon_l$, $\forall \, l \, \in \, [1, M]$, and k_{fl}^* denotes the representative large reaction rate constant (evaluated at T_0) in the lth fast reaction set. $\mathcal{V}_{fl}(x)$ is a $(n \times P_{fl})$ generalized stoichiometric matrix and $\bar{r}_{fl}(x)$ represents a P_{fl}-dimensional vector of scaled reaction rates. $f(x)$ is a vector field containing the terms associated with mass flow/heat transfer and slow reactions which do not contain large parameters. The following assumptions are also assumed to hold at each step l of the sequential application of singular perturbations:

(i) the $(n \times \sum_{j=l}^{M} P_f j)$ matrix $[\,\mathcal{V}_{fl}(x) \,|\ldots|\, \mathcal{V}_{fM}(x)]$ has full column rank, i.e., the stoichiometric vectors of reactions faster than the $(l-1)$th fast reactions are linearly independent.

(ii) the $(\sum_{j=l}^{M} P_{fj} \times n)$ matrix $\left[\left(\dfrac{\partial \bar{r}_{fl}(x)}{\partial x} \right)^T \,|\, \cdots \,|\, \left(\dfrac{\partial \bar{r}_{fM}(x)}{\partial x} \right)^T \right]^T$ has full row rank.

The above assumptions have a very concrete physical interpretation in this case. The first captures the need for *stoichiometric independence* of the reactions associated with each time-scale (and across time-scales). The second captures the requirement of *kinetic independence* of reaction rates which avoids redundancy in the algebraic constraints generated.

Proceeding similar to the method described in the previous section, in the limit $\epsilon_l \rightarrow 0$, we obtain the following constraints:

$$0 = \bar{r}_{fj}(x), \quad j = l, \ldots, M \tag{4.4}$$

These constraints must be satisfied after the lth boundary layer and they define a low-dimensional manifold $\mathcal{M}_l = \{x \in X : \bar{r}_{fj}(x), \quad j = l, \ldots, M\}$ of dimension $(n - \sum\limits_{j=l}^{M} P_{fj})$, where the slow dynamics after the lth boundary layer evolve.

Reduced-order models of the slow dynamics after the lth boundary layer can be obtained by considering the limit $\epsilon_l \to 0$, in which case the term $\dfrac{\bar{r}_{fl}}{\epsilon_l}$ becomes indeterminate, resulting in no explicit expression to evaluate the reaction rates of the lth fast reaction set in the subsequent slow time scales. Let $z_{l_i} = \lim\limits_{\epsilon_l \to 0} \dfrac{\bar{r}_{fl_i}}{\epsilon_l}$ denote the vector of indeterminate (but finite) reaction rates in the subsequent slow time-scales. Then the slow dynamics of the system in Eq. 4.3 after the lth boundary layer is described by the following system:

$$\dot{x} = f(x) + \sum_{j=1}^{l-1} \frac{1}{\epsilon_j} \mathcal{V}_{fj}(x)\, \bar{r}_{fj}(x) + \sum_{j=l}^{M} \mathcal{V}_{fj}(x)\, z_j$$

$$0 = \bar{r}_{fj}(x) \quad j = l, \ldots, M \tag{4.5}$$

The system in Eq. 4.5 is a DAE system of nontrivial index. As before, we assume that the variables $z_j \ \forall j \in [l, M]$ are obtained after one differentiation of the constraints $0 = \bar{r}_{fj}(x)$. Then a state-space realization of the DAE system can be readily obtained as:

$$\dot{x} = f(x) + \sum_{j=1}^{l-1} \frac{1}{\epsilon_j} \mathcal{V}_{fj}(x)\, \bar{r}_{fj}(x) - [\mathcal{V}_{fl}(x) \mid \ldots \mid \mathcal{V}_{fM}(x)]$$

$$\begin{bmatrix} L_{\mathcal{V}_{fl}}\bar{r}_{fl} & \cdots & L_{\mathcal{V}_{fM}}\bar{r}_{fl} \\ \vdots & \ddots & \vdots \\ L_{\mathcal{V}_{fl}}\bar{r}_{fM} & \cdots & L_{\mathcal{V}_{fM}}\bar{r}_{fM} \end{bmatrix}^{-1} \left\{ \begin{bmatrix} L_f\bar{r}_{fl} \\ \vdots \\ L_f\bar{r}_{fM} \end{bmatrix} + \sum_{j=1}^{l-1}\frac{1}{\epsilon_j} \begin{bmatrix} L_{\mathcal{V}_{fj}}\bar{r}_{fl} \\ \vdots \\ L_{\mathcal{V}_{fj}}\bar{r}_{fM} \end{bmatrix} \bar{r}_{fj} \right\}$$

$$0 = \bar{r}_{fj}(x) \quad j = l, \ldots, M \tag{4.6}$$

4.2 Application

We consider a non-isothermal isobaric batch reactor system with the hydrogen oxidation mechanism (see Table 1) taking place at one standard atmosphere [63]. The reaction system consists of eight species, specifically, $O, O_2, H, H_2, OH, H_2O, HO_2, H_2O_2$. The reaction system comprises of 20 reactions, and the reaction rate r_i for the ith reaction is given as, $r_i = k_i(T)\, \bar{r}_i(C)$,

Table 1 Reaction mechanism of the hydrogen oxidation System

Number (R_i)	Reaction	k^0	β	E
1	$H_2 + O_2 \rightarrow OH + OH$	0.170×10^{14}	0.0	47,780
2	$OH + H_2 \rightarrow H_2O + H$	0.117×10^{10}	1.3	3,626
3	$O + OH \rightarrow O_2 + H$	0.400×10^{15}	−0.5	0.0
4	$O + H_2 \rightarrow OH + H$	0.506×10^{05}	2.67	6,290
5	$H + O_2 + M \rightarrow HO_2 + M$	0.361×10^{18}	−0.72	0.0
6	$OH + HO_2 \rightarrow H_2O + O_2$	0.750×10^{13}	0.0	0.0
7	$H + HO_2 \rightarrow OH + OH$	0.140×10^{15}	0.0	1,073
8	$O + HO_2 \rightarrow O_2 + OH$	0.140×10^{14}	0.0	1,073
9	$OH + OH \rightarrow O + H_2O$	0.600×10^{09}	1.3	0.0
10	$H + H + M \rightarrow H_2 + M$	0.100×10^{19}	−1.0	0.0
11	$H + H + H_2 \rightarrow H_2 + H_2$	0.920×10^{17}	−0.6	0.0
12	$H + H + H_2O \rightarrow H_2 + H_2O$	0.600×10^{20}	−1.25	0.0
13	$H + OH + M \rightarrow H_2O + M$	0.160×10^{23}	−2.0	0.0
14	$H + O + M \rightarrow OH + M$	0.620×10^{17}	−0.6	0.0
15	$O + O + M \rightarrow O_2 + M$	0.189×10^{14}	0.0	−1788
16	$H + HO_2 \rightarrow H_2 + O_2$	0.125×10^{14}	0.0	0.0
17	$HO_2 + HO_2 \rightarrow H_2O_2 + O_2$	0.200×10^{13}	0.0	0.0
18	$H_2O_2 + M \rightarrow OH + OH + M$	0.130×10^{18}	0.0	45,500
19	$H_2O_2 + H \rightarrow HO_2 + H_2$	0.160×10^{13}	0.0	3,800
20	$H_2O_2 + OH \rightarrow H_2O + HO_2$	0.100×10^{14}	0.0	1,800

where the reaction rate constant k_i is given by a modified Arrhenius equation, $k_i = k_i^o\, T^{\beta_i}\, exp(-\dfrac{E_i}{RT})$, and $\bar{r}_i(C)$ is expressed as a product of reactant concentrations. The reaction rate constant data are in SI units, specifically the activation energy E is expressed in KJ/mole, and the pre-exponential factor k^o is expressed in mole-m-sec-K. The enthalpy and specific heat capacity data are obtained from [37]. The system is studied in the temperature range of 1,000–2,000 K. The full order ODE model comprising of the material and energy balances can be readily obtained in the form of Eq. 4.1. The reactions $5, 7, 10, 11, 12, 13, 14, 16$ are classified as faster reactions, the reactions $3, 8, 9, 15$ are classified as fast reactions, and the reactions $1, 2, 4, 6, 17, 18, 19, 20$ are classified as slow reactions in the temperature range of interest. Note that although a finer classification (in more than three groups) is possible, this will make the model reduction procedure more complex. We proceed by defining $k_{f1}^* = k_9(T^o)$ and $k_{f2}^* = k_7(T^o)$, where $T^o = 1,000$ K.

In the faster time scale for the eight faster reactions, the requirement of stoichiometrically independent reactions with kinetically independent reaction rates results in the identification of a single constraint:

$$\bar{r}_{f2}(x) = C_H = 0 \qquad (4.7)$$

Note that this corresponds to the condition of complete conversion of fast radical species H and is equivalent to the PSSH for the species H. The corresponding (9×1) stoichiometric matrix is:

$$
\mathcal{V}_{f2}(x) =
\begin{bmatrix}
-\bar{k}_{14}\, C_O\, M \\
-\bar{k}_5\, C_{O_2}\, M + \bar{k}_{16}\, C_{HO_2} \\
-\{\bar{k}_5\, C_{O_2}\, M + \bar{k}_7\, C_{HO_2} + \bar{k}_{10}\, 2\, C_H\, M + \bar{k}_{11}\, 2\, C_{H_2}\, C_H \\
+\bar{k}_{12}\, 2\, C_{H_2O}\, C_H + \bar{k}_{13}\, C_{OH}\, M + \bar{k}_{14}\, C_O\, M + \bar{k}_{16}\, C_{HO_2}\} \\
\bar{k}_{10}\, 2\, C_H\, M + \bar{k}_{11}\, 2\, C_{H_2}\, C_H + \bar{k}_{12}\, 2\, C_{H_2O}\, C_H + \bar{k}_{16}\, C_{HO_2} \\
2\, \bar{k}_7\, C_{HO_2} + \bar{k}_{14}\, C_O\, M - \bar{k}_{13}\, C_{OH}\, M \\
\bar{k}_{13}\, C_{OH}\, M \\
\bar{k}_5\, C_{O_2}\, M - \bar{k}_7\, C_{HO_2} - \bar{k}_{16}\, C_{HO_2} \\
0 \\
\beta(x)
\end{bmatrix}
\tag{4.8}
$$

where $\bar{k}_i = \epsilon_2\, k_i$, $i = 5, 7, 10, 11, 12, 13, 14, 16$. $\beta(x)$ denotes a function of x and is omitted for brevity. The algebraic variable z_{f2} can be readily obtained as:

$$
z_{f2} = (k_2\, C_{OH}\, C_{H_2} + k_3\, C_O\, C_{OH} + k_4\, C_O\, C_{H_2})\, /
$$
$$
(\bar{k}_5\, C_{O_2}\, M + \bar{k}_7\, C_{HO_2} + \bar{k}_{10}\, 2\, C_H\, M + \bar{k}_{11}\, 2\, C_{H_2}\, C_H
$$
$$
+\bar{k}_{12}\, 2\, C_{H_2O}\, C_H + \bar{k}_{13}\, C_{OH}\, M + \bar{k}_{14}\, C_O\, M + \bar{k}_{16}\, C_{HO_2}) \tag{4.9}
$$

Similarly, in the fast time-scale, the requirement of stoichiometrically independent reactions with kinetically independent reaction rates results in the following constraints:

$$
\bar{r}_{f1} =
\begin{bmatrix}
C_O \\
C_{OH}
\end{bmatrix}
= 0
\tag{4.10}
$$

Note that this corresponds to the condition of complete conversion of fast radical species OH and O and is equivalent to the PSSH for the species OH and O. The corresponding (9×2) stoichiometric matrix is:

$$
\mathcal{V}_{f1}(x) =
\begin{bmatrix}
-\bar{k}_8\, C_{HO_2} - 2\, \bar{k}_{15}\, C_O\, M - \bar{k}_3\, C_{OH} & \bar{k}_9 \\
\bar{k}_8\, C_{HO_2} + \bar{k}_{15}\, C_O\, M + \bar{k}_3\, C_{OH} & 0 \\
\bar{k}_3\, C_{OH} & 0 \\
0 & 0 \\
\bar{k}_8\, C_{HO_2} - \bar{k}_3\, C_{OH} & -2\, \bar{k}_9 \\
0 & \bar{k}_9 \\
-\bar{k}_8\, C_{HO_2} & 0 \\
0 & 0 \\
\alpha_1(x) & \alpha_2(x)
\end{bmatrix}
\tag{4.11}
$$

where $\bar{k}_i = \epsilon_1 k_i$, $i = 3, 8, 9, 15$. $\alpha_1(x)$ denotes a function of x and is omitted for brevity.

Simulations confirming the three-time-scale behavior of the reaction system and the excellent match between the full- and reduced-order models are reported in [69].

5 Dynamics and Control of Mass-Integrated Process Networks

Process integration through material and energy recycle is a key feature of modern chemical plants. Integration allows efficient utilization of material and energy resources, leading to leaner plant configurations and improved economics. At the same time, it gives rise to core plant dynamics that are usually slow and can be highly nonlinear owing to the feedback interconnections among the different process units. The dynamic analysis and control of process networks with mass and energy integration has received considerable attention in the last two decades. Our research in this area has established that in the limit of tight process integration, exemplified by large material and/or energy recycle streams, multiple-time-scale dynamics emerge, with the core network dynamics evolving in a slow time-scale and being of very low order [3, 5, 6, 8, 9, 18, 31–35, 51, 52]. Singular perturbations can be used to derive DAE models of this slow network dynamics.

5.1 Simple Mass-Integrated Networks

Initial research [35, 51] focused on simple reactor-separator networks, with a large recycle stream (compared to the throughput streams) from the separation to the reaction section. This is often used to enhance conversion or selectivity of desired reactions or species, by keeping a low single-pass conversion in the reactor and recycling large quantities of unreacted species. It is also used routinely in networks which involve recovery and recycle of organic solvents whose environmental impact can be severe. The large rate of recycle implies similarly large flowrates in the internal network streams and leads to a highly structured form in the material balance equations of such networks. Specifically, it was shown that the underlying models of such networks have the form:

$$\dot{x} = f(x) + g^s(x)u^s + \frac{1}{\epsilon}g^l(x)u^l \qquad (5.1)$$

where u^s is a vector of scaled input variables that correspond to the small inlet and outlet flowrates, u^l is a vector of scaled input variables that correspond to large internal and recycle flowrates, $f(x)$ is a smooth vector field containing reaction terms (which are assumed to be small), $g^s(x)$ and $g^l(x)$ are smooth matrices of appropriate dimensions, and ϵ is the small singular perturbation parameter which corresponds to the ratio of the nominal (steady-state) values of the feed and recycle flowrates.

5.1.1 Illustrative Example

We consider the network of a CSTR and a distillation column (see [51]). Reactant A is fed at a molar flowrate F_0 to the CSTR, where the first-order irreversible reactions $A \rightarrow B \rightarrow C$ produce the desired product B and the undesired byproduct C. The outlet stream from the reactor is fed to the distillation column with N trays (numbered from top to bottom) on tray f at a flowrate F. The light unconverted reactant A is distilled at the top of the column and recycled completely to the reactor at a flowrate D, while the heavier product B and byproduct C are withdrawn at the bottom from the reboiler at a flowrate B. For simplicity we assume isothermal operation of the reactor, constant molar overflow and relative volatilities $\alpha_A > \alpha_B > \alpha_C = 1$, and equal latent heat of vaporization for all components.

A dynamic model for the reactor-distillation column network takes then the form:

$$
\left.
\begin{aligned}
\dot{M}_R &= F_0 + D - F \\
\dot{x}_{A,R} &= \frac{F_0(1 - x_{A,R}) + D(x_{A,D} - x_{A,R})}{M_R} - k_1 x_{A,R} \\
\dot{x}_{B,R} &= \frac{-F_0 x_{B,R} + D(x_{B,D} - x_{B,R})}{M_R} + k_1 x_{A,R} - k_2 x_{B,R}
\end{aligned}
\right\} \text{reactor}
$$

$$
\left.
\begin{aligned}
\dot{M}_D &= V_B - R - D \\
\dot{x}_{A,D} &= \frac{V_B(y_{A,1} - x_{A,D})}{M_D} \\
\dot{x}_{B,D} &= \frac{V_B(y_{B,1} - x_{B,D})}{M_D}
\end{aligned}
\right\} \text{condenser}
$$

$$
\left.
\begin{aligned}
\dot{x}_{A,i} &= \frac{1}{M_i}[V_B(y_{A,i+1} - y_{A,i}) + (R + F)(x_{A,i-1} - x_{A,i})] \\
\dot{x}_{B,i} &= \frac{1}{M_i}[V_B(y_{B,i+1} - y_{B,i}) + (R + F)(x_{B,i-1} - x_{B,i})]
\end{aligned}
\right\} \text{tray } i
$$

$$
\left.
\begin{aligned}
\dot{M}_B &= R + F - V_B - B \\
\dot{x}_{A,B} &= \frac{1}{M_B}[(R + F)(x_{A,N} - x_{A,B}) - V_B(y_{A,B} - x_{A,B})] \\
\dot{x}_{B,B} &= \frac{1}{M_B}[(R + F)(x_{B,N} - x_{B,B}) - V_B(y_{B,B} - x_{B,B})]
\end{aligned}
\right\} \text{reboiler} \quad (5.2)
$$

where M_R, M_D, M_B, M_i denote the molar liquid holdups in the reactor, condenser, reboiler, and trays i, $x_{A,i}, x_{B,i}$ etc. denote the corresponding mole fractions of

A and B, and $y_{A,i}$, $y_{B,i}$ denote the vapor phase mole fractions given by the following relations:

$$y_{A,i} = \frac{\alpha_A x_{A,i}}{1 + (\alpha_A - 1)x_{A,i} + (\alpha_B - 1)x_{B,i}}, \quad y_{B,i} = \frac{\alpha_B x_{B,i}}{1 + (\alpha_A - 1)x_{A,i} + (\alpha_B - 1)x_{B,i}}$$

(For brevity we included only the equations for the trays of the distillation column after the feed in the above model.) The network model comprises of a total of $(2N+9)$ ODEs.

In this process, the two reactions with rate constants $k_1 \approx k_2$ compete in series to produce the main product B and byproduct C. It is desired to have high conversion of the reactant A and high product selectivity for B. This is achieved by keeping the single-pass conversion in the reactor low, and using a recycle flowrate D much larger than the reactant feed flowrate F_0 (where F_{0s} is assumed to be of $O(1)$ for simplicity). Note that a large recycle flowrate D implies that the flowrates F and V_B are also equally large, while the flowrates R and B are comparable to F_0.

Defining the small parameter $\epsilon = \dfrac{F_{0s}}{D_s}$, the ratios: $\kappa_1 = \dfrac{F_s}{D_s}$ and $\kappa_2 = \dfrac{V_{Bs}}{D_s}$, which are of $O(1)$, and the scaled input functions $u_1 = \dfrac{F}{F_s}$, $u_2 = \dfrac{V_B}{V_{Bs}}$ $u_3 = \dfrac{B}{B_s}$, $u_4 = \dfrac{R}{R_s}$ and $u_R = \dfrac{D}{D_s}$, which are also of $O(1)$, the model of Eq. 5.2 takes the form:

$$\dot{M}_R = F_0 + \frac{1}{\epsilon}F_{0s}(u_R - \kappa_1 u_1)$$

$$\dot{x}_{A,R} = \frac{F_0(1 - x_{A,R}) + \dfrac{1}{\epsilon}F_{0s}(x_{A,D} - x_{A,R})u_R}{M_R} - k_1 x_{A,R}$$

$$\dot{x}_{B,R} = \frac{-F_0 x_{B,R} + \dfrac{1}{\epsilon}F_{os}(x_{B,D} - x_{B,R})u_R}{M_R} + k_1 x_{A,R} - k_2 x_{B,R}$$

$$\dot{M}_D = \frac{1}{\epsilon}F_{0s}(\kappa_2 u_2 - u_R) - R_s u_4$$

$$\dot{x}_{A,D} = \frac{\dfrac{1}{\epsilon}F_{0s}\kappa_2 u_2(y_{A,1} - x_{A,D})}{M_D}$$

$$\dot{x}_{B,D} = \frac{\dfrac{1}{\epsilon}F_{0s}\kappa_2 u_2(y_{B,1} - x_{B,D})}{M_D}$$

$$\dot{x}_{A,i} = \frac{1}{M_i}[\frac{1}{\epsilon}F_{0s}\kappa_2 u_2(y_{A,i+1} - y_{A,i}) + R_s u_4(x_{A,i-1} - x_{A,i})$$

$$+\frac{1}{\epsilon}F_{0s}\kappa_1 u_1(x_{A,i-1} - x_{A,i})]$$

$$\dot{x}_{B,i} = \frac{1}{M_i}[\frac{1}{\epsilon}F_{0s}\kappa_2 u_2(y_{B,i+1} - y_{B,i}) + R_s u_4(x_{B,i-1} - x_{B,i})$$

$$+\frac{1}{\epsilon}F_{0s}\kappa_1 u_1(x_{B,i-1} - x_{B,i})]$$

$$\dot{M}_B = R_s u_4 - B_s u_3 + \frac{1}{\epsilon}F_{0s}(\kappa_1 u_1 - \kappa_2 u_2)$$

$$\dot{x}_{A,B} = \frac{1}{M_B}[R_s u_4(x_{A,N} - x_{A,B}) + \frac{1}{\epsilon}F_{0s}\kappa_1 u_1(x_{A,N} - x_{A,B})$$

$$-\frac{1}{\epsilon}F_{0s}\kappa_2 u_2(y_{A,B} - x_{A,B})]$$

$$\dot{x}_{B,B} = \frac{1}{M_B}[R_s u_4(x_{B,N} - x_{B,B}) + \frac{1}{\epsilon}F_{0s}\kappa_1 u_1(x_{B,N} - x_{B,B})$$

$$-\frac{1}{\epsilon}F_{0s}\kappa_2 u_2(y_{B,B} - x_{B,B})] \tag{5.3}$$

which is in the form of Eq. 5.1.

5.1.2 Model Reduction

The system of Eq. 5.1 is in a nonstandard singularly perturbed form which does not conform to the ones studied previously [44, 58]. The decomposition of the above system in non-stiff models of the fast and slow dynamics can be addressed following an approach similar to the one highlighted previously for systems in nonstandard singularly perturbed form.

Defining the fast time-scale $\tau = t/\epsilon$ (which is in the order of magnitude of the residence time in an individual process unit) and considering the limit $\epsilon \to 0$, we obtain the following description of the fast dynamics of the system of Eq. 5.1 (corresponding to the dynamics of the individual process units):

$$\frac{dx}{d\tau} = g^l(x)u^l \tag{5.4}$$

The dynamics in this fast time-scale involves only the large recycle and internal flowrates that are included in u^l and does not involve the small feed/product flowrates u^s. Therefore, any control objectives for the individual process units in the fast time-scale τ must be addressed with the large inputs u^l.

Turning now to the slow dynamics, multiplying Eq. 5.1 by ϵ and considering the limit $\epsilon \to 0$, the constraints $g^l(x)u^l = 0$ must be satisfied in the slow time-scale. In this limit, the term $(g^l(x)u^l)/\epsilon$ becomes indeterminate. Defining $z = lim_{\epsilon\to0}(g^l(x)u^l)/\epsilon$ as this finite but unknown term, the system of Eq. 5.1 takes the form:

$$\dot{x} = f(x) + g^s(x)u^s + z$$
$$0 = g^l(x)u^l \tag{5.5}$$

which represents the model of the slow dynamics of the process network, induced by the large recycle flowrate. Notice that the small flowrates u^s are available as manipulated inputs to address the control objectives for the overall recycle network in this slow time-scale. Note also that the above DAE system has a nontrivial index, since the solution for the algebraic variables z cannot be obtained directly from the algebraic equations.

6 General Mass-Integrated Networks

A more general prototype of a mass-integrated network was considered in [5, 9] and includes generic reaction-separation networks with a large recycle stream *and* a small purge stream used to remove small amounts of feed impurities or unwanted reaction byproducts (accumulation of such impurities in the recycle stream can be detrimental to the operation or economics of the plant). Such networks can be found in most integrated plants and their dynamics influence strongly the product purity and production rate in a plant-wide setting.

A modeling framework was developed in the general case of $S + 1$ units, a total of C components, with the recycle stream connecting units S and 1, and the product stream leaving the network through unit $S+1$ (see Fig. 1). Specifically, let F_o denote the feed flowrate to the first unit, F_{Io} the rate at which the impurity is input to the network, F_j, $j = 1, \ldots S + 1$, the outlet flowrate from the jth unit, F_R the recycle flowrate, and F_P the purge flowrate. Also, let N_I denote the net rate at which the impurity is separated from the recycle loop. Assuming that:

- the nominal flowrate of the recycle stream, $F_{R,s}$, is much larger than that of the network feed stream $F_{o,s}$:

$$\frac{F_{o,s}}{F_{R,s}} = \epsilon_1 \ll 1 \tag{6.1}$$

Fig. 1 Generic
reactor-separator process
network with large recycle
and purge

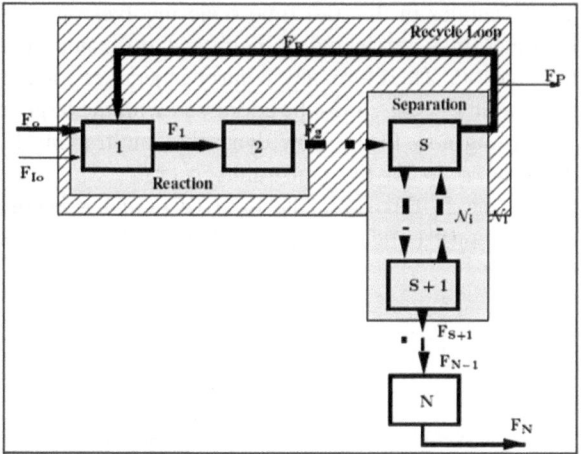

- the nominal inlet impurity flowrate $F_{Io,s}$ and the flowrate of the purge stream $F_{P,s}$ are much smaller than the flowrate of the reactant feed stream $F_{o,s}$, i.e.,

$$\frac{F_{P,s}}{F_{o,s}} = \epsilon_2 \ll 1 \tag{6.2}$$

and

$$F_{Io,s}/F_{o,s} = \beta_1 \epsilon_2$$

- the net rate of impurity removal from the recycle loop in the separation unit is much smaller than the rate at which the impurity is input to the network:

$$\mathcal{N}_{I,s}/F_{o,s} = \beta_2 \epsilon_2^2$$

and defining $u_j = F_j/F_{j,s}$, $j = 0, 1, \ldots, S + 1, R, P$ to be the scaled inputs that correspond to the flowrates $F_o, \ldots, F_j, \ldots, F_R$, and F_P, the mathematical model that describes the overall and component material balances of the network was shown to take the form:

$$\frac{d\mathbf{x}}{dt} = \bar{\mathbf{f}}(\mathbf{x}, \mathbf{u}^s) + \frac{1}{\epsilon_1}\mathbf{G}^l(\mathbf{x})\mathbf{u}^l + \epsilon_2\left[\mathbf{g}^{lo}(\mathbf{x}) + \epsilon_2\mathbf{g}^l(\mathbf{x}) + \mathbf{g}^P(\mathbf{x})u_p\right] \tag{6.3}$$

with $\mathbf{u}^l \in \mathbb{R}^{m^l}$ being the vector of scaled input variables corresponding to the "large" flowrates $F_1, \ldots, F_{S-1}, F_R$, $\mathbf{u}^s \in \mathbb{R}^{m^s}$ being the vector of scaled input variables corresponding to the "small" flowrates F_o and F_{S+1}. $\bar{\mathbf{f}}(\mathbf{x}, \mathbf{u}^s), \mathbf{g}^{lo}(\mathbf{x}), \mathbf{g}^l(\mathbf{x})$, and $\mathbf{g}^P(\mathbf{x})$ are n-dimensional vector functions, and $\mathbf{G}^l(\mathbf{x}) \in \mathbb{R}^{n \times m^l}$.

Equation 6.3 contains $\mathcal{O}(1)$ terms, along with small ($\mathcal{O}(\epsilon_2)$) and very large ($\mathcal{O}(\frac{1}{\epsilon_1})$) terms. Following a singular perturbation approach similar to the one discussed earlier in the paper, we have established that the systems in Eq. 6.3 evolve over a fast, an intermediate, and a slow time scale, and derived reduced-order, non-stiff DAE models in each time scale. These results are summarized below.

6.1 Nonlinear Model Reduction

Defining the fast time-scale $\tau = 1/\epsilon_1$ (which is in the order of magnitude of the residence time in an individual process unit) and considering the limit $\epsilon_1 \to 0$ results in the following description of the fast dynamics of the system:

$$\frac{d\mathbf{x}}{d\tau} = \mathbf{G}^l(\mathbf{x})\mathbf{u}^l \tag{6.4}$$

The above system involves only the large recycle and internal flowrates $\mathbf{u}^l$. Also note that the large flowrates $\mathbf{u}^l$ do not affect the *total holdup* in the recycle loop of any of the components $1, \ldots, C - 1$—which is only influenced by the small flowrates $\mathbf{u}^s$—or the total holdup of I, influenced exclusively by the inflow F_{Io}, the transfer rate $\mathcal{N}_I$ in the separator, and the purge stream u^P. Hence, the differential equations in Eq. 6.4 are not independent. Equivalently, the quasi-steady-state condition $\mathbf{0} = \mathbf{G}^l(\mathbf{x})\mathbf{u}^l$ does not specify a set of isolated equilibrium points, but rather, a low-dimensional equilibrium manifold. The vector function $\mathbf{G}^l$ can be reformulated [18,51] as:

$$\mathbf{G}^l(\mathbf{x}) = \mathbf{B}(\mathbf{x})\tilde{\mathbf{G}}^l(\mathbf{x}) \tag{6.5}$$

with $\mathbf{B}(\mathbf{x}) \in \mathbb{R}^{n\times(n-C-m)}$ being a full column rank matrix and the matrix $\tilde{\mathbf{G}}^l(\mathbf{x}) \in \mathbb{R}^{(n-C-m)\times m^l}$, with $m < n$ being the number of states associated with unit(s) $S + 1$, having linearly independent rows, which suggests that the dimension of the equilibrium manifold of the fast dynamics of the units within the recycle loop has an upper bound in $C + m$.

Taking into account Eq. 6.5, the quasi-steady-state constraints that must be satisfied in the slower time-scales are:

$$\mathbf{0} = \tilde{\mathbf{G}}^l(\mathbf{x})\mathbf{u}^l \tag{6.6}$$

In this limit, the term $(\tilde{\mathbf{G}}^l(\mathbf{x})\mathbf{u}^l)/\epsilon_1$ becomes indeterminate. By defining $\mathbf{z} = \lim_{\epsilon\to0}(\tilde{\mathbf{G}}^l(\mathbf{x})\mathbf{u}^l)/\epsilon_1$, $\mathbf{z} \in \mathbb{R}^{n-C-m}$ as this finite but unknown term, the slow dynamics of the network take the form:

$$\begin{aligned}\frac{d\mathbf{x}}{dt} &= \bar{\mathbf{f}}(\mathbf{x}, \mathbf{u}^s) + \mathbf{B}(\mathbf{x})\mathbf{z} + \epsilon_2[\mathbf{g}^{lo}(\mathbf{x}) + \epsilon_2\mathbf{g}^l(\mathbf{x}) + \mathbf{g}^P(\mathbf{x})u_p] \\ \mathbf{0} &= \tilde{\mathbf{G}}^l(\mathbf{x})\mathbf{u}^l\end{aligned} \tag{6.7}$$

Thus, the model of the dynamics after the fast boundary layer consists of a DAE system of nontrivial index. The algebraic variables $\mathbf{z}$ physically correspond to the net material flows of the recycle loop in the slower time-scales. Also, note that the DAE model of Eq. 6.7 has a well-defined index only if the flowrates $\mathbf{u}^l$ which appear in the algebraic constraints that determine the constrained state-space are specified as functions of the state variables $\mathbf{x}$, via a control law $\mathbf{u}^l(\mathbf{x})$. It is then possible to differentiate the constraints in Eq. 6.7 to obtain (typically after one differentiation) a solution for the algebraic variables $\mathbf{z}$. In this case, the index of the DAE system (6.7) is two and the dimension of the underlying ODE system describing the dynamics after the fast boundary layer is $\mathcal{C} + m$. An ODE description of this dynamics can be obtained by substituting $\mathbf{z}$ in Eq. 6.7, to obtain:

$$\frac{d\mathbf{x}}{dt} = \tilde{\mathbf{f}}(\mathbf{x}, \mathbf{u}^s) + \epsilon_2[\tilde{\mathbf{g}}^{lo}(\mathbf{x}) + \epsilon_2\tilde{\mathbf{g}}^I(\mathbf{x}) + \tilde{\mathbf{g}}^P(\mathbf{x})u_p]$$
$$0 = \tilde{\mathbf{G}}^l(\mathbf{x})\mathbf{u}^l(\mathbf{x}) \tag{6.8}$$

A minimal-order ODE representation of the system in Eq. 6.8 can be subsequently obtained by employing a coordinate change of the form:

$$\begin{bmatrix} \zeta \\ \eta \end{bmatrix} = \mathbf{T}_1(\mathbf{x}) = \begin{bmatrix} \boldsymbol{\phi}(\mathbf{x}) \\ \tilde{\mathbf{G}}^l(\mathbf{x})\mathbf{u}^l(\mathbf{x}) \end{bmatrix} \tag{6.9}$$

Specifically, the dynamics after the fast boundary layer will be of the form:

$$\frac{d\zeta}{dt} = \hat{\mathbf{f}}(\zeta, \mathbf{u}^s) + \epsilon_2[\hat{\mathbf{g}}^{lo}(\zeta) + \epsilon_2\hat{\mathbf{g}}^I(\zeta) + \hat{\mathbf{g}}^P(\zeta)u_P]$$
$$\eta \equiv 0 \tag{6.10}$$

This is the core dynamics of the network, present due to the large recycle flowrate. Observe, however, that the model of Eq. 6.10 still contains both $\mathcal{O}(1)$ and $\mathcal{O}(\epsilon_2)$ terms and is, therefore, stiff. The time evolution of the network after the fast boundary layer has thus the potential to feature two time-scales. We can proceed in a similar way, by considering the limiting case of the purge flowrate and the impurity feed being set to zero, i.e., $\epsilon_2 \to 0$. In this limit, we obtain a description of the dynamics in the intermediate time-scale:

$$\frac{d\zeta}{dt} = \hat{\mathbf{f}}(\zeta, \mathbf{u}^s) \tag{6.11}$$

The description of the intermediate dynamics in Eq. 6.11 only involves the flowrates $\mathbf{u}^s$. However, these flowrates do not affect the total holdup of the impurity in the recycle loop. Consequently, one of the differential equations describing the intermediate dynamics is redundant, and the Eq. (6.11) are not independent. Correspondingly, the steady-state conditions:

$$0 = \hat{\mathbf{f}}(\zeta, \mathbf{u}^s) \tag{6.12}$$

specify a one-dimensional sub-manifold in which a slower dynamics will evolve. Following the approach taken in the beginning of this subsection, we rewrite the vector function $\hat{\mathbf{f}}(\zeta, \mathbf{u}^s)$ as:

$$\hat{\mathbf{f}}(\zeta, \mathbf{u}^s) = \hat{\mathbf{B}}(\mathbf{x})\check{\mathbf{f}}(\zeta, \mathbf{u}^s) \tag{6.13}$$

where the matrix $\hat{\mathbf{B}}(\mathbf{x}) \in \mathbb{R}^{(C+m)\times(C+m-1)}$ has full column rank, and the vector $\check{\mathbf{f}}(\zeta, \mathbf{u}^s) \in \mathbb{R}^{(C+m-1)}$ has linearly independent rows.

Next, in order to obtain a description of the slow dynamics, we define the slow, compressed, time-scale $\theta = \epsilon_2 t$, and we consider the limit $\epsilon_2 \to 0$, in which the constraints $0 = \hat{\mathbf{f}}(\zeta, \mathbf{u}^s)$, or, equivalently, the linearly independent constraints $0 = \check{\mathbf{f}}(\zeta, \mathbf{u}^s)$ are obtained. These constraints must be satisfied in the slow time-scale. Dividing Eq. 6.10 by ϵ_2, and considering the same limiting case under the constraints above, we obtain a description of the slow dynamics of the system. Note that, in this limit, the term $\check{\mathbf{f}}(\zeta, \mathbf{u}^s)/\epsilon_2$ becomes indeterminate. By defining $\hat{\mathbf{z}} = \lim_{\epsilon_2 \to 0} \check{\mathbf{f}}(\zeta, \mathbf{u}^s)/\epsilon_2$, $\hat{\mathbf{z}} \in \mathbb{R}^{C+m-1}$, the slow dynamics of the network in Eq. 6.3 takes the form:

$$\begin{aligned} \frac{d\mathbf{x}}{d\tau} &= \hat{\mathbf{g}}^{lo}(\zeta) + \hat{\mathbf{g}}^P(\zeta)u_p + \hat{\mathbf{B}}(\zeta)\hat{\mathbf{z}} \\ 0 &= \check{\mathbf{f}}(\zeta, \mathbf{u}^s) \end{aligned} \tag{6.14}$$

In the DAE system in Eq. 6.14, the variables $\hat{\mathbf{z}} \in \mathbb{R}^{C+m-1}$ are again implicitly fixed by the algebraic constraints, and thus the index of the system is again nontrivial. Also, note that similarly to the previous reduction step, the index of Eq. 6.14 is well-defined only if the flowrates $\mathbf{u}^s$ are specified as a function of the state variables (in this case, expressed in the new coordinates ζ), i.e., $\mathbf{u}^s = \mathbf{u}^s(\zeta)$. Specifying these flowrates via feedback control laws allows for the determination of $\hat{\mathbf{z}}$ through the differentiation of the algebraic constraints in Eq. 6.14. In this case, the DAE model describing the slow dynamics (6.14) is of index two, and the underlying dimension of the ODE system describing the evolution of the network in the slow time-scale is one. An explicit state-space realization of the slow dynamics can then be obtained via an appropriate coordinate change. This will be the dynamics associated with the small amount of feed impurity, removed by the small purge stream.

6.2   Hierarchical Controller Design

Each of the derived reduced-order models for the fast, intermediate and slow dynamics only involve one group of manipulated inputs, namely, the large internal flowrates $\mathbf{u}^l$, the small flowrates $\mathbf{u}^s$, and the purge flowrate u^p, respectively. Furthermore, each set of inputs starting from the faster time-scale has to be specified through a control law in order for the model in the subsequent time scale to be well-posed. This leads to a hierarchical controller design framework described below. **Distributed Control at the Unit Level**: In a fast time scale, the large flowrates $\mathbf{u}^l$ are available for addressing regulatory control objectives at the unit level, such as liquid level/holdup control, as well as for the rejection of fast disturbances. Typically, the above control objectives are fulfilled using simple linear controllers, possibly with integral action, depending on the stringency of the control objectives. **Supervisory Control at the Network Level**: In the intermediate time scale, the small flowrates $\mathbf{u}^s$ are available for addressing control objectives at the network level, such as controlling the product purity, stabilizing the total material holdup, and setting the production rate. Very often (especially as they also serve for the regulatory control for the units outside the recycle loop), the number of available manipulated inputs $\mathbf{u}^s$ is exceeded by the number of network level control objectives. In this case, it is possible to use the set points $\mathbf{y}_{sp}^l$ of the controllers in the fast time-scale as manipulated inputs in the intermediate time-scale, which leads to cascaded control configurations. However, the constrained state-space of the DAE description of the dynamics after the fast dynamics (Eq. 6.7) becomes control dependent (i.e., $\tilde{\mathbf{G}}^l(\mathbf{x})\mathbf{u}^l = \tilde{\mathbf{G}}^l(\mathbf{x})\mathbf{u}^l(\mathbf{x}, \mathbf{y}_{sp}^l),$) and the derivation of a corresponding ODE representation of the type (Eq. 6.8) is nontrivial. For the class of systems under consideration it was shown that a dynamic pre-compensation (adding integrators in certain input channels) suffices to overcome this problem [18, 19].
Control of Impurity Levels: The presence of impurities in the feed stream can lead to the accumulation of the impurities in the recycle loop over a slow time-scale, with detrimental effects on the operation of the network and the process economics. To avoid these adverse effects the control of the impurity levels should be addressed using the flowrate of the purge stream, u^P, as a manipulated input.

The above analysis and control framework, with nonlinear controllers embedded in the supervisory control level, was applied successfully to several prototype reactor-separator networks and high-purity distillation columns, and was shown to lead to superior performance and robustness characteristics compared with linear decentralized control [5, 9, 51]. It was also combined with a self-optimizing control approach for the selection of controlled outputs and was shown to be an effective strategy for control structure selection with well-characterized steady-state optimality and dynamic response characteristics [4].

Finally, a similar dynamic structure in the process model was documented for process networks with large energy (rather than material) recycle [31–34]. Examples of such tightly integrated processes include reactor-heat exchanger networks, heat-integrated distillation columns, thermally coupled distillation columns, etc. The time-scale separation in these models leads to a similar, hierarchical control structure where energy management can be systematically addressed in the slow time-scale using the underlying DAE models (see e.g. [7]).

7 Conclusions

In this paper we provided a tutorial overview of chemical process models described by DAE systems of high index. These correspond to slow, quasi-equilibrium models of fast-rate chemical processes modeled in a nonstandard singularly perturbed form. The resulting DAE systems are in semi-explicit form, with the algebraic equations being singular with respect to algebraic variables defined as the finite (but indeterminated) rates in the slow time-scale. There is a broad variety of chemical processes that give rise to such models. These include reaction and separation units with fast rates of reaction, heat transfer or mass transfer, as well as networks of chemical processes connected through large rates of material or energy flows. Control of such processes is logically addressed through a combination of separate controllers in the fast and slow time-scales, designed on the basis of the corresponding fast and slow subsystems. Thus, the high-index DAE models play a critical role in designing well-conditioned nonlinear feedback controllers for improved operation of such processes.

References

1. Aris, R., Gavalas, G.R.: On the theory of reactions in continuous mixtures. Philos. Trans. R. Soc. **269**(9A260), 351–393 (1966)
2. Bailey, J.E.: Lumping analysis of reactions in continuous mixtures. Chem. Eng. J. **3**, 52–71 (1972)
3. Baldea, M., Daoutidis, P.: Model reduction and control of reactor–heat exchanger networks. J. Process Control **16**, 265–274 (2006)
4. Baldea, M., Araujo, A., Skogestad, S., Daoutidis, P.: Dynamic considerations in the synthesis of self-optimizing control structures. AIChE J. **54**, 1830 (2008)
5. Baldea, M., Daoutidis. P.: Dynamics and control of integrated process networks - a multi-time perspective. Comput. Chem. Eng. **31**, 426 (2007)
6. Baldea, M., Daoutidis, P.: Modeling, dynamics and control of process networks with high energy throughput. Comput. Chem. Eng. **32**(9), 1964–1983 (2008)
7. Baldea, M., Daoutidis, P.: Control of integrated chemical process systems using underlying DAE models. In: Control and Optimization with Differential-Algebraic Constraints, pp. 281–300. SIAM, Philadelphia (2012)
8. Baldea, M., Daoutidis, P.: Dynamics and nonlinear control of integrated process systems. In: Chemical Engineering Series. Cambridge University Press, Cambridge (2012)
9. Baldea, M., Daoutidis, P., Kumar, A.: Dynamics and control of integrated networks with purge streams. AIChE J. **52**, 1460–1472 (2006)
10. Bauer, I., Bock, H.G., Korkel, S., Schloder, J.P.: Numerical methods for optimum experimental design in dae systems. J. Comput. Appl. Math. **120**, 1–25 (2000)

11. Biegler, L., Campbell, S.L., Mehrmann, V. (eds.): Control and Optimization with Differential-Algebraic Constraints. SIAM, Philadelphia (2012)
12. Bowen, J.R., Acrivos, A., Oppenheim, A.K.: Singular perturbation refinement to quasi-steady-state approximation in chemical kinetics. Chem. Eng. Sci. **18**, 177–188 (1963)
13. Brenan, K.E., Campbell, S.L., Petzold, L.R.: Numerical solution of initial-value problems in differential-algebraic equations. In: Classics in Applied Mathematics. SIAM, Philadelphia (1996)
14. Byrne, G.D., Ponzi, P.R.: Differential-algebraic systems, their applications and solutions. Comput. Chem. Eng. **12**, 377–382 (1988)
15. Campbell, S.L.: Singular systems of differential equations II. In: Research Notes in Mathematics, vol. 61, Pitman Books, San Francisco, London, Melbourne (1982)
16. Cervantes, A.M., Wachter, A., Tutuncu, R.H., Biegler, L.T.: A reduced space interior point strategy for optimization of differential algebraic systems. Comput. Chem. Eng. **24**, 39–51 (2000)
17. Christofides, P.D., Daoutidis, P.: Feedback control of two-time-scale nonlinear systems. Int. J. Control **63**, 965–994 (1996)
18. Contou-Carrère, M.N., Baldea, M., Daoutidis, P.: Dynamic precompensation and output feedback control of integrated process networks. Ind. Eng. Chem. Res. **43**, 3528–3538 (2004)
19. Contou-Carrère, M.N., Daoutidis, P.: An output feedback precompensator for nonlinear differential-algebraic-equation systems with control-dependent state-space. IEEE Trans. Automat. Control **50**, 1831 (2005)
20. Contou-Carrère, M.N., Daoutidis, P.: Model reduction and control of multi-scale reaction-convection processes. Chem. Eng. Sci. **63**, 4012 (2007)
21. Dai, L.: Singular Control Systems. In: Lecture Notes in Control and Information Sciences, vol. 118, Springer, Heidelberg (1989)
22. Diehl, M., Bock, H.G., Schloder, J.P., Findeisen, R., Nagy, Z., Allgower, F.: Real-time optimization and nonlinear model predictive control of processes governed by differential-algebraic equations. J. Process Control **12** 577–585 (2002)
23. Feehery, W.F., Tolsma, J.E., Barton. P.I.: Efficient sensitivity analysis of large-scale differential-algebraic systems. Appl. Numer. Math. **25**, 41–54 (1997)
24. Fenichel, N.: Geometric singular perturbation theory for ordinary differential equations. J. Differ. Equ. **31**, 53 (1979)
25. Findeisen, R., Allgower, F.: Nonlinear model predictive control for index-one DAE systems In: Progress in Systems and Control Theory, vol. 26, pp. 145–161. Birkhauser, Basel (2000)
26. Fraser, S.J.: The steady state and equilibrium approximations: A geometric picture. J. Chem. Phys. **88**, 4732–4738 (1988)
27. Gani, R., Cameron, I.T.: Modeling for dynamic simulation of chemical processes: The index problem. Chem. Eng. Sci. **47**, 1311–1315 (1992)
28. Gerdts, M.: Direct shooting methods for the numerical solution of higher-index dae optimal control problems. J. Optim. Theory Appl. **117**, 267–294 (2003)
29. Gerdtzen, Z.P., Daoutidis, P., Hu, W.S.: Non-linear reduction for kinetic models of metabolic reaction networks. Metab. Eng. **6**(2), 140–154 (2004)
30. Hoppensteadt, F.: Properties of solutions of ordinary differential equations with small parameters. Commun. Pure Appl. Math. **XXIV**(6), 807–840 (1971)
31. Jogwar, S.S., Baldea, M., Daoutidis, P.: Dynamics and control of process networks with large energy recycle. Ind. Eng. Chem. Res. **48**(13), 6087–6097 (2009)
32. Jogwar, S.S., Baldea, M., Daoutidis, P.: Tight energy integration: Dynamic impact and control advantages. Comput. Chem. Eng. **34**(9), 1457–1466 (2010)
33. Jogwar, S.S., Daoutidis, P.: Dynamics and control of vapor recompression distillation. J. Process Control **19**(10), 1737–1750 (2009)
34. Jogwar, S.S., Daoutidis, P.: Energy flow patterns and control implications for integrated distillation networks. Ind. Eng. Chem. Res. **49**(17), 8048–8061 (2010)
35. Jogwar, S.S., Torres, A.I., Daoutidis, P.: Networks with large solvent recycle: Dynamics, hierarchical control and a biorefinery application. AIChE J. **58**(6), 1764–1777 (2012)

36. Kameswaran, S., Biegler, L.T.: Simultaneous dynamic opimization strategies: Recent advances and challenges. Comput. Chem. Eng. **30**, 1560–1575 (2006)
37. Kee, R.J., Rupley, F.M., Miller, J.A.: Chemkin-II: A fortran chemical kinetics package for the analysis of gas phase chemical kinetics. In: Sandia Report SAND89-8009B, UC-706, Livermore, CA (1989)
38. Khalil, H., Kokotovic, P.V.: Control of linear systems with multiparameter singular perturbations. Automatica **15** 197–207 (1979)
39. Khalil, H.K.: Nonlinear Systems. 2nd edn. Prentice-Hall, Upper Saddle River (1996)
40. Kokotovic, P.V., Bensoussan, A., Blankenship, G.: Singular perturbations and asymptotic analysis in control systems. In: Lecture Notes in Control and Information Sciences, vol. 90. Springer, Heidelberg (1987)
41. Kokotovic, P.V., Khalil, H.K., O'Reilly, J.: Singular Perturbations in Control: Analysis and Design. Academic Press, London (1986)
42. Kokotovic, P.V., O'Malley, R.E., Sannuti, P.: Singular perturbations and order reduction in control theory - an overview. Automatica **12**, 123–132 (1976)
43. Kroner, A., Marquardt, W., Gilles, E.D.: Getting around consistent initialization of dae systems. Comput. Chem. Eng. **21**, 145–158 (1997)
44. Kumar, A., Christofides, P.D., Daoutidis, P.: Singular perturbation modeling of nonlinear processes with non-explicit time-scale separation. Chem. Eng. Sci. **53**, 1491–1504 (1998)
45. Kumar, A., Daoutidis, P.: Control of nonlinear differential-algebraic-equation systems with disturbances. Ind. Eng. Chem. Res. **34**, 2060–2076 (1995)
46. Kumar, A., Daoutidis, P.: Feedback control of nonlinear differential-algebraic-equation systems. AIChE J. **41**(3), 619–636 (1995)
47. Kumar, A., Daoutidis, P.: Dynamic feedback regularization and control of nonlinear differential-algebraic-equation systems. AIChE J. **42**, 2175–2198 (1996)
48. Kumar, A., Daoutidis, P.: High-index dae systems in modeling and control of chemical processes. In: Preprints of IFAC Conference on Control of Industrial Systems, vol. 1, pp. 518–523. Belfort, France (1997)
49. Kumar, A., Daoutidis, P.: Control of nonlinear differential equation systems. In: Research Notes in Mathematics, vol. 397, Chapman & Hall/CRC, Boca Raton, London, New York, Washington DC (1999)
50. Kumar, A., Daoutidis, P.: Modeling, analysis and control of ethylene glycol reactive distillation column. AIChE J. **45**, 51 (1999)
51. Kumar, A., Daoutidis, P.: Dynamics and control of process networks with recycle. J. Process Control **12**, 475–484 (2002)
52. Kumar, A., Daoutidis, P.: Nonlinear model reduction and control for high-purity distillation columns. Ind. Eng. Chem. Res. **42**, 4495–4505 (2003)
53. Lam, S.H., Goussis, D.A.: The csp method for simplifying kinetics. J. Chem. Kin. **26**, 461–486 (1994)
54. Lefkopoulos, A., Stadherr, M.A.: Index analysis of unsteady state chemical process systems - ii. strategies for determining the overall flowsheet index. Comput. Chem. Eng. **17**, 415–430 (1993)
55. Li, G., Tomlin, A.S., Rabitz, H., Toth. J.: A general analysis of approximate nonlinear lumping in chemical kinetics. J. Chem. Phys. **101**, 1172–1187 (1994)
56. Li, S., Petzold, L.: Software and algorithms for sensitivity analysis of large-scale differential-algebraic systems. J. Comput. Appl. Math. **125**, 131–145 (2000)
57. Maas U., Pope, S.B.: Simplifying chemical kinetics: Intrinsic low-dimensional manifolds in composition space. Combust. Flame **88**, 239–264 (1992)
58. Marino R., Kokotovic, P.V.: A geometric approach to nonlinear singular perturbed control systems. Automatica **24**, 31–41 (1988)
59. Martinson, W.S., Barton, P.I.: A diffferentiation index for partial differential algebraic equations. SIAM J. Sci. Comput. **21**, 2295–2315 (2000)
60. O'Malley Jr. R.E.: Singular Perturbation Methods for Ordinary Differential Equations. Springer, New York (1991)

61. Pantelides, C.C., Gritsis, D., Morison, K.R., Sargent, R.W.H.: The mathematical modeling of transient systems using differential-algebraic equations. Comput. Chem. Eng. **12**, 449–454 (1988)
62. Peters, N.: Reduced Kinetic Mechanisms and Asymptotic Approximations for Methane-Air Flames. Springer, Berlin (1991)
63. Petzold, L., Zhu, W.: Model reduction for chemical kinetics: An optimization approach. AIChE J. **45**, 869–886 (1999)
64. Rabitz, H., Kramer, M., Dacol. D.: Sensitivity analysis of chemical kinetics. Annu. Rev. Phys. Chem. **34**, 419–461 (1983)
65. Rhem, A., Allgower, F.: General quadratic performance analysis and synthesis of differential algebraic equation (dae) systems. J. Process Control **12**, 467–474 (2002)
66. Tichonov, A.N.: Systems of differential equations containing a small parameter multiplying the derivative. Mat. Sb. **31**, 575–586 (1952)
67. Turanyi, T., Berces, T., Vajda, S.: Reaction rate analysis of complex kinetic systems. Int. J. Chem. Kinet. **21**, 83–99 (1989)
68. Vieira, R.C., Biscaia, E.C.: Direct methods for consistent initialization of dae systems. Comput. Chem. Eng. **25**, 1299–1311 (2001)
69. Vora, N.P.: Nonlinear model reduction and control of multiple time scale chemical processes: chemical reaction systems and reactive distillation columns. PhD thesis, University of Minnesota - Twin Cities (2000)
70. Vora, N.P., Contou-Carrère, M.N., Daoutidis, P.: Model reduction of multiple time scale processes in non-standard singularly perturbed form. In: Coarse Graining and Model Reduction Approaches for Multiscale Phenomena. Lecture Notes Series, pp. 99–116. Springer, New York (2006)
71. Vora, N.P., Daoutidis, P.: Dynamics and control of an ethyl acetate reactive distillation column. Ind. Eng. Chem. Res. **40**, 833–849 (2001)
72. Vora, N.P., Daoutidis, P.: Nonlinear model reduction of chemical reaction systems. AIChE J. **47**, 2320–2332 (2001)
73. Wei, J., Kuo, J.C.W.: A lumping analysis in monomolecular reaction systems. Ind. Eng. Chem. Fund. **8**, 114–123 (1969)
74. Williams, F.A.: The fundamental theory of chemically reactive systems. In: Combustion Theory. Benjamin/Cummings, Menlo Park (1985)

A Survey on Optimal Control Problems with Differential-Algebraic Equations

Matthias Gerdts

Abstract The paper provides an overview on necessary and, whenever available, on sufficient conditions of optimality for optimal control problems with differential-algebraic equations (DAEs) and on numerical approximation techniques. Local and global minimum principles of Pontryagin type are discussed for convex linear-quadratic optimal control problems and for non-convex problems. The main steps for the derivation of such conditions will be explained. The basic working principles of different approaches towards the numerical solution of DAE optimal control problems are illustrated for direct shooting methods, full discretization techniques, projected gradient methods, and Lagrange–Newton methods in a function space setting.

Keywords Differential-algebraic equations • Direct discretization method • Lagrange–Newton method • Necessary conditions • Projected gradient method • Optimal control • Sufficient conditions

Subject Classifications: 49-02, 49K15, 49M05, 49M15, 49M25, 49N10, 34A09, 65L80

Contents

M. Gerdts (✉)
Department of Aerospace Engineering, Institute of Mathematics and Applied Computing,
Universität der Bundeswehr München, Werner-Heisenberg-Weg 39, 85577 Neubiberg, Germany
e-mail: matthias.gerdts@unibw.de

© Springer International Publishing Switzerland 2015   103
A. Ilchmann, T. Reis (eds.), *Surveys in Differential-Algebraic Equations II*,
Differential-Algebraic Equations Forum, DOI 10.1007/978-3-319-11050-9_3

1 Introduction

Differential-algebraic equations (DAEs) are composite systems of ordinary differential equations and algebraic equations. Such systems are frequently used in mechanical engineering, process engineering, electrical engineering, and in other disciplines as well, to describe the motion of a time dependent process. In an industrial environment, DAEs are appealing because such systems can be set up automatically by software packages, whereas an equivalent formulation in terms of explicit ordinary differential equations (ODEs) is typically more difficult to achieve and depends on a proper choice of coordinates, e.g., minimal coordinates for mechanical multi-body systems. However, the formulation as a DAE, although simple from the modeling point of view, has implications in view of numerical treatment and solution properties in general. More specifically, the existence of a solution is not guaranteed in general and depends on the structure of the DAE and on external inputs, e.g., control inputs. Moreover, not all initial values are suitable and thus only so-called consistent initial conditions are permitted. A survey on solution theory for linear DAEs can be found in the recent survey paper [106]. A comprehensive structural analysis of linear and nonlinear DAEs can be found in the monographs [68] and [80]. Finally, DAEs differ in their stability properties from ODEs. While ODEs can be viewed as well-behaved systems (with respect to small perturbations and subject to a Lipschitz condition of the right hand-side), DAEs are inherently ill-posed. The degree of ill-posedness depends on the structure and can be measured by the so-called perturbation index, compare [59, Def. 1.1]. This inherent ill-posedness of higher index DAEs requires suitable numerical integration schemes, see [21, 58, 59], [68, Chap. 5], and index reduction techniques, see [68, Chap. 6], as well as stabilization techniques, see e.g., [42, 43].

In this paper we focus on DAEs, which can be controlled by some vector-valued control input $u : I \longrightarrow \mathbb{R}^{n_u}$, $n_u \in \mathbb{N}$, on some fixed compact time interval $I = [t_0, t_f]$ with initial time t_0 and final time $t_f > t_0$. In its most general form, the DAE is defined by the implicit ordinary differential equation

$$F(t, x(t), x'(t), u(t)) = 0, \qquad (1.1)$$

where $x : I \longrightarrow \mathbb{R}^{n_x}$, $n_x \in \mathbb{N}$, denotes the state, x' its derivative w.r.t. time t, and $F : I \times \mathbb{R}^{n_x} \times \mathbb{R}^{n_x} \times \mathbb{R}^{n_u} \longrightarrow \mathbb{R}^{n_x}$ is a given function. Throughout it is assumed that F is sufficiently smooth, i.e., it possesses continuous partial derivatives up to a requested order. If the Jacobian $F'_{x'}$ is nonsingular in a solution x, then Eq. (1.1) can be solved for x' by the implicit function theorem and an explicit ODE is obtained. We are particularly interested in the case where $F'_{x'}$ is singular, in which case (1.1) contains ODEs as well as algebraic equations. Particular examples with singular Jacobian are semi-explicit DAEs of type

$$F(t, x, x', u) = \begin{pmatrix} M(t, x_d)x'_d - f(t, x_d, x_a, u) \\ g(t, x_d, x_a, u) \end{pmatrix}, \qquad x := (x_d, x_a)^\top, \qquad (1.2)$$

with a nonsingular matrix M and the so-called differential state vector x_d and the algebraic state vector x_a. Such systems occur, e.g., in process engineering and mechanical multi-body systems, whereas electric circuits can be modeled by quasi linear DAEs of type

$$F(t, x, x', u) = Q(t, x)x' - f(t, x, u) \qquad (1.3)$$

with a possibly singular matrix function Q. Linear DAEs of type

$$F(t, x, x', u) = E(t)x' - A(t)x - B(t)u - q(t) \qquad (1.4)$$

with time dependent matrices E, A, B, and inhomogeneity q are of particular interest in controller design. This form is referred to as descriptor form.

Now the task is to find an optimal control u such that a given objective function is minimized subject to the DAE and further control and state constraints. To this end let $\varphi : \mathbb{R}^{n_x} \times \mathbb{R}^{n_x} \longrightarrow \mathbb{R}$, $f_0 : I \times \mathbb{R}^{n_x} \times \mathbb{R}^{n_u} \longrightarrow \mathbb{R}$, $c : I \times \mathbb{R}^{n_x} \times \mathbb{R}^{n_u} \longrightarrow \mathbb{R}^{n_c}$, $n_c \in \mathbb{N}$, $\psi : \mathbb{R}^{n_x} \times \mathbb{R}^{n_x} \longrightarrow \mathbb{R}^{n_\psi}$, $n_\psi \in \mathbb{N}$, be given functions and $\mathscr{U} \subseteq \mathbb{R}^{n_u}$ a set. A general prototype optimal control problem reads as follows:

Problem 1.1 (OCP) Minimize

$$\varphi(x(t_0), x(t_f)) + \int_{t_0}^{t_f} f_0(t, x(t), u(t))dt$$

subject to the DAE (1.1), the control-state constraints

$$c(t, x(t), u(t)) \leq 0,$$

the boundary condition

$$\psi(x(t_0), x(t_f)) = 0,$$

and the set constraints

$$u(t) \in \mathcal{U}.$$

It is important to point out that OCP is merely a generic container for different types of DAE optimal control problems, but its formulation is yet too vague for a thorough analysis. Most importantly, the function spaces from which x and u shall be chosen have to be defined appropriately. Typically, in the nonlinear setting of OCP, u is supposed to be an essentially bounded function from the Banach space $L^\infty(I, \mathbb{R}^{n_u})$, or a square integrable function from the Hilbert space $L^2(I, \mathbb{R}^{n_u})$ for linear-quadratic problems. The choice of a proper space for the state x is more subtle since the components of x may have different smoothness properties depending on the structure of the DAE. In classic ODE optimal control theory the natural space for x would be the Banach space $W^{1,\infty}(I, \mathbb{R}^{n_x})$ of essentially bounded functions with essentially bounded first derivatives in the nonlinear case or the Hilbert space $W^{1,2}(I, \mathbb{R}^{n_x})$ of L^2-functions with L^2-derivative in the linear-quadratic case. In the presence of DAEs, however, the DAE might not possess a solution in these classic spaces. This immediately becomes clear for the semi-explicit DAE (1.2). Suppose that the algebraic constraint $0 = g(t, x_d, x_a, u)$ can be solved for x_a, i.e., $x_a = x_a(t, x_d, u)$. In this case the smoothness of x_a depends on the smoothness of the control input u. Thus, for an L^∞-input u one can only expect the algebraic variable x_a to be an L^∞-function rather than a $W^{1,\infty}$-function. One can easily imagine that the situation for general DAEs (1.1) is even more complicated. For this reason the discussion in the following sections is restricted to appropriate subclasses of (1.1) and of OCP whenever necessary.

Once appropriate spaces have been identified, the key issue in the analysis of OCP is necessary optimality conditions. To this end OCP and the special cases in Sects. 2 and 3 will be considered as nonlinear optimization problems in a Banach space setting of type NLP in Problem 1.2 below. Then, the necessary conditions in Theorems 1.3 and 1.4 will be applied to the optimal control problems in Sects. 2 and 3.

Let Z, V, W be Banach spaces equipped with norms $\|\cdot\|_Z$, $\|\cdot\|_V$, $\|\cdot\|_W$, respectively, and let $J : Z \longrightarrow \mathbb{R}$, $G : Z \longrightarrow W$, $H : Z \longrightarrow V$ be given mappings. Let $K \subseteq W$ be a non-empty closed convex cone with vertex at zero and $S \subseteq Z$ a non-empty set.

Problem 1.2 (NLP) Minimize $J(z)$ subject to the constraints

$$z \in S, \qquad G(z) \in K, \qquad H(z) = 0.$$

Herein, $z \in S$ is referred to as set constraint, $G(z) \in K$ as inequality (or cone) constraint, and $H(z) = 0$ as equality constraint. Note that the cone K induces a partial ordering on W according to the relation $x \leq_K y$ which holds if and only if $x - y \in K$. With this the cone constraint $G(z) \in K$ reads $G(z) \leq_K 0$.

First order necessary conditions of optimality are provided by the following theorem, compare [48, Theorem 2.3.24], which can be found in similar form in [83] and [75, Theorems 3.1,4.1]. Herein,

$$K^- := \{k^* \in W^* \mid k^*(k) \leq 0 \; \forall k \in K\}$$

denotes the negative dual cone of K and W^* denotes the dual space of W.

Theorem 1.3 (Fritz John Conditions) *Let Banach spaces* $(Z, \|\cdot\|_Z)$, $(V, \|\cdot\|_V)$, $(W, \|\cdot\|_W)$ *be given.*

(a) Let $S \subseteq Z$ be a closed convex set with non-empty interior and $K \subseteq W$ a closed convex cone with vertex at zero and non-empty interior.

(b) Let $J : Z \longrightarrow \mathbb{R}$ and $G : Z \longrightarrow W$ be Fréchet-differentiable and let $H : Z \longrightarrow V$ be continuously Fréchet-differentiable.

(c) Let $\hat{z}$ be a local minimum of NLP.

(d) Let the image of $H'(\hat{z})$ be closed in V.

Then there exist nontrivial multipliers $(\ell_0, \mu^*, \lambda^*) \in \mathbb{R} \times W^* \times V^*$, $(\ell_0, \mu^*, \lambda^*) \neq 0$, *such that*

$$\ell_0 \geq 0, \tag{1.5}$$

$$\mu^* \in K^-, \tag{1.6}$$

$$\mu^*(G(\hat{z})) = 0, \tag{1.7}$$

$$\ell_0 J'(\hat{z})(d) + \mu^*(G'(\hat{z})(d)) + \lambda^*(H'(\hat{z})(d)) \geq 0, \quad \text{for all } d \in S - \{\hat{z}\}. \tag{1.8}$$

Every point $(z, \ell_0, \mu^*, \lambda^*) \in Z \times \mathbb{R} \times W^* \times V^*$ with $(\ell_0, \mu^*, \lambda^*) \neq 0$ and (1.5)–(1.8) is called Fritz John point of NLP. Every Fritz John point $(z, \ell_0, \mu^*, \lambda^*)$ with $\ell_0 \neq 0$ is called Karush-Kuhn-Tucker (KKT) point.

Assumption (d) is satisfied, if $H'(\hat{z})$ is surjective. It may happen that Theorem 1.3 only holds with $\ell_0 = 0$. In this case the Fritz John conditions (1.5)–(1.8) are not very useful since the objective function J does not appear. To exclude this degenerated situation, a constraint qualification is required.

Theorem 1.4 (KKT Conditions) *Let the assumptions of Theorem 1.3 be satisfied. Theorem 1.3 holds with $\ell_0 = 1$ if one of the following constraint qualifications is satisfied:*

(a) Linear independence constraint qualification (LICQ):
   $\hat{z} \in int(S)$ and the operator $(G'(\hat{z}), H'(\hat{z}))$ is surjective in $W \times V$.

(b) Constraint qualification of Mangasarian-Fromowitz:
   $H'(\hat{z})$ is surjective and there exists some $\hat{d} \in int(S - \{\hat{z}\})$ with

$$H'(\hat{z})(\hat{d}) = 0, \qquad G(\hat{z}) + G'(\hat{z})(\hat{d}) \in int(K).$$

For a proof see [48, Sect. 2.3.5], [75, Theorems 3.1,4.1].

The purpose of the paper is to provide an overview on different aspects in optimal control with DAEs and to summarize the literature in this field to the best of the author's knowledge. Naturally many aspects occur and not all of them can be covered in depth, compare the recent monograph [16]. The reader will find more detailed references for the different topics in the subsequent sections. An outline of the paper is as follows. Necessary and sufficient conditions for linear-quadratic DAE optimal control problems will be investigated in Sect. 2. In Sect. 3 we will focus on necessary conditions in terms of local (or weak) minimum principles and global (or strong) minimum principles for a class of nonlinear DAE optimal control problems. Herein, the local minimum principles are derived by application of Theorem 1.3. Direct discretization methods are discussed in Sect. 4, while Sect. 5 is devoted to two function space methods, the projected gradient method and the Lagrange–Newton method. Conclusions and suggestions for further research in the section "Conclusions and Future Directions" conclude the paper.

Notation

We use the convention that the dimension of a real-valued vector x is denoted by $n_x \in \mathbb{N}$, that is $x \in \mathbb{R}^{n_x}$.

In order to simplify notation, we use the abbreviation $f[t]$ for a function of type $f(t, z(t))$, which depends on time and on one or more time dependent functions.

The partial derivatives of a function $f : \mathbb{R}^{n_x} \times \mathbb{R}^{n_y} \longrightarrow \mathbb{R}^{n_f}$, $(x, y) \mapsto f(x, y)$, at (x, y) are denoted by $f'_x(x, y)$ and $f'_y(x, y)$.

The Lebesque space of all measurable vector valued functions $f : I \longrightarrow \mathbb{R}^n$, $I = [t_0, t_f]$, which are bounded in the L^p-norm $\|f\|_p := (\int_I \|f(t)\|^p \, dt)^{1/p}$, is denoted by $L^p(I, \mathbb{R}^n)$, $1 \le p < \infty$. $L^\infty(I, \mathbb{R}^n)$ with norm $\|\cdot\|_\infty$ denotes the space of essentially bounded functions on I. $C(I, \mathbb{R}^n)$ denotes the space of continuous vector valued functions $f : I \longrightarrow \mathbb{R}^n$, $BV(I, \mathbb{R}^n)$ the space of measurable vector valued functions of bounded variation on I, and $NBV(I, \mathbb{R}^n)$ the space of normalized measurable vector valued functions of bounded variation on I, i.e., measurable vector valued functions of bounded variation that are continuous from the right and zero at t_0.

For $1 \le p \le \infty$, $W^{1,p}(I, \mathbb{R}^n)$ denotes the Sobolev space of absolutely continuous vector valued functions f that are bounded w.r.t. the norm $\|f\|_{1,p} := \max\{\|f\|_p, \|f'\|_p\}$.

Throughout, X^* denotes the dual space of a Banach space X.

2   Optimality Conditions for Linear-Quadratic DAE Optimal Control Problems

Linear-quadratic optimal control problems play an important role in the design of feedback controllers for control problems, see, e.g., [77, 87]. Typically the linear dynamics are obtained by linearization of the nonlinear dynamics (1.1) at an equilibrium point or along a given reference trajectory. The task is to minimize a convex quadratic objective function with the aim to minimize the control effort and the deviation of the state from a reference solution, which is given w.l.o.g. by the zero function.

Linear-quadratic optimal control problems are commonly approached as follows:

(a) A boundary value problem (BVP) ("optimality system") is formulated that consists of the original DAE with initial condition, of an adjoint DAE with terminal condition, and of a stationarity condition.
(b) It is typically straightforward to show with minimal assumptions that solvability of the BVP is actually sufficient for optimality.
(c) The solvability of the BVP is analyzed and conditions are formulated under which a solution of the BVP exists.

This approach differs to some extend from the standard approach in optimization, where necessary conditions are investigated first and afterwards it is shown that the necessary conditions in the convex case are sufficient as well. In the DAE context however, it may happen that the "optimality system" in (a) is not necessary for optimality since existence of a solution in (b) is not guaranteed automatically.

In order to motivate the adjoint DAE and to illustrate the difficulties in deriving necessary conditions we try to derive them by exploitation of Theorem 1.4. To this end we need to properly formulate the function spaces in which the optimal control problem lives. Instead of the descriptor form (1.4) of a linear DAE we follow [10] and consider linear DAEs with the special leading term formulation

$$A(t)(D(t)x(t))' + B(t)x(t) + P(t)u(t) = q(t) \qquad (2.1)$$

with continuous matrix functions $A \in L^{\infty}(I, \mathbb{R}^{n_x \times m})$, $m \in \mathbb{N}$, $D \in L^{\infty}(I, \mathbb{R}^{m \times n_x})$, $B \in L^{\infty}(I, \mathbb{R}^{n_x \times n_x})$, $P \in L^{\infty}(I, \mathbb{R}^{n_x \times n_u})$, $q \in L^{\infty}(I, \mathbb{R}^{n_x})$, and ker $A \oplus$ im $D = \mathbb{R}^m$. If D in (2.1) or E in (1.4) are smooth, both forms can be transformed into each other, compare [8, 10]. As we shall see, the leading term formulation reveals a nice symmetry in the original DAE and its adjoint DAE.

The leading term $A(Dx)'$ has the advantage that it indicates more appropriately from which space the state x and the control u should be chosen. For linear DAEs in descriptor form (1.4) transformations are necessary to reveal the spaces, compare Sect. 2.3.

To this end, the matrix D can be understood as a filter that filters out the differentiable components of x. A natural choice is to use the spaces

$$X := W_D^{1,2}(I, \mathbb{R}^{n_x}) := \{x \in L^2(I, \mathbb{R}^{n_x}) \mid Dx \in W^{1,2}(I, \mathbb{R}^m)\},$$
$$U := L^2(I, \mathbb{R}^{n_u}),$$

both being Hilbert spaces, compare Theorem 7.1 in Sect. 7.

For essentially bounded symmetric matrices $Q \in L^\infty(I, \mathbb{R}^{n_x \times n_x})$ and $R \in L^\infty(I, \mathbb{R}^{n_u \times n_u})$ and a given consistent vector $x_0 \in \mathbb{R}^{n_x}$ with x_0 in the range of $D(t_0)$ consider the following linear-quadratic optimal control problem.

Problem 2.1 (LQOCP) Minimize

$$\frac{1}{2} \int_I x(t)^\top Q(t) x(t) + u(t)^\top R(t) u(t) dt$$

with respect to $(x, u) \in X \times U$ subject to the constraints

$$A(t)(D(t)x(t))' + B(t)x(t) + P(t)u(t) = q(t),$$
$$D(t_0)x(t_0) = x_0.$$

The following subsections are devoted to the analysis of necessary and sufficient conditions for LQOCP.

2.1 Necessary Conditions

Define the functional $J : X \times U \longrightarrow \mathbb{R}$ by

$$J(x, u) := \frac{1}{2} \int_I x(t)^\top Q(t) x(t) + u(t)^\top R(t) u(t) dt$$

and the linear operator $H : X \times U \longrightarrow L^2(I, \mathbb{R}^{n_x}) \times \mathbb{R}^m$ by

$$H(x, u) := \begin{pmatrix} A(\cdot)(D(\cdot)x(\cdot))' + B(\cdot)x(\cdot) + P(\cdot)u(\cdot) - q(\cdot) \\ D(t_0)x(t_0) - x_0 \end{pmatrix}.$$

J and H are continuously Fréchet-differentiable and LQOCP reads

$$\text{Minimize} \quad J(x, u) \quad \text{subject to} \quad H(x, u) = 0,$$

i.e., LQOCP is of type NLP with $z = (x, u)$, $Z = X \times U$, $S = Z$, and we like to apply Theorem 1.4. To this end, let $\hat{z} = (\hat{x}, \hat{u})$ be a minimum of LQOCP and let Assumption 2.2 below hold.

Assumption 2.2 $H'(\hat{x}, \hat{u})$ *is surjective, that is, the initial value problem (IVP)*

$$A(t)(D(t)x(t))' + B(t)x(t) + P(t)u(t) = r(t),$$

$$D(t_0)x(t_0) = r_0$$

possesses a solution $(x, u) \in X \times U$ *for any* $(r, r_0) \in L^2(I, \mathbb{R}^{n_x}) \times \mathbb{R}^m$.

Under these assumptions, Theorem 1.4 yields the existence of a multiplier $(\lambda^*, \sigma) \in L^2(I, \mathbb{R}^{n_x})^* \times \mathbb{R}^m$ with

$$0 = L'_{(x,u)}(\hat{x}, \hat{u}, \lambda^*, \sigma)(x, u) \qquad \forall (x, u) \in X \times U, \tag{2.2}$$

where $L : X \times U \times L^2(I, \mathbb{R}^{n_x})^* \times \mathbb{R}^m \longrightarrow \mathbb{R}$,

$$L(x, u, \lambda^*, \sigma) := J(x, u) + \lambda^*(H(x, u)) + \sigma^\top (D(t_0)x(t_0) - x_0)$$

denotes the Lagrange function of LQOCP. Since $L^2(I, \mathbb{R}^{n_x})$ is a Hilbert space, the multiplier λ^* can be represented as $\lambda^*(h) = \langle \lambda, h \rangle_{L^2(I, \mathbb{R}^{n_x})}$ with some $\lambda \in L^2(I, \mathbb{R}^{n_x})$ by the theorem of Riesz. Hence,

$$L(x, u, \lambda^*, \sigma) = J(x, u) + \int_I \lambda(t)^\top \left(A(t)(D(t)x(t))' + B(t)x(t) + P(t)u(t) - q(t) \right) dt$$

$$+ \sigma^\top (D(t_0)x(t_0) - x_0)$$

and the variational equation (2.2) implies

$$0 = \int_I \hat{x}(t)^\top Q(t)x(t) + \lambda(t)^\top \left(A(t)(D(t)x(t))' + B(t)x(t) \right) dt$$

$$+ \sigma^\top D(t_0)x(t_0), \tag{2.3}$$

$$0 = \int_{t_0}^{t_f} \left(R(t)^\top \hat{u}(t) + P(t)^\top \lambda(t) \right)^\top u(t) dt \tag{2.4}$$

for all $x \in X$ and all $u \in U$. The variational equation (2.4) implies the condition

$$0 = R(t)^\top \hat{u}(t) + P(t)^\top \lambda(t) \qquad \text{a.e. in } I. \tag{2.5}$$

For a further manipulation of the variational equation (2.3) we assume that $A(\cdot)^\top \lambda(\cdot)$ is in $W^{1,2}(I, \mathbb{R}^m)$ or equivalently, $\lambda \in W_{A^\top}^{1,2}(I, \mathbb{R}^{n_x}) := \{\lambda \in L^2(I, \mathbb{R}^{n_x}) \mid A^\top \lambda \in W^{1,2}(I, \mathbb{R}^m)\}$. Then partial integration yields for all $x \in X$,

$$0 = \left[\left(A(t)^\top \lambda(t) \right)^\top (D(t)x(t)) \right]_{t_0}^{t_f} + \sigma^\top D(t_0)x(t_0)$$
$$+ \int_{t_0}^{t_f} \left(\hat{x}(t)^\top Q(t) + \lambda(t)^\top B(t) - \left((A(t)^\top \lambda(t))' \right)^\top D(t) \right) x(t)\, dt.$$

Application of a variation lemma, see [48, Lemma 3.1.9], yields the adjoint DAE

$$- D(t)^\top (A(t)^\top \lambda(t))' + Q(t)\hat{x}(t) + B(t)^\top \lambda(t) = 0 \qquad \text{a.e. in } I, \qquad (2.6)$$

and the transversality conditions

$$D(t_0)^\top \left(A(t_0)^\top \lambda(t_0) - \sigma \right) = 0, \qquad D(t_f)^\top A(t_f)^\top \lambda(t_f) = 0. \qquad (2.7)$$

We summarize the findings.

Theorem 2.3 (Necessary Conditions for LQOCP) *Let $(\hat{x}, \hat{u})$ be a minimum of LQOCP and let Assumption 2.2 hold.*

Then there exists $\lambda \in L^2(I, \mathbb{R}^{n_x})$ and $\sigma \in \mathbb{R}^m$ such that (2.5) and (2.3) hold for every $x \in X$.

Moreover, if $\lambda \in W_{A^\top}^{1,2}(I, \mathbb{R}^{n_x})$, then (2.6) and (2.7) hold as well.

Theorem 2.3 is based on two crucial assumptions, namely Assumption 2.2 and the assumption $\lambda \in W_{A^\top}^{1,2}(I, \mathbb{R}^{n_x})$, which will not be satisfied for arbitrary data. A further analysis of the DAE (2.1) and the adjoint DAE (2.6) becomes necessary.

For index-1 tractable problems, the surjectivity assumption in Assumption 2.2 follows with [10, Theorem 3.2] with $D(t_0)$ being surjective, but for index-2 tractable problems surjectivity cannot be satisfied in general, since the term $r(t) - P(t)u(t)$ in Assumption 2.2 has to satisfy additional smoothness requirements, compare [10, Theorem 3.4] and [80, Theorem 2.52].

Since the adjoint DAE (2.6) has a leading term structure as well, the existence results in [10, Theorems 3.2, 3.4] can be applied to the adjoint DAE in order to guarantee that the adjoint DAE admits a solution $\lambda \in W_{A^\top}^{1,2}(I, \mathbb{R}^{n_x})$. In fact, it was shown in [80, Proposition 11.6, Theorem 11.9] that properties of the DAE (2.1) like the "properly stated leading term" and index-1 and index-2 tractability are inherited by the adjoint DAE.

2.2 Sufficient Conditions for Linear-Quadratic DAE Optimal Control Problems

The derivation of necessary conditions was subject to additional assumptions that needed to be imposed. However, if Q and R are symmetric and positive semi-definite, it is straightforward to show that the solvability of the "optimality system" (2.1), (2.5), (2.6), and (2.7) is already sufficient for optimality without any further assumptions.

Theorem 2.4 (Sufficient Condition) *Consider the linear-quadratic optimal control problem LQOCP. Let Q and R be symmetric and uniformly positive semi-definite on I. Let $(\hat{x}, \hat{u}) \in X \times U$ and $\lambda \in W_{A^\top}^{1,2}(I, \mathbb{R}^{n_x})$ on I satisfy*

$$A(t)(D(t)\hat{x}(t))' + B(t)\hat{x}(t) + P(t)\hat{u}(t) = q(t), \quad D(t_0)\hat{x}(t_0) = x_0, \tag{2.8}$$

$$-D(t)^\top (A(t)^\top \lambda(t))' + Q(t)\hat{x}(t) + B(t)^\top \lambda(t) = 0, \quad D(t_f)^\top A(t_f)^\top \lambda(t_f) = 0, \tag{2.9}$$

$$R(t)^\top \hat{u}(t) + P(t)^\top \lambda(t) = 0. \tag{2.10}$$

Then $(\hat{x}, \hat{u})$ is a global minimizer of LQOCP.

Proof Let (x, u) be feasible for LQOCP. Exploitation of the assumptions and neglecting the explicit time dependence for notational convenience yield

$$
\begin{aligned}
J(x, u) - J(\hat{x}, \hat{u}) &= \int_I \frac{1}{2} x^\top Q x + \frac{1}{2} u^\top R u - \frac{1}{2} \hat{x}^\top Q \hat{x} - \frac{1}{2} \hat{u}^\top R \hat{u} \, dt \\
&= \int_I \frac{1}{2} x^\top Q x + \frac{1}{2} u^\top R u - \frac{1}{2} \hat{x}^\top Q \hat{x} - \frac{1}{2} \hat{u}^\top R \hat{u} \\
&\quad + \lambda^\top \left(A(Dx)' + Bx + Pu - q \right) \\
&\quad - \lambda^\top \left(A(D\hat{x})' + B\hat{x} + P\hat{u} - q \right) \, dt \\
&= \int_I \frac{1}{2} x^\top Q x + \frac{1}{2} u^\top R u - \frac{1}{2} \hat{x}^\top Q \hat{x} - \frac{1}{2} \hat{u}^\top R \hat{u} \\
&\quad + \lambda^\top B(x - \hat{x}) + \lambda^\top P(u - \hat{u}) + (A^\top \lambda)^\top \left((Dx)' - (D\hat{x})' \right) \, dt \\
&= \int_I \frac{1}{2} x^\top Q x + \frac{1}{2} u^\top R u - \frac{1}{2} \hat{x}^\top Q \hat{x} - \frac{1}{2} \hat{u}^\top R \hat{u} \\
&\quad + \lambda^\top B(x - \hat{x}) + \lambda^\top P(u - \hat{u}) - \left((A^\top \lambda)' \right)^\top (Dx - D\hat{x}) \, dt \\
&= \int_I \frac{1}{2} x^\top Q x + \frac{1}{2} u^\top R u - \frac{1}{2} \hat{x}^\top Q \hat{x} - \frac{1}{2} \hat{u}^\top R \hat{u} \\
&\quad + \lambda^\top B(x - \hat{x}) + \lambda^\top P(u - \hat{u}) - \hat{x}^\top Q(x - \hat{x}) - \lambda^\top B(x - \hat{x}) \, dt
\end{aligned}
$$

$$= \int_I \frac{1}{2} x^\top Q x + \frac{1}{2} u^\top R u - \frac{1}{2} \hat{x}^\top Q \hat{x} - \frac{1}{2} \hat{u}^\top R \hat{u} - \hat{u}^\top R (u - \hat{u})$$

$$- \hat{x}^\top Q (x - \hat{x}) \, dt$$

$$= \frac{1}{2} \int_I (x - \hat{x})^\top Q (x - \hat{x}) + (u - \hat{u})^\top R (u - \hat{u}) \, dt$$

$$\geq 0,$$

since Q and R are supposed to be uniformly positive semi-definite. □

The theorem states the following: If the optimality system (2.8)–(2.10) has a solution, then this solution is optimal. It is important to point out, that in this context the solvability of the overall optimality system (2.8)–(2.10) viewed as one single DAE is important, not the solvability of the individual DAEs (2.8) and (2.9) appearing in the optimality system. In general it is not guaranteed that the optimality system has a solution, see [8, Beispiel 3.16] for a counterexample. Sufficient conditions for solvability of the optimality system (2.8)–(2.10) can be found in [76, Theorem 5.5] or [8, Satz 3.22].

Let us finally state a sufficient condition for LQOCP in the presence of control-state constraints of type

$$S(t) x(t) + F(t) u(t) - r(t) \leq 0 \tag{2.11}$$

with matrix functions $S \in L^\infty(I, \mathbb{R}^{\ell \times n_x})$, $F \in L^\infty(I, \mathbb{R}^{\ell \times n_u})$, and $r \in L^\infty(I, \mathbb{R}^\ell)$. The investigation of necessary conditions subject to restrictions similar to those in Sect. 2.1 leads to the following optimality system:

$$A(t)(D(t)\hat{x}(t))' + B(t)\hat{x}(t) + P(t)\hat{u}(t) = q(t), \tag{2.12}$$

$$-D(t)^\top (A(t)^\top \lambda(t))' + Q(t)\hat{x}(t) + B(t)^\top \lambda(t) + S(t)^\top \eta(t) = 0, \tag{2.13}$$

$$R(t)^\top \hat{u}(t) + P(t)^\top \lambda(t) + F(t)^\top \eta(t) = 0, \tag{2.14}$$

$$D(t_0)\hat{x}(t_0) = x_0, \qquad D(t_f)^\top A(t_f)^\top \lambda(t_f) = 0, \tag{2.15}$$

$$0 \leq \eta(t) \quad \perp \quad -(S(t)\hat{x}(t) + F(t)\hat{u}(t) - r(t)) \geq 0, \tag{2.16}$$

where $0 \leq a \perp b \geq 0$ is an abbreviation for the complementarity conditions $a \geq 0, b \geq 0, a^\top b = 0$. Solvability of this complementarity system is again sufficient for optimality:

Theorem 2.5 (Sufficient Condition in the Presence of Control-State Constraints) *Consider the linear-quadratic optimal control problem LQOCP subject to the control-state constraint (2.11). Let Q and R be symmetric and uniformly positive*

semi-definite on I. Let $(\hat{x}, \hat{u}) \in X \times U$, $\lambda \in W_{A^\top}^{1,2}(I, \mathbb{R}^{n_x})$, and $\eta \in L^2(I, \mathbb{R}^\ell)$ satisfy (2.12)–(2.16) on I.
  Then $(\hat{x}, \hat{u})$ is a global minimizer of LQOCP subject to (2.11).

The proof can be found in Appendix 7. The investigation of sufficient conditions for the solvability of (2.12)–(2.16) is the subject of future research.

2.3 Problems in Descriptor Form

Let the DAE in LQOCP be given in descriptor form (1.4) with a singular matrix function E on $I = [t_0, t_f]$:

Problem 2.6 (LQOCP-DF) Minimize

$$\frac{1}{2} \int_I x(t)^\top Q(t) x(t) + u(t)^\top R(t) u(t) dt$$

subject to the constraints

$$E(t) x'(t) = A(t) x(t) + B(t) u(t) + q(t), \tag{2.17}$$

$$x(t_0) = x_0.$$

Remark 2.1 The assignment $x(t_0) = x_0$ in LQOCP-DF, i.e., fixing the entire initial state vector, is used for notational convenience only. In fact it is misleading since in general it is not possible to fix the entire initial state vector $x(t_0)$ in advance (if continuous controls are considered). Some components may depend implicitly on the control. For illustration consider the DAE

$$x_1'(t) = x_2(t), \qquad x_2'(t) = x_3(t) + u(t), \qquad 0 = x_2(t) - t.$$

The initial value of x_1 can be arbitrary whereas $x_2(t_0)$ and $x_3(t_0)$ have to be consistent, that is $x_2(t_0) = t_0$ and $x_3(t_0) = 1 - u(t_0)$. To this end, if the problem was considered in the space of continuous controls, then LQOCP-DF might fail to have a solution if the entire vector $x(t_0)$ is fixed a priori. However, if LQOPC-DF is considered in the L^2-space, then LQOCP-DF may still have a solution but with a discontinuous control at t_0.

Owing to the singularity of E not all components of x need to be differentiable. But for the correct definition of the function spaces in LQOCP-DF it is necessary to identify the differentiable components and those with lower differentiability requirements. A concept that is capable to identify such components is the behavior approach and the strangeness index, see [68, Definition 3.15]. Herein, states and controls are unified in one state vector $z := (x, u)^\top$ without imposing a priori assumptions on the smoothness of its components. The DAE (2.17) is then equivalently written as the DAE

$$\mathscr{E}(t)z'(t) = \mathscr{A}(t)z(t) + q(t) \tag{2.18}$$

with $\mathscr{E}(t) := \big(E(t) \big| 0 \big)$, $\mathscr{A}(t) := \big(A(t) \big| B(t) \big)$.

Assuming the existence of the strangeness index, a canonical form of type

$$z_1'(t) = \mathscr{A}_{13}(t)z_3(t) + q_1(t),$$
$$0 = z_2(t) + q_2(t),$$
$$0 = q_3(t)$$

can be derived using suitable projectors for a suitable partition of z and q, compare [68, Theorem 3.17]. This canonical form allows to identify the smoothness requirements of the components of z. The components of z_2 are fixed by the inhomogeneity q_2 and thus they possess the same smoothness properties as q_2. An initial value $z_2(t_0)$ is consistent if and only if $z_2(t_0) = -q_2(t_0)$. The smoothness of z_1 is determined by the smoothness of z_3 and q_1, that is, if z_3 and q_1 are L^1-integrable function, then z_1 is absolutely continuous. The last equation $0 = q_3(t)$ is a measure of inconsistency of the descriptor system, i.e., a solution exists if and only if $q_3 \equiv 0$.

Once this partitioning of the descriptor system is known, it is possible to consider optimization problems with the appropriate function spaces. However, this approach does not distinguish states and controls and may require a higher smoothness of a control than it is actually realistic from an application point of view. Hence, for control problems a refined analysis as in [68, Sect. 3.6] is necessary. Moreover, the strangeness index requires E and A to be sufficiently smooth, which might be a restriction if the linear DAE (2.17) results from a linearization of the nonlinear DAE (1.1) along an optimal solution.

We omit the details and summarize necessary and sufficient conditions for LQOCP-DF. It was shown in [67, Theorems 2,3] (for a more general setting but with $q \equiv 0$) that the solvability of the linear BVP

$$\begin{pmatrix} E(t) & 0 & 0 \\ 0 & -E(t)^\top & 0 \\ 0 & 0 & 0 \end{pmatrix} \begin{pmatrix} x'(t) \\ \lambda'(t) \\ u'(t) \end{pmatrix} = \begin{pmatrix} A(t) & 0 & B(t) \\ Q(t) & A(t)^\top + E'(t)^\top & 0 \\ 0 & B(t)^\top & R(t) \end{pmatrix} \begin{pmatrix} x(t) \\ \lambda(t) \\ u(t) \end{pmatrix},$$

$$x(t_0) = x_0, \qquad E(t_f)^\top \lambda(t_f) = 0,$$

is a necessary condition and a sufficient condition for optimality, if Q and R are symmetric and positive semi-definite. A characterization for the strangeness index to be zero is provided in [67, Proposition 4]. Further solution properties are discussed as well. Optimal control problems with autonomous descriptor systems are considered in [87].

3 Necessary Optimality Conditions for Nonlinear DAE Optimal Control Problems

The Fritz John conditions and the KKT conditions in Theorems 1.3 and 1.4, respectively, are the basic tools for the derivation of necessary conditions for nonlinear control and state constrained DAE optimal control problems. In order to avoid technical matters with regard to the proper identification of function spaces, as it was outlined for the linear-quadratic case in Sect. 2, we restrict the discussion to semi-explicit DAEs of index two and consider the following autonomous problem on the compact time interval $I = [t_0, t_f]$ with fixed time points $t_0 < t_f$, the control set $\mathscr{U} \subseteq \mathbb{R}^{n_u}$, and functions

$$\varphi : \mathbb{R}^{n_x} \times \mathbb{R}^{n_x} \longrightarrow \mathbb{R},$$

$$f_0 : \mathbb{R}^{n_x} \times \mathbb{R}^{n_y} \times \mathbb{R}^{n_u} \longrightarrow \mathbb{R},$$

$$f : \mathbb{R}^{n_x} \times \mathbb{R}^{n_y} \times \mathbb{R}^{n_u} \longrightarrow \mathbb{R}^{n_x},$$

$$g : \mathbb{R}^{n_x} \longrightarrow \mathbb{R}^{n_y},$$

$$s : \mathbb{R}^{n_x} \longrightarrow \mathbb{R}^{n_s},$$

$$\psi : \mathbb{R}^{n_x} \times \mathbb{R}^{n_x} \longrightarrow \mathbb{R}^{n_\psi}.$$

φ, f_0, f, s, ψ are supposed to be continuously differentiable and g is supposed to be twice continuously differentiable.

Problem 3.1 (OCP-SE) Minimize

$$\varphi(x(t_0), x(t_f)) + \int_I f_0(x(t), y(t), u(t)) \, dt$$

w.r.t. $x \in W^{1,\infty}(I, \mathbb{R}^{n_x})$, $y \in L^\infty(I, \mathbb{R}^{n_y})$, $u \in L^\infty(I, \mathbb{R}^{n_u})$ subject to the constraints

$$\begin{aligned}
x'(t) &= f(x(t), y(t), u(t)) & &\text{a.e. in } I, \\
0 &= g(x(t)) & &\text{in } I, \\
s(x(t)) &\leq 0 & &\text{in } I, \\
0 &= \psi(x(t_0), x(t_f)), \\
u(t) &\in \mathscr{U} & &\text{a.e. in } I.
\end{aligned}$$

This sufficiently simple problem class already exhibits the main difficulties in deriving necessary conditions. Moreover, OCP-SE contains practically important problem classes like mechanical multi-body systems in Gear-Gupta-Leimkuhler formulation, see [43]. For extensions towards more general systems with properly stated leading terms we refer to [8, Kapitel 5] and [80, Sect. 11.2] and for general nonlinear descriptor systems to [69].

Let $X := W^{1,\infty}(I, \mathbb{R}^{n_x})$, $Y := L^{\infty}(I, \mathbb{R}^{n_y})$, $U := L^{\infty}(I, \mathbb{R}^{n_u})$, $V := L^{\infty}(I, \mathbb{R}^{n_x}) \times W^{1,\infty}(I, \mathbb{R}^{n_y}) \times \mathbb{R}^{n_\psi}$, $W := C(I, \mathbb{R}^{n_s})$, $K := \{k \in C(I, \mathbb{R}^{n_s}) \mid k(t) \leq 0 \text{ in } I\}$, $Z := X \times Y \times U$, and $S := \{(x, y, u) \in Z \mid u(t) \in \mathscr{U} \text{ a.e. in } I\}$. Define $J : Z \longrightarrow \mathbb{R}$, $H : Z \longrightarrow V$, and $G : Z \longrightarrow W$ by

$$J(x, y, u) := \varphi(x(t_0), x(t_f)) + \int_I f_0(x(t), y(t), u(t)) \, dt,$$

$$H(x, y, u) := \begin{pmatrix} x'(\cdot) - f(x(\cdot), y(\cdot), u(\cdot)) \\ -g(x(\cdot)) \\ \psi(x(t_0), x(t_f)) \end{pmatrix},$$

$$G(x, y, u) := s(x(\cdot)).$$

With these definitions and $z = (x, y, u)$, OCP-SE fits into the problem class of Problem 1.2, i.e.,

$$\text{Minimize} \quad J(z) \quad \text{s.t.} \quad z \in S, \; G(z) \in K, \; H(z) = 0.$$

In the sequel, $\hat{z} = (\hat{x}, \hat{y}, \hat{u})$ denotes a local minimum of OCP-SE. Throughout we assume that the DAE has index two:

Assumption 3.2 *Let the matrix*

$$M(t) := g_x'(\hat{x}(t)) f_y'(\hat{x}(t), \hat{y}(t), \hat{u}(t))$$

be non-singular almost everywhere in I and let $M(\cdot)^{-1}$ be essentially bounded in I.

3.1 A Local (or Weak) Minimum Principle

The local minimum principle (or weak minimum principle) requires a convex control set $\mathscr{U}$ with non-empty interior and it provides a necessary condition for a local (or weak) minimum $\hat{z}$ of OCP-SE. Local minimality means that $\hat{z} = (\hat{x}, \hat{y}, \hat{u})$ minimizes J w.r.t. to all feasible $z = (x, y, u)$ with

$$\|z - \hat{z}\|_Z = \max\{\|x - \hat{x}\|_{1,\infty}, \|y - \hat{y}\|_\infty, \|u - \hat{u}\|_\infty\} < \varepsilon$$

for some $\varepsilon > 0$. Note that this neighborhood contains comparatively few functions. For instance, if $\hat{u}$ is a bang–bang control, then the above (weak) neighborhood only contains controls u with the same points of discontinuity as of $\hat{u}$. This weak neighborhood will be enlarged in Sect. 3.2, which eventually results in stronger necessary conditions.

In the sequel we only outline the main steps to prove a local minimum principle. We omit technical difficulties and technical assumptions, which can be found in [48, Chap. 4] and [46,47]. We emphasize that the same steps essentially can be followed for more general DAEs, see [69]. It turns out that the linearized DAE along the local minimum plays an important role and the existence and uniqueness results mentioned in Sect. 2 re-enter the scene.

The first step towards necessary conditions is to apply Theorem 1.3 to OCP-SE, which under appropriate assumptions yields the existence of multipliers $\ell_0 \geq 0$, $\lambda^* \in V^*$, and $\mu^* \in W^*$ such that the complementarity conditions

$$\mu^* \in K^- \quad \text{and} \quad \mu^*(G(\hat{x}, \hat{y}, \hat{u})) = 0, \tag{3.1}$$

and the variational inequality

$$0 \leq \ell_0 J'(\hat{x}, \hat{y}, \hat{u})(x, y, u) + \mu^*(G'(\hat{x}, \hat{y}, \hat{u})(x, y, u)) + \lambda^*(H'(\hat{x}, \hat{y}, \hat{u})(x, y, u)) \tag{3.2}$$

hold for all $(x, y, u) \in S - \{(\hat{x}, \hat{y}, \hat{u})\}$.

The second step is to obtain useful representations of the multipliers $\lambda^* \in V^* = L^\infty(I, \mathbb{R}^{n_x})^* \times W^{1,\infty}(I, \mathbb{R}^{n_y})^* \times \mathbb{R}^{n_\psi}$ and $\mu^* \in W^* = C(I, \mathbb{R}^{n_s})^*$. Without such representations, conditions (3.1) and (3.2) are of little practical use. The derivation of suitable representation (especially for λ^*) is rather technical and we omit the details, which can be found in [47] or [48, Chap. 3]. The analysis exploits the linearized DAE along the local minimum and results in the following representations with functions $\lambda_f \in BV(I, \mathbb{R}^{n_x})$, $\lambda_g \in L^\infty(I, \mathbb{R}^{n_y})$, $\zeta \in \mathbb{R}^{n_y}$, and $\mu \in NBV(I, \mathbb{R}^{n_s})$, compare [48, Corollary 3.2.4]:

$$\lambda_f^*(h_1) = -\int_I \left(\lambda_f(t)^\top + \lambda_g(t)^\top g_x'[t]\right) h_1(t)\,dt \qquad (h_1 \in L^\infty(I, \mathbb{R}^{n_x})),$$

$$\lambda_g^*(h_2) = -\zeta^\top h_2(t_0) - \int_I \lambda_g(t)^\top h_2'(t)\,dt \qquad (h_2 \in W^{1,\infty}(I, \mathbb{R}^{n_y})),$$

$$\mu^*(h) = \int_I h(t)^\top d\mu(t) \qquad (h \in C(I, \mathbb{R}^{n_s})).$$

Introducing these representations into (3.1)–(3.2) allows to deduce three variational inequalities and equalities for x, y, and u, respectively. Application of a variation lemma, compare [48, Lemma 3.1.9], eventually yields the following local minimum principle, compare [48, Theorem 3.2.7], which uses the Hamilton function

$$\mathscr{H}(x, y, u, \lambda_f, \lambda_g, \ell_0) := \ell_0 f_0(x, y, u) + \lambda_f^\top f(x, y, u) + \lambda_g^\top g_x'(x) f(x, y, u).$$

Theorem 3.3 (Local Minimum Principle for OCP-SE) *Let $(\hat{x}, \hat{y}, \hat{u})$ be a local minimum of OCP-SE. Let $\mathscr{U}$ be a closed and convex set with non-empty interior. Let Assumption 3.2 be valid and let the functions in OCP-SE be sufficiently regular, see [48, Assumption 2.2.8].*

Then there exist multipliers

$$\ell_0 \in \mathbb{R}, \ \lambda_f \in BV(I, \mathbb{R}^{n_x}), \ \lambda_g \in L^\infty(I, \mathbb{R}^{n_y}), \ \mu \in NBV(I, \mathbb{R}^{n_s}), \ \zeta \in \mathbb{R}^{n_y}, \ \sigma \in \mathbb{R}^{n_\psi}$$

such that the following conditions are satisfied:

(a) $\ell_0 \geq 0$, $(\ell_0, \zeta, \sigma, \lambda_f, \lambda_g, \mu) \neq 0$,
(b) Adjoint equations: Almost everywhere in I we have

$$\lambda_f(t) = \lambda_f(t_f) + \int_t^{t_f} \mathscr{H}_x'(\hat{x}(\tau), \hat{y}(\tau), \hat{u}(\tau), \lambda_f(\tau), \lambda_g(\tau), \ell_0)^\top d\tau,$$

$$+ \int_t^{t_f} s_x'(\hat{x}(\tau))^\top d\mu(\tau) \tag{3.3}$$

$$0 = \mathscr{H}_y'(\hat{x}(t), \hat{y}(t), \hat{u}(t), \lambda_f(t), \lambda_g(t), \ell_0)^\top. \tag{3.4}$$

(c) Transversality conditions:

$$\lambda_f(t_0)^\top = -\Big(\ell_0 \varphi_{x_0}'(\hat{x}(t_0), \hat{x}(t_f)) + \sigma^\top \psi_{x_0}'(\hat{x}(t_0), \hat{x}(t_f))$$

$$+ \zeta^\top g_x'(\hat{x}(t_0))\Big), \tag{3.5}$$

$$\lambda_f(t_f)^\top = \ell_0 \varphi_{x_f}'(\hat{x}(t_0), \hat{x}(t_f)) + \sigma^\top \psi_{x_f}'(\hat{x}(t_0), \hat{x}(t_f)). \tag{3.6}$$

(d) Stationarity of Hamilton function: Almost everywhere in I we have

$$\mathscr{H}_u'(\hat{x}(t), \hat{y}(t), \hat{u}(t), \lambda_f(t), \lambda_g(t), \ell_0)(u - \hat{u}(t)) \geq 0. \tag{3.7}$$

for all $u \in \mathscr{U}$.
(e) Complementarity condition: μ_i, $i \in \{1, \ldots, n_s\}$, is non-decreasing on I and constant on every interval (t_1, t_2) with $t_1 < t_2$ and $s_i(\hat{x}(t)) < 0$ for all $t \in (t_1, t_2)$.

Normality of the multiplier ℓ_0 (meaning that $\ell_0 = 1$ can be chosen) requires an additional constraint qualification, for instance the linear independence constraint qualification or the Mangasarian-Fromowitz constraint qualification, compare Theorem 1.4. In practice, it is often assumed that $\ell_0 = 1$ holds and with this assumption one tries to satisfy the necessary conditions either numerically or analytically.

The adjoint integral equation (3.3) with the Riemann–Stieltjes integral implies that the adjoint λ_f in general is of bounded variation only and hence λ_f may be discontinuous. The complementarity condition in (e) implies that the discontinuities may occur on active arcs of the state constraint only and the discontinuities are triggered by the multiplier μ. One can show that λ_f in between two discontinuities satisfies the differential equation

$$\lambda'_f(t) = -\mathscr{H}'_x(\hat{x}(t), \hat{y}(t), \hat{u}(t), \lambda_f(t), \lambda_g(t), \ell_0)^\top - s'_x(\hat{x}(t))^\top \mu'(t),$$

and at every point of discontinuity $t \in (t_0, t_f)$, λ_f satisfies the jump condition

$$\lambda_f(t) - \lambda_f(t-) = -s'_x(\hat{x}(t))^\top (\mu(t) - \mu(t-)),$$

where $\lambda_f(t-)$ and $\mu(t-)$ denote the left-sided limits at t.

Further local minimum principles for smooth problems with mixed control-state constraints and a combination of pure state constraints and mixed control-state constraints are derived in [46, Theorem 3.2].

A weak (or local) minimum principle for non-smooth optimal control problems with mixed control-state constraints and semi-explicit DAEs is proved in [33, Theorems 3.1,3.2].

A constructive way to prove a local minimum principle for higher index DAEs by means of first-order approximations using adjoint equations is described in [93,94].

Remark 3.1 The local minimum principle contains some hidden index reduction which is visible in the Hamilton function through the term $g'_x f$.

This raises the question whether an analog local minimum principle holds for the Hamilton function

$$\tilde{\mathscr{H}}(x, y, u, \lambda_f, \lambda_g, \ell_0) := \ell_0 f_0(x, y, u) + \lambda_f^\top f(x, y, u) + \lambda_g^\top g(x) \ ?$$

Using $\tilde{\mathscr{H}}$ instead of $\mathscr{H}$ in Theorem 3.3 yields the so-called formal necessary conditions. In general, these conditions do not constitute necessary optimality conditions, hence the terminology "formal necessary condition" is used. However, there are cases where the formal necessary conditions are actually necessary optimality conditions and can be related to the true necessary conditions, compare [70] for details in the linear case.

3.2   A Global (or Strong) Minimum Principle

In contrast to the local (or weak) minimum principle, the global minimum principle (or strong minimum principle) states necessary conditions for a strong local minimum $\hat{z}$ of J, which means that $\hat{z} = (\hat{x}, \hat{y}, \hat{u})$ minimizes J w.r.t. to all feasible $z = (x, y, u)$ with $\|x - \hat{x}\|_\infty < \varepsilon$ for some $\varepsilon > 0$. This neighborhood is comparatively large, as the distances of u and y to $\hat{u}$ and $\hat{y}$ are not taken into account explicitly. This allows to relax the assumptions on the control set $\mathscr{U}$, which in the sequel is merely supposed to be a bounded measurable subset of $\mathbb{R}^{n_u}$. $\mathcal{U}$ may even be a discrete set.

The following theorem is taken from [48, Theorem 7.1.6] and applies to OCP-SE without pure state constraints. An analog result holds for semi-explicit index-1 problems, see [52, Theorem 9.7]. Similar results are obtained in [97, Propositions 2–4] for semi-explicit Hessenberg DAEs up to index 3. The proof technique in [97] applies the minimum principle from ODE theory to the underlying ODE of the Hessenberg DAEs, while the proof technique in [48, 52] exploits a variable time transformation technique in combination with the local minimum principle.

Theorem 3.4 (Global Minimum Principle) *Let $(\hat{x}, \hat{y}, \hat{u})$ be a strong local minimum of OCP-SE (without pure state constraints $s(x(t)) \le 0$). Let Assumption 3.2 be valid and let the functions in OCP-SE be sufficiently regular, see [48, Assumption 2.2.8].*

Then there exist multipliers

$$\ell_0 \in \mathbb{R}, \ \lambda_f \in W^{1,\infty}(I, \mathbb{R}^{n_x}), \ \lambda_g \in L^{\infty}(I, \mathbb{R}^{n_y}), \ \zeta \in \mathbb{R}^{n_y}, \ \sigma \in \mathbb{R}^{n_\psi}$$

such that the following conditions are satisfied:

(a) $\ell_0 \ge 0$, $(\ell_0, \zeta, \sigma, \lambda_f, \lambda_g) \ne 0$

(b) Adjoint equations: Almost everywhere in I we have

$$\lambda'_f(t) = -\mathcal{H}'_x(\hat{x}(t), \hat{y}(t), \hat{u}(t), \lambda_f(t), \lambda_g(t), \ell_0)^\top,$$

$$0 = \mathcal{H}'_y(\hat{x}(t), \hat{y}(t), \hat{u}(t), \lambda_f(t), \lambda_g(t), \ell_0)^\top.$$

(c) Transversality conditions:

$$\lambda_f(t_0)^\top = -\left(\ell_0 \varphi'_{x_0}(\hat{x}(t_0), \hat{x}(t_f)) + \sigma^\top \psi'_{x_0}(\hat{x}(t_0), \hat{x}(t_f)) + \zeta^\top g'_x(\hat{x}(t_0))\right),$$

$$\lambda_f(t_f)^\top = \ell_0 \varphi'_{x_f}(\hat{x}(t_0), \hat{x}(t_f)) + \sigma^\top \psi'_{x_f}(\hat{x}(t_0), \hat{x}(t_f)).$$

(d) Optimality condition: Almost everywhere in I we have

$$\mathcal{H}(\hat{x}(t), \hat{y}(t), \hat{u}(t), \lambda_f(t), \lambda_g(t), \ell_0) \le \mathcal{H}(\hat{x}(t), y, u, \lambda_f(t), \lambda_g(t), \ell_0)$$

for all $(u, y) \in \Omega(\hat{x}(t))$, where

$$\Omega(x) = \{(u, y) \in \mathcal{U} \times \mathbb{R}^{n_y} \mid g'_x(x) f(x, y, u) = 0\}.$$

(e) The Hamilton function is constant with respect to time:

$$\mathcal{H}(\hat{x}(t), \hat{y}(t), \hat{u}(t), \lambda_f(t), \lambda_g(t), \ell_0) \equiv const.$$

Note that the stationarity condition (3.7) in the local minimum principle can be seen as a necessary condition for the optimality condition (d) in the global minimum principle (if $\mathcal{U}$ is convex). Hence, the global minimum principle is a much stronger statement as it requires global minimality of the Hamilton function on Ω. The set $\Omega(x)$ contains feasible controls and algebraic variables that obey not only the explicit control constraints $u \in \mathcal{U}$ but also the hidden algebraic constraint $g'_x(x) f(x, y, u) = 0$.

The minimization on the set Ω is essential and it was observed in [34, Example, page 495] that a global minimum principle without this restriction may fail to hold, while a local minimum principle still holds, see [34, Theorem 3.2]. A global minimum principle for non-smooth optimal control problems was obtained in [34, Theorem 3.1] under a convexity assumption for the velocity set of a semi-explicit index-1 DAE.

In [36] a global minimum principle is derived for optimal control problems subject to implicit differential inequalities and control and state constraints under an analog of the Mangasarian-Fromowitz constraint qualification.

3.3 Indirect Methods and Boundary Value Problems

The indirect approach for optimal control problems is based on the semi-analytic exploitation of the necessary optimality conditions and its transformation to a multi-point boundary value problem. For simplicity we neglect state constraints in order to illustrate the basic approach and discuss an example first.

Example 3.5 Let constants $\alpha_1, \ldots, \alpha_4 > 0$ be given.
Minimize

$$\alpha_1 x_1(1)^2 + \alpha_2 (x_2(1) + 1)^2 + \int_0^1 \alpha_3 u(t)^2 + \alpha_4 x_1(t)^2 \, dt$$

subject to the constraints

$$\begin{aligned}
x_1'(t) &= u(t) - y(t), & x_1(0) &= 0, \\
x_2'(t) &= u(t), & x_2(0) &= 1, \\
x_3'(t) &= -x_2(t), & x_3(0) &= 0, \\
0 &= x_1(t) + x_3(t).
\end{aligned}$$

Differentiation of the algebraic constraint yields $0 = u(t) - y(t) - x_2(t)$ and the DAE has index two. The Hamilton function (with $\ell_0 = 1$, $x = (x_1, x_2, x_3)^{\mathsf{T}}$, $\lambda_f = (\lambda_1, \lambda_2, \lambda_3)^{\mathsf{T}}$) is defined by

$$\mathcal{H}(x, y, u, \lambda_f, \lambda_g, \ell_0) = \alpha_3 u^2 + \alpha_4 x_1^2 + \lambda_1 (u - y) + \lambda_2 u - \lambda_3 x_2 + \lambda_g (u - y - x_2).$$

Let $(\hat{x}, \hat{u})$ be a local minimum. Then the local minimum principle in Theorem 3.3 yields the conditions

$$0 = \mathcal{H}_u' = 2\alpha_3 \hat{u} + \lambda_1 + \lambda_2 + \lambda_g,$$

$$\lambda_1' = -\mathcal{H}_{x_1}' = -2\alpha_4 \hat{x}_1, \qquad\qquad \lambda_1(1) = 2\alpha_1 \hat{x}_1(1),$$

$$\lambda_2' = -\mathcal{H}_{x_2}' = \lambda_3 + \lambda_g, \qquad\qquad \lambda_2(1) = 2\alpha_2(\hat{x}_2(1) + 1),$$

$$\lambda_3' = -\mathcal{H}_{x_3}' = 0, \qquad\qquad\qquad \lambda_3(1) = 0,$$

$$0 = \mathcal{H}_y' = -\lambda_1 - \lambda_g.$$

The first equation can be solved for $\hat{u}$:

$$\hat{u} = -\frac{1}{2\alpha_3}\left(\lambda_1 + \lambda_2 + \lambda_g\right).$$

Introducing this expression into the DAE yields the following optimality system, which is a linear DAE boundary value problem with the index-2 algebraic variable $\hat{y}$ and the index-1 variable λ_g:

$$\hat{x}_1' = -\frac{1}{2\alpha_3}\left(\lambda_1 + \lambda_2 + \lambda_g\right) - \hat{y}, \qquad \hat{x}_1(0) = 0,$$

$$\hat{x}_2' = -\frac{1}{2\alpha_3}\left(\lambda_1 + \lambda_2 + \lambda_g\right), \qquad \hat{x}_2(0) = 1,$$

$$\hat{x}_3' = -\hat{x}_2, \qquad\qquad\qquad\qquad \hat{x}_3(0) = 0,$$

$$\lambda_1' = -2\alpha_4 \hat{x}_1, \qquad\qquad\qquad \lambda_1(1) = 2\alpha_1 \hat{x}_1(1),$$

$$\lambda_2' = \lambda_3 + \lambda_g, \qquad\qquad\qquad \lambda_2(1) = 2\alpha_2(\hat{x}_2(1) + 1),$$

$$\lambda_3' = 0, \qquad\qquad\qquad\qquad \lambda_3(1) = 0,$$

$$0 = \hat{x}_1 + \hat{x}_3,$$

$$0 = -\lambda_1 - \lambda_g.$$

In this case the linear BVP can be solved analytically, and we leave the details to the reader. In general, however, the BVP will be nonlinear and numerical methods are required to solve it. Observe that the index of the overall optimality system is still 2 with y being the index-2 algebraic variable and λ_g being an index-1 algebraic variable.   □

The key step in Example 3.5 was to express the control $\hat{u}$ not as a function of time but in feedback form $\hat{u} = u(\hat{x}, \hat{y}, \lambda_f, \lambda_g, \ell_0)$ (with $\ell_0 = 1$). Typically, this feedback form is obtained by solving the stationarity condition

$$\mathcal{H}_u'(\hat{x}, \hat{y}, \hat{u}, \lambda_f, \lambda_g, \ell_0) = 0$$

for $\hat{u}$, which is possible according to the implicit function theorem, if the Hessian matrix $\mathcal{H}_{uu}''$ is uniformly positive definite along the optimal solution. Note that the introduction of $\hat{u} = u(\hat{x}, \hat{y}, \lambda_f, \lambda_g, \ell_0)$ into the state and adjoint DAEs might change the index of these equations. Hence, it makes sense to consider and to analyze the overall optimality system as a coupled DAE with the additional algebraic constraint $\mathcal{H}_u' \equiv 0$ and u being the corresponding algebraic variable, see e.g., [9, 10, 67] for optimality systems arising in linear-quadratic optimal control problems.

If $\mathcal{H}_u' \equiv 0$ cannot be solved for u, the situation is more complicated. We illustrate the complications by a modification of Example 3.5.

Example 3.6 Consider again the optimal control problem in Example 3.5, but assume that $\alpha_3 = 0$.

In this case, we need to add constraints on u in order to obtain a meaningful optimization problem. So, suppose u is bound to the set

$$\mathcal{U} := [u_{min}, u_{max}] \qquad (u_{min} < u_{max}).$$

The global minimum principle in Theorem 3.4 states that the optimal control $\hat{u}$ minimizes the Hamilton function on the set

$$\Omega(\hat{x}(t)) = \{(u, y) \mid u \in [u_{min}, u_{max}], u - y - \hat{x}_2(t) = 0\}$$

for a.e. t. Minimizing the Hamilton function on $\Omega(\hat{x}(t))$ (observe $u - y = \hat{x}_2(t)$) yields:

$$\hat{u}(t) = \begin{cases} u_{min}, & \text{if } \lambda_2(t) > 0, \\ u_{max}, & \text{if } \lambda_2(t) < 0, \\ \text{singular}, & \text{if } \lambda_2(t) = 0 \text{ in some interval } [t_1, t_2] \text{ with } t_1 < t_2 \end{cases}$$

This determines the optimal control in feedback form, i.e., as a function of the adjoint λ_2. In the singular case we obtain the algebraic constraint $\lambda_2(t) = 0$ on the interval $[t_1, t_2]$. Differentiation yields $0 = \lambda_3(t) + \lambda_g(t)$. Since $\lambda_3 \equiv 0$ and $\lambda_g = -\lambda_1$ we find $\lambda_1(t) = 0$. Differentiation implies $\hat{x}_1(t) = 0$ and another differentiation yields $0 = \hat{u}(t) - \hat{y}(t) = \hat{x}_2(t)$ on $[t_1, t_2]$. By differentiation the latter holds if $\hat{u}(t) = 0$ on $[t_1, t_2]$, i.e., the control is zero on singular arcs. In order to obtain this singular control law on the interval $[t_1, t_2]$, we needed 4 differentiations of the algebraic constraint $\lambda_2 = 0$ and some algebraic manipulations to finally obtain the condition $\hat{u} = 0$ on $[t_1, t_2]$. Hence, the (differentiation) index is five on a singular subarc. On non-singular subarcs, $\hat{u}$ is explicitly given by u_{min} or u_{max}. Introducing this into the state DAE and the adjoint DAE yields a DAE of index two on non-singular subarcs, with y being the index-two algebraic variable and λ_g being the index-one algebraic variable, but an index-5 DAE on singular subarcs. □

Example 3.6 illustrates the difficulties that arise in the presence of control or state constraints. It is easy to imagine that the situation can be very involved if many control or state constraints are present in a nonlinear problem. Moreover, we have analyzed singular and non-singular arcs in Example 3.6, but we still do not know the time intervals on which the constraints are active or inactive/singular. Singular arcs may not even exist in Example 3.6. Apparently, singular subarcs cannot occur in Example 3.6 if $0 \notin [u_{min}, u_{max}]$.

In general a good intuition, a sound knowledge of the underlying physical problem, or good homotopy methods, see [35], are needed to recognize the switching structure of the optimal solution, i.e., the sequence of active, inactive, or singular arcs of the inequality constraints. On active arcs the optimality system can be interpreted as a DAE which consists of the original DAE, the adjoint DAE, the active control-state constraint (viewed as an algebraic equation), and the stationarity condition $\mathscr{H}'_u = 0$ (if $\mathscr{U} = \mathbb{R}^{n_u}$). Herein, the control vector u and the multipliers for the constraints serve as algebraic variables. On inactive arcs the optimality system can be viewed as a DAE as well but without the algebraic constraints resulting from the active inequality constraints. Hence, the overall optimality system can be viewed as a piecewise defined DAE with appropriate coupling conditions at the junction points of the different arcs. The index of the piecewise defined DAE may switch at the junction points. The overall system is a multi-point DAE boundary value problem. Its numerical solution requires a very good initial guess for the switching structure and the adjoints. For complicated problems this is hard to realize and for that reason other methods like direct discretization methods or function space methods are usually preferred. Note that also for the simpler problems in [8, 80] it may happen, that the overall optimality DAE fails to have index one around the solution, see e.g., [80, Example 11.19, Sect. 11.3].

A detailed overview on collocation methods and shooting methods for boundary value problems is provided by the monographs [6] and [68, Chap. 7]. Collocation methods for DAE boundary value problems can be found in [7, 65, 71, 73, 104]. Shooting methods are discussed in [72, 78, 79]. A parallel multiple shooting algorithm is developed in [63]. A damped Newton method and finite difference approximations are used in [13] to solve the BVP, which is formulated for mechanical multi-body systems using symbolic computations. [100] use a piecewise defined boundary value problem to solve optimal control problems with miniaturized manipulators. An algorithm for 2-point boundary value problems arising in economic problems with an infinite time horizon is discussed in [74].

4   Direct Discretization Methods for DAE Optimal Control Problems

Direct discretization methods have a long tradition in solving optimal control problems subject to ODEs and many extensions to DAEs exist as well, see, e.g., [12, 17, 20, 25, 27, 39, 40, 57, 64, 84, 88, 90–92, 95, 102, 103, 111]. Direct discretization methods can be grouped into collocation methods, compare [14, 60, 112], and direct (single or multiple) shooting methods, see, e.g., [19,23,44,53,105]. Both approaches have in common that the infinite dimensional optimal control problem is discretized and transformed into a finite dimensional nonlinear optimization problem. The approaches differ in the way the optimal control problem is discretized. Depending on the discretization the resulting optimization problems exhibit a small but dense structure or a large-scale but sparse structure. The sparse structure needs to be exploited in order to obtain an efficient numerical method. We do not focus on these algorithmic details since state-of-the-art nonlinear programming solvers are able to handle all these issues.

Consider the following optimal control problem on $I = [t_0, t_f]$ and note that more general problems like in Problem 1.1 can be transformed to this form by standard techniques.

Problem 4.1 Minimize

$$\varphi(x(t_0), x(t_f))$$

subject to the constraints

$$F(t, x(t), x'(t), u(t)) = 0 \qquad\qquad \text{a.e. in } I,$$
$$c(t, x(t), u(t)) \leq 0 \qquad\qquad \text{a.e. in } I,$$
$$\psi(x(t_0), x(t_f)) = 0.$$

4.1 The Full Discretization (Full Transcription)

Define the (for simplicity) equidistant grid

$$\mathbb{G}_N := \{t_i \mid t_i = t_0 + ih, i = 0, 1, \ldots, N\} \qquad (N \in \mathbb{N}),$$

with step-size $h = (t_f - t_0)/N$ and approximate the DAE on $\mathbb{G}_N$ by a suitable discretization method. Essentially any suitable discretization scheme for DAEs can be used for discretization, see [21, 58, 59] for an overview on BDF methods and Runge–Kutta methods. For notational simplicity we use the implicit Euler method

$$F\left(t_{i+1}, x_{i+1}, \frac{x_{i+1} - x_i}{h}, u_{i+1}\right) = 0, \qquad i = 0, 1, \ldots, N - 1, \qquad (4.1)$$

with approximations $x_i \approx x(t_i), i = 0, \ldots, N$, and $u_i \approx u(t_i), i = 1, \ldots, N$.

If the initial value x_0 is not fixed but contains degrees of freedom that are to be exploited in the optimization, a procedure is necessary to ensure consistency of the initial value. Consistent initial values can be achieved numerically, e.g., by adding algebraic constraints and hidden constraints at x_0 explicitly to the discretized problem, by projection, see [49] or [48, Sect. 4.5.1], or by relaxation, see [103] or [48, Sect. 4.5.2]. Alternate methods have been suggested by [5, 22, 28, 54, 82, 89].

The projection method projects a possibly inconsistent initial value x_0 onto algebraic constraints and yields a consistent value $\tilde{x}_0 = X_0(x_0, u_h)$ with a differentiable function X_0 and the control parameterization $u_h = (u_1, \ldots, u_N)^\top$. We impose the constraint $x_0 = X_0(x_0, u_h)$ to ensure that x_0 in the optimal solution is actually consistent. For semi-explicit Hessenberg-DAEs the function X_0 can be realized by solving a least squares problem.

In contrast to the projection method, the relaxation technique does not modify the initial value x_0 within each iteration of the numerical procedure to solve the optimization problem but instead it modifies the DAE such that x_0 becomes consistent for the modified DAE. Therefore additional consistency constraints have to be added to the optimization problem to ensure equivalence of the original DAE and the modified DAE in the optimal solution. This technique is mainly used in shooting methods.

An approximation of Problem 4.1 with the projection method for consistent initialization then reads as follows:

Problem 4.2 (DOCP) Minimize

$$\varphi(x_0, x_N)$$

with respect to $x_h := (x_0, \ldots, x_N)^\top$ and $u_h := (u_1, \ldots, u_N)^\top$ subject to the constraints (4.1) and

$$x_0 - X_0(x_0, u_h) = 0,$$
$$c(t_i, x_i, u_i) \leq 0, \qquad\qquad i = 0, \ldots, N, \text{ with } u_0 := u_1,$$
$$\psi(x_0, x_N) = 0.$$

DOCP is a finite dimensional optimization problem of type NLP, which is large-scale but exhibits a sparse structure in the Jacobian of the constraints and the Hessian matrix of the Lagrange function. It can be solved by any optimization method that is able to handle large-scale and sparse structures. The most frequently used paradigms are sequential quadratic programming (SQP), interior-point methods, or multiplier methods. The sparsity patterns have to be exploited in the computation of derivatives and on linear algebra level, where linear equations with saddle point matrices (KKT matrices) need to be solved, compare [14] or [102] for a reduced SQP method.

The accuracy of the solution of DOCP can be increased by grid refinement techniques. These techniques often use the local discretization error of the discretization (implicit Euler in our case) to measure accuracy, compare [15]. A grid refinement strategy conceptionally works as follows:

Algorithm 4.3 (Full Discretization Method)

(0) Init: Choose $N \in \mathbb{N}$, $tol > 0$,
(1) Solve DOCP by a suitable optimization method.

(2) Estimate the local discretization error of the discrete solution.
(3) If the local discretization error is less than *tol*, STOP.
(4) Refine the grid based on the local discretization error and go to (1) with the refined grid.

It was observed in [17, 39] that full discretization methods are often able to solve constrained optimal control problems whose index on active state constraints, if viewed as a DAE, is high (in other words: the order of the state constraint is high). An example is a state constrained optimal control problem subject to a method-of-lines discretization of a heat conduct equation, where the index on active arcs depends on the PDE discretization parameters and can be arbitrarily high. Full discretization methods are still able to solve such problems. This effect was analyzed further in [27, 64]. It is argued that shooting methods are likely to fail in this case. This is certainly true, if one attempts to solve the high index DAE on active arcs explicitly, which can be the case in the classic indirect method. Direct shooting methods, however, would not do this, since they work with discretized state constraints as well and impose them as inequality constraints in the optimization problem. Hence, direct shooting methods are able to solve many problems with higher order state constraints as well.

4.2   The Reduced Discretization (Direct Shooting Methods)

The reduced discretization approach (or shooting method) aims to reduce the size of Problem DOCP by elimination of the equations in (4.1) from DOCP. Suppose Eq. (4.1) can be solved for x_{i+1} step by step. This is guaranteed by the implicit function theorem (provided a solution exists), if the so-called iteration matrix

$$F'_x + \frac{1}{h} F'_{x'}$$

is non-singular (this requires regular local pencils). By solving the equations step by step we obtain

$$\tilde{x}_0 = X_0(x_0, u_h),$$

$$x_1 = \bar{X}_1(\tilde{x}_0, u_1) = \bar{X}_1(X_0(x_0, u_h), u_1) =: X_1(x_0, u_h),$$

$$\vdots$$

$$x_{i+1} = \bar{X}_{i+1}(x_i, u_{i+1}) = \bar{X}_{i+1}(X_i(x_0, u_h), u_{i+1}) =: X_{i+1}(x_0, u_h),$$

$$\vdots$$

$$x_N = X_N(x_0, u_h),$$

where the value x_0 is projected onto a consistent initial value $\tilde{x}_0$ using the function X_0. Clearly, solving Eq. (4.1) in the above way is just the single shooting idea, where the initial value problem $F(t, x(t), x'(t), u(t)) = 0$, $x(t_0) = \tilde{x}_0$ is solved by the implicit Euler method for a consistent initial value $\tilde{x}_0$. In case of unique solvability the solution will depend on the initial value and the control input only. That is what the functions $X_i(x_0, u_h)$ indicate.

Additional intermediate shooting nodes could be introduced in order to obtain a multiple shooting method but we prefer to focus on the single shooting case for simplicity and leave the details to the reader.

Introducing the implicitly defined values $x_i = X_i(x_0, u_h)$ into DOCP yields an optimization problem of reduced size with considerably fewer constraints and optimization variables.

Problem 4.4 (R-DOCP) Minimize

$$\varphi(X_0(x_0, u_h), X_N(x_0, u_h))$$

with respect to $z := (x_0, u_h)^\top$, $u_h = (u_1, \ldots, u_N)^\top$ subject to the constraints

$$c(t_i, X_i(x_0, u_h), u_i) \le 0, \qquad i = 0, \ldots, N,, \text{ with } u_0 := u_1,$$
$$\psi(X_0(x_0, u_h), X_N(x_0, u_h)) = 0.$$

Let

$$J(z) := \varphi(X_0(z), X_N(z)),$$
$$H(z) := \psi(X_0(z), X_N(z)),$$
$$G(z) := \begin{pmatrix} c(t_0, X_0(z), u_1) \\ c(t_1, X_1(z), u_1) \\ \vdots \\ c(t_N, X_N(z), u_N) \end{pmatrix}.$$

With these definitions R-DOCP fits into the problem class NLP. Compared to DOCP the reduced problem R-DOCP is comparatively small but it has a dense structure (apart from some minor sparsity introduced by control approximations with local support). Hence standard nonlinear programming methods using dense linear algebra are sufficient to solve R-DOCP. However, most nonlinear programming solvers require to evaluate the gradient of the objective function $J'(z^k)$ and the Jacobians of the constraints $H'(z^k)$ and $G'(z^k)$ at some intermediate iterate z^k. These derivatives can be computed in various ways. A simple but often inefficient and inaccurate approach is to use finite difference approximations.

Instead the expressions

$$J'(z^k) = \varphi'_{x_0}[z^k]X'_0(z^k) + \varphi'_{x_f}[z^k]X'_N(z^k),$$

$$H'(z^k) = \psi'_{x_0}[z^k]X'_0(z^k) + \psi'_{x_f}[z^k]X'_N(z^k),$$

$$G'(z^k) = \begin{pmatrix} c'_x[t_0]X'_0(z^k) + c'_u[t_0]\frac{\partial u_1(z^k)}{\partial z} \\ c'_x[t_1]X'_1(z^k) + c'_u[t_1]\frac{\partial u_1(z^k)}{\partial z} \\ \vdots \\ c'_x[t_N]X'_N(z^k) + c'_u[t_N]\frac{\partial u_N(z^k)}{\partial z} \end{pmatrix},$$

are useful, where the notations $[z^k]$ and $[t_i]$ indicate that the respective functions are evaluated at $X_i(z^k)$ and at the grid points t_i, $i = 0, \ldots, N$. Herein, it is convenient to view the control value u_i as a function of z^k, i.e., $u_i = u_i(z^k)$. The sensitivities $S_i := X'_i(z^k)$, $i = 0, 1, \ldots, N$ need to be computed. That is, the numerical solution of the initial value problem needs to be differentiated w.r.t. the optimization vector z. This can be achieved by internal numerical differentiation (IND), compare [18]. The idea is to differentiate the numerical discretization scheme w.r.t. the optimization variables. Application of this idea to the implicit Euler discretization in (4.1) yields the linearized equations

$$\left(F'_x[z^k] + \frac{1}{h}F'_{x'}[z^k]\right) X'_{i+1}(z^k) - \frac{1}{h}F'_{x'}[z^k]X_i(z^k) + F'_u[z^k]\frac{\partial u_{i+1}(z^k)}{\partial z} = 0$$

for $i = 0, 1, \ldots, N - 1$. The IND approach coincides with the sensitivity equation approach, if the sensitivity equation (or variational equation)

$$F'_x[t]S(t) + F'_{x'}[t]S'(t) + F'_u[t]\frac{\partial u(t; z^k)}{\partial z} = 0$$

with a control parameterization $u = u(t; z^k)$ (e.g., a B-spline representation) is discretized by the same numerical method with the same step-sizes as for the nonlinear DAE $F(t, x(t), x'(t), u(t; z^k)) = 0$. Extensions of this approach can be found in [11, 30, 41, 86]. These numerical procedures often re-use the iteration matrix $F'_x + \frac{1}{h}F'_{x'}$ for several integration steps in order to save computation time. A drawback of these sensitivity equation approaches is that the dimension of the sensitivity matrix $S_i = X'_i(z^k)$ grows with the dimension of the optimization vector z. Hence, the approach is not very efficient if many optimization variables but few constraints are present.

In this case the adjoint equation approach in [29] is more efficient. Herein, an adjoint equation has to be solved for each constraint of the optimization problem. The dimension of the adjoint equation does not depend on the number of optimization variables and hence the adjoint equation approach is more efficient if few constraints but many optimization variables are present. The adjoint method for gradient computations will be used in Sect. 5.1 to compute the gradient of a reduced objective functional in a projected gradient method.

Finally, an accurate and efficient approach with provable efficiency bounds is to use tools from algorithmic differentiation in combination with checkpointing strategies, compare [55, 56]. Checkpointing is a way to reduce the storage requirements. Note that the adjoint equation requires either to store the state at all grid points or to recompute it whenever it is needed in the adjoint system. The idea of a checkpointing strategy is to introduce a suitable number of intermediate time points at which the state is stored and from which the state is recomputed by forward integration if the state is needed by the adjoint scheme at intermediate time points. This technique allows to balance storage requirements and computing effort.

The following numerical experiments illustrate the reduced discretization approach. We used the software OCPID-DAE1 by the author, which is available for academic use on the webpage http://www.optimal-control.de\.

Example 4.5 We revisit Example 3.5:
Minimize

$$\alpha_1 x_1(1)^2 + \alpha_2(x_2(1) + 1)^2 + \int_0^1 \alpha_3 u(t)^2 + \alpha_4 x_1(t)^2 \, dt$$

subject to the constraints

$$
\begin{aligned}
x_1'(t) &= u(t) - y(t), & x_1(0) &= 0, \\
x_2'(t) &= u(t), & x_2(0) &= 1, \\
x_3'(t) &= -x_2(t), & x_3(0) &= 0, \\
0 &= x_1(t) + x_3(t).
\end{aligned}
$$

The direct (single) shooting method with tolerance $tol = 10^{-8}$, $\alpha_1 = \alpha_2 = \alpha_4 = 1$, $\alpha_3 = 0.5 \cdot 10^{-2}$, $N = 100$, and initial guess $u_h^0 \equiv -3$ produces the following iterates:

```
=== sqpfiltertoolbox  (C) MATTHIAS GERDTS, UNIVERSITAET DER BUNDESWEHR   ====
----------------------------------------------------------------------------
 ITER   QPIT   ALPHA               OBJ                   CV            KKT
----------------------------------------------------------------------------
 0       0    0.0000E+00   0.1328333044250002E+01   0.0000E+00    0.3103E-01
 1       1    0.1000E+01   0.1261996137706250E+01   0.0000E+00    0.3018E-01
 2       1    0.1000E+01   0.9645290326151866E+00   0.0000E+00    0.2600E-01
 3       1    0.1000E+01   0.1626445121083435E+00   0.0000E+00    0.7893E-02
 4       1    0.1000E+01   0.5933569494404990E-01   0.0000E+00    0.1340E-02
 5       1    0.1000E+01   0.5929137763126855E-01   0.0000E+00    0.1340E-02
...
 74      1    0.1000E+01   0.4105117130358388E-01   0.0000E+00    0.2923E-07
 75      1    0.1000E+01   0.4105117130166638E-01   0.0000E+00    0.2956E-07
 76      1    0.1000E+01   0.4105117129766830E-01   0.0000E+00    0.2720E-07
 77      1    0.1000E+01   0.4105117129130897E-01   0.0000E+00    0.1935E-07
 78      1    0.1000E+01   0.4105117128571807E-01   0.0000E+00    0.8277E-08
============================================================================
END OF SQP METHOD
============================================================================
KKT CONDITIONS SATISFIED TO THE REQUESTED ACCURACY (IER=          0  )!
```

The column "ITER" shows the major iteration index of a filter and linesearch SQP method, "QPIT" indicates the iterations needed to solve a linear-quadratic optimization problem within each major iteration, "ALPHA" is the step-size used in a globalization procedure, "OBJ" is the objective function value, "CV" denotes the constraint violation, and "KKT" contains the error in the optimality conditions (KKT condition).

The Following Pictures Show The Converged Solution Of The Direct Shooting Method: □

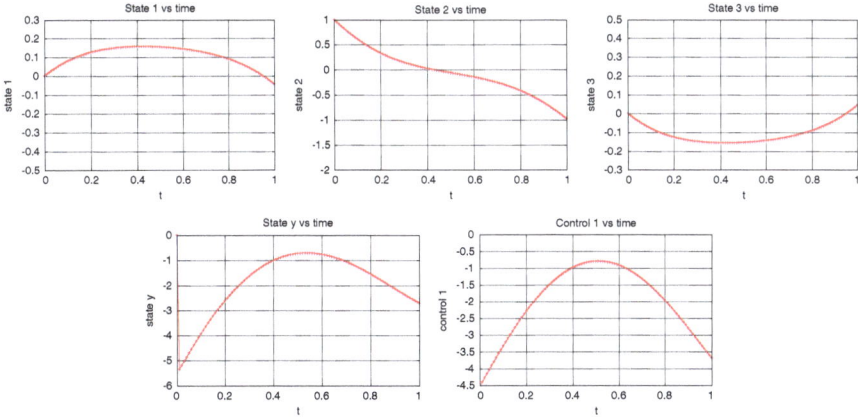

Example 4.6 We revisit Example 3.5 with additional control constraints and $\alpha_3 > 0$:

Minimize

$$\alpha_1 x_1(1)^2 + \alpha_2(x_2(1) + 1)^2 + \int_0^1 \alpha_3 u(t)^2 + \alpha_4 x_1(t)^2 \, dt$$

subject to the constraints

$$
\begin{aligned}
x_1'(t) &= u(t) - y(t), & x_1(0) &= 0, \\
x_2'(t) &= u(t), & x_2(0) &= 1, \\
x_3'(t) &= -x_2(t), & x_3(0) &= 0, \\
0 &= x_1(t) + x_3(t), \\
u(t) &\in [-3, -1].
\end{aligned}
$$

The direct (single) shooting method with optimality and feasibility tolerance $tol = 10^{-8}$, $\alpha_1 = \alpha_2 = \alpha_4 = 1$, $\alpha_3 = 0.5 \cdot 10^{-2}$, $N = 100$, and initial guess $u_h^0 \equiv -3$ produces the following iterates. Please note the high number 202 of QP iterations needed in iteration 1. The QP is being solved by a primal active set method and the high number of iterations indicates that the initial guess corresponds to an active set which is far away from the optimal active set. Hence, the QP needs quite a lot of iterations to identify the active constraints correctly.

```
=== sqpfiltertoolbox  (C) MATTHIAS GERDTS, UNIVERSITAET DER BUNDESWEHR  ====
-----------------------------------------------------------------------------
ITER    QPIT   ALPHA              OBJ                   CV           KKT
-----------------------------------------------------------------------------
0        0    0.0000E+00    0.1328333044250002E+01   0.0000E+00   0.3103E-01
1      202    0.1000E+01    0.1261996137706250E+01   0.0000E+00   0.3018E-01
2        1    0.1000E+01    0.9645290326151866E+00   0.0000E+00   0.2600E-01
3        1    0.1000E+01    0.1626445121083435E+00   0.0000E+00   0.7893E-02
4        1    0.1000E+01    0.5933569494404990E-01   0.0000E+00   0.1340E-02
5        1    0.1000E+01    0.5929137763126855E-01   0.0000E+00   0.1340E-02
...
270      1    0.1000E+01    0.4202430002584630E-01   0.0000E+00   0.2972E-07
271      1    0.1000E+01    0.4202430002537009E-01   0.0000E+00   0.2943E-07
272      1    0.1000E+01    0.4202430002306404E-01   0.0000E+00   0.2807E-07
273      1    0.1000E+01    0.4202430001330967E-01   0.0000E+00   0.2154E-07
274      1    0.1000E+01    0.4202429999601267E-01   0.0000E+00   0.8365E-08
=============================================================================
END OF SQP METHOD
=============================================================================
KKT CONDITIONS SATISFIED TO THE REQUESTED ACCURACY (IER=        0 )!
```

The following pictures show the converged solution of the direct shooting method: □

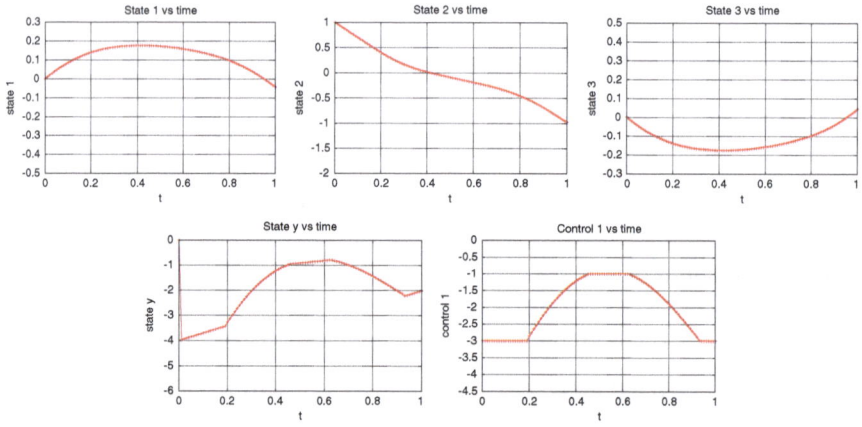

Example 4.7 We revisit Example 3.6 and choose $\alpha_3 = 0$. In this case, the optimal solution is a bang–bang-solution and singular subarcs cannot appear owing to the box constraints for u.

Minimize

$$\alpha_1 x_1(1)^2 + \alpha_2 (x_2(1) + 1)^2 + \int_0^1 \alpha_3 u(t)^2 + \alpha_4 x_1(t)^2 \, dt$$

subject to the constraints

$$
\begin{aligned}
x_1'(t) &= u(t) - y(t), & x_1(0) &= 0, \\
x_2'(t) &= u(t), & x_2(0) &= 1, \\
x_3'(t) &= -x_2(t), & x_3(0) &= 0, \\
0 &= x_1(t) + x_3(t), & & \\
u(t) &\in [-3, -1]. & &
\end{aligned}
$$

The direct (single) shooting method with optimality and feasibility tolerance $tol = 10^{-8}$, $\alpha_1 = \alpha_2 = \alpha_4 = 1$, $\alpha_3 = 0$, $N = 100$, and initial guess $u_h^0 \equiv -3$ produces the following iterates:

```
=== sqpfiltertoolbox   (C) MATTHIAS GERDTS, UNIVERSITAET DER BUNDESWEHR   ====
-------------------------------------------------------------------------------
ITER   QPIT   ALPHA              OBJ                     CV          KKT
-------------------------------------------------------------------------------
0        0    0.0000E+00    0.1283333044250000E+01    0.0000E+00    0.3073E-01
1      202    0.1000E+01    0.1218503970294439E+01    0.0000E+00    0.2989E-01
2        1    0.1000E+01    0.9276782329801730E+00    0.0000E+00    0.2577E-01
3        1    0.1000E+01    0.1415878077993118E+00    0.0000E+00    0.7918E-02
4        1    0.1000E+01    0.3865667516225475E-01    0.0000E+00    0.1287E-02
5        1    0.1000E+01    0.3861717769109438E-01    0.0000E+00    0.1287E-02
...
437      1    0.1000E+01    0.1817325950042694E-01    0.0000E+00    0.3984E-05
438      1    0.1000E+01    0.1817323931406664E-01    0.0000E+00    0.3878E-05
439      1    0.1000E+01    0.1817314890395615E-01    0.0000E+00    0.3359E-05
440      1    0.1000E+01    0.1817290710543137E-01    0.0000E+00    0.1110E-05
441      1    0.1000E+01    0.1817287743385263E-01    0.0000E+00    0.2496E-10
===============================================================================
END OF SQP METHOD
===============================================================================
KKT CONDITIONS SATISFIED TO THE REQUESTED ACCURACY (IER=         0 )!
```

The following pictures show the converged solution of the direct shooting method: □

Finally, a more challenging problem with control and state constraints is discussed.

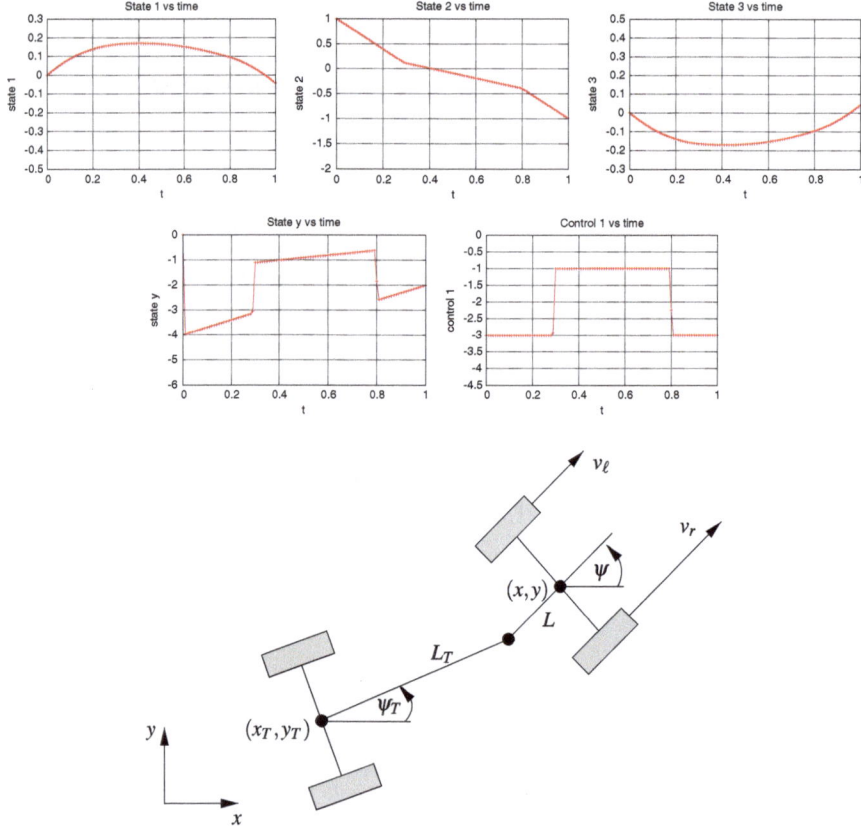

Fig. 1 Configuration of a vehicle and a trailer

Example 4.8 Consider a vehicle with two wheels and a trailer, which is attached to the vehicle, see Fig. 1.

Herein, (x, y) denotes the center of gravity of the vehicle, (x_T, y_T) the center of gravity of the trailer, ψ the yaw angle of the vehicle, ψ_T the yaw angle of the trailer, $B = 0.11$ the width of the vehicle and the trailer, $L = 0.1$ the distance of the center of gravity of the vehicle to the attachment point for the trailer, and $L_T = 0.2$ the distance of the center of gravity of the trailer to the attachment point of the vehicle. The velocities of the vehicle's wheels are denoted by v_ℓ (left) and v_r (right) and are subject to state constraints $v_\ell, v_r \in [-0.3, 0.3]$. The acceleration of each wheel of the vehicle can be controlled by controls u_ℓ and u_r subject to box constraints $u_\ell, u_r \in [-0.5, 0.5]$. The velocities of the wheels of the trailer serve as algebraic variables y_ℓ and y_r in the following optimal control problem. The task is to perform a reverse parking maneuver from a given initial position to a given terminal position in minimal time. This leads to the following optimal control problem.

Minimize t_f subject to the index-2 DAE

$$x'(t) = \frac{v_\ell(t) + v_r(t)}{2} \cos \psi(t),$$

$$y'(t) = \frac{v_\ell(t) + v_r(t)}{2} \sin \psi(t),$$

$$\psi'(t) = \frac{v_r(t) - v_\ell(t)}{B},$$

$$v_\ell(t) = u_\ell(t),$$

$$v_r(t) = u_r(t),$$

$$x_T'(t) = \frac{y_\ell(t) + y_r(t)}{2} \cos \psi_T(t),$$

$$y_T'(t) = \frac{y_\ell(t) + y_r(t)}{2} \sin \psi_T(t),$$

$$\psi_T'(t) = \frac{y_r(t) - y_\ell(t)}{B},$$

$$0 = x_T(t) + L_T \cos \psi_T(t) - (x(t) - L \cos \psi(t)),$$

$$0 = y_T(t) + L_T \sin \psi_T(t) - (y(t) - L \sin \psi(t)),$$

the control and state constraints

$$u_\ell(t), u_r(t) \in [-0.5, 0.5], \qquad v_\ell(t), v_r(t) \in [-0.3, 0.3],$$

the initial conditions

$$x(0) = y(0) = \psi(0) = y_T(0) = \psi_T(0) = v_\ell(0) = v_r(0) = 0, x_T(0) = -0.3,$$

and the terminal conditions

$$x(t_f) = -1, y(t_f) = -0.4, \psi(t_f) = v_\ell(t_f) = v_r(t_f)) = \psi_T(t_f) = 0.$$

The direct (single) shooting method with optimality and feasibility tolerance $tol = 10^{-8}$, $N = 100$, and initial guess $u_\ell, u_r \equiv -0.3$ produces the following iterates (note that the objective function was scaled by a factor of 10):

```
=== sqpfiltertoolbox   (C) MATTHIAS GERDTS, UNIVERSITAET DER BUNDESWEHR   ====
-------------------------------------------------------------------------------
ITER   QPIT   ALPHA               OBJ                  CV           KKT
-------------------------------------------------------------------------------
0        0    0.0000E+00    0.4000000000000000E+02   0.9000E+00    0.1000E+02
1      503    0.1000E+01    0.4035136156884651E+02   0.1477E+00    0.4500E+01
2      323    0.1000E+01    0.4319188910480073E+02   0.1478E+00    0.6860E+01
3       38    0.1000E+01    0.4617600877472075E+02   0.5227E-01    0.6438E+01
4       78    0.1000E+01    0.4691915302534527E+02   0.5860E-02    0.2459E+01
```

```
   5     111   0.1000E+01   0.4691471552645523E+02   0.1142E-03   0.2211E+01
 ...
 298       8   0.1000E+01   0.4683007498066451E+02   0.3274E-09   0.1572E-06
 299       5   0.1000E+01   0.4683007498043936E+02   0.2709E-09   0.3665E-07
 300       7   0.1000E+01   0.4683007498037029E+02   0.5909E-09   0.3183E-07
 301       7   0.1000E+01   0.4683007498046145E+02   0.2047E-09   0.5966E-08
========================================================================
END OF SQP METHOD
========================================================================
KKT CONDITIONS SATISFIED TO THE REQUESTED ACCURACY (IER=            0 )!
```

The following pictures show the converged solution of the direct shooting method on a normalized time scale. The final time amounts to $t_f \approx 4.683$: ☐

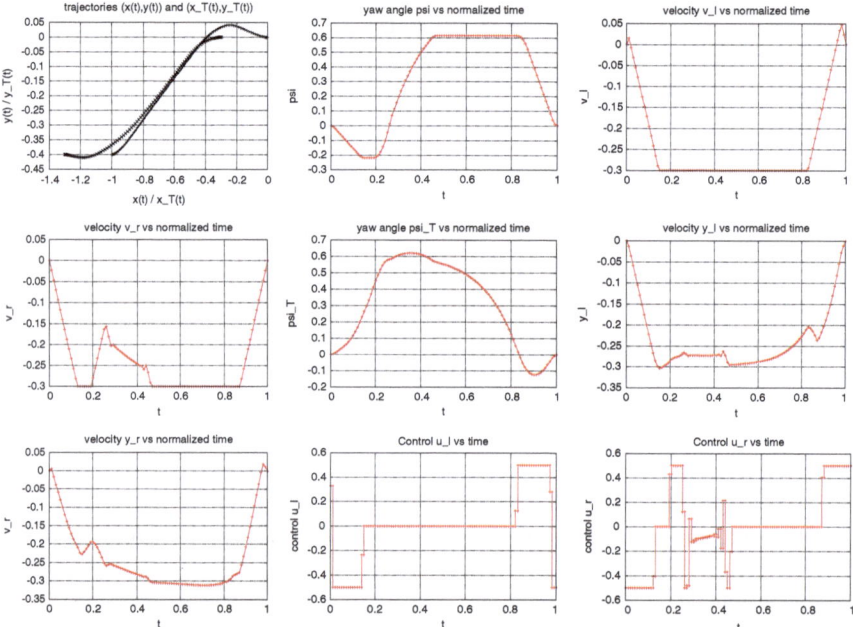

Direct discretization methods are frequently applied in model-predictive control, compare [37, 38], and extensions towards real-time optimal control by means of a parametric sensitivity analysis can be found in [24].

Direct discretization methods have been used successfully to solve mixed-integer optimal control problems with a partially discrete control set $\mathcal{U}$ in Problem 1.1. To this end the mixed-integer optimal control problem is either relaxed and combined with an appropriate rounding strategy as in [98, 99] or transformed by a time transformation technique as in [45, 81].

5   Function Space Methods

Function space methods apply solution methods in the same function spaces where the optimal control problem lives. This has the advantage that discretization errors are not introduced immediately. Moreover, structures of the optimal control problem can be exploited efficiently. However, a detailed functional analytic background is necessary to set up the methods and the analysis becomes particularly challenging in the presence of pure state constraints as the Lagrange multipliers are measures in this case. Function space methods are frequently used in PDE constrained optimal control, see the monographs [61, 62, 107] for the state-of-the-art in this field. However, function space methods are less frequently used in ODE or DAE constrained optimal control problems since these problems typically exhibit strong nonlinearities, non-convexity, and complicated state and control constraints. Direct discretization methods and especially the underlying state-of-the-art nonlinear programming solvers are able to handle these difficulties since they use sophisticated globalization strategies and measures to deal with inconsistencies, ill-conditioning, rank deficiencies etc. Without these additional measures, conceptual function space methods like gradient methods, interior-point methods, or sequential-quadratic programming (SQP) methods are often less robust for highly nonlinear problems. Nevertheless, they can perform very well for well-behaved problems. In order to illustrate the working principle of function space methods we discuss a gradient method and a Lagrange–Newton method.

5.1   Gradient Method

The gradient method and its extension, the projected gradient method, are the most simple gradient-based optimization methods for unconstrained problems or problems with sufficiently simple constraints like box constraints.

The projected gradient method is demonstrated for the following DAE optimal control problem subject to simple bounds, where $I := [t_0, t_f]$ is a compact interval, $\bar{x} \in \mathbb{R}^{n_x}$ is a given vector, $\varphi : \mathbb{R}^{n_x} \longrightarrow \mathbb{R}$, $f : \mathbb{R}^{n_x} \times \mathbb{R}^{n_y} \times \mathbb{R}^{n_u} \longrightarrow \mathbb{R}^{n_x}$ are continuously differentiable functions, $g : \mathbb{R}^{n_x} \longrightarrow \mathbb{R}^{n_y}$ is a twice continuously differentiable function, and $\mathcal{U} \subseteq \mathbb{R}^{n_u}$ is a closed convex set.

Problem 5.1 (OCP-SB)   Minimize

$$\Gamma(x, y, u) := \varphi(x(t_f))$$

with respect to $x \in W^{1,\infty}(I, \mathbb{R}^{n_x})$, $y \in L^\infty(I, \mathbb{R}^{n_y})$, $u \in L^\infty(I, \mathbb{R}^{n_u})$ subject to the constraints

$$x'(t) = f(x(t), y(t), u(t)) \qquad \text{a.e. in } I, \qquad (5.1)$$

$$0 = g(x(t)) \qquad \text{in } I, \qquad (5.2)$$

$$x(t_0) = \bar{x}, \tag{5.3}$$

$$u(t) \in \mathcal{U} \qquad\qquad\qquad \text{a.e. in } I. \tag{5.4}$$

The DAE (5.1)–(5.2) is supposed to be of index two, see Assumption 3.2, and $\bar{x}$ is supposed to be consistent, i.e., $g(\bar{x}) = 0$. Moreover, in order to allow for the elimination of the equality constraints we assume the following:

Assumption 5.2

(a) *The initial value problem (5.1)–(5.3) possesses a unique solution $(x(u), y(u)) \in W^{1,\infty}(I, \mathbb{R}^{n_x}) \times L^{\infty}(I, \mathbb{R}^{n_y})$ for every control input $u \in L^{\infty}(I, \mathbb{R}^{n_u})$.*

(b) *The control-to-state mapping $L^{\infty}(I, \mathbb{R}^{n_u}) \ni u \mapsto (x(u), y(u)) \in W^{1,\infty}(I, \mathbb{R}^{n_x}) \times L^{\infty}(I, \mathbb{R}^{n_y})$ is continuously Fréchet-differentiable.*

Remark 5.1 The idea of considering the control-to-state mapping in Assumption 5.2 is exactly the single shooting idea. The realization requires to solve the initial value problem (5.1)–(5.3) for a given control input u by some suitable discretization scheme, amongst them are the implicit Euler method, BDF methods, and implicit Runge–Kutta methods. The existence of solutions in Assumption 5.2 (a) can be shown for ODEs (Carathéodory solutions) or under standard assumptions in the case that f and g in (5.1)–(5.2) are linear. General existence and uniqueness results can be found in [96], but for L^{∞}-inputs existence and uniqueness is not fully understood up to the knowledge of the author. Hence we merely assume that a unique solution exists for a given control input.

Exploitation of Assumption 5.2 allows to eliminate the equality constraints in Problem 5.1 and to consider an equivalent minimization problem for the reduced functional

$$J(u) := \Gamma(x(u), y(u), u) \tag{5.5}$$

subject to the simple bounds $u(t) \in \mathcal{U}$ only.

Problem 5.3 (Reduced Problem)

Minimize $J(u)$ with respect to $u \in L^{\infty}(I, \mathbb{R}^{n_u})$ subject to $u(t) \in \mathcal{U}$ a.e. in I.

Let $\hat{u}$ be a local minimum of Problem 5.3. Then the first order necessary optimality condition reads

$$J'(\hat{u})(u - \hat{u}) \geq 0 \qquad \forall u \in \mathcal{U}_{ad}, \tag{5.6}$$

where the admissible set $\mathcal{U}_{ad}$ is defined by

$$\mathcal{U}_{ad} := \{u \in L^{\infty}(I, \mathbb{R}^{n_u}) \mid u(t) \in \mathcal{U} \text{ a.e. in } I\}.$$

The variational inequality (5.6) with $\nabla J(\hat{u})$ being the Riesz representation of the functional $J'(\hat{u})$) (in a Hilbert space) is equivalent with the condition

$$\hat{u} = \Pi_{\mathcal{U}_{ad}} \left(\hat{u} - \nu \nabla J(\hat{u})\right) \quad (\nu > 0)$$

that has to hold a.e. in I, where $\Pi_{\mathcal{U}_{ad}} : L^\infty(I, \mathbb{R}^{n_u}) \longrightarrow \mathcal{U}_{ad}$ is the projection onto the admissible set $\mathcal{U}_{ad}$. For box constraints

$$\mathcal{U} = \{u \in \mathbb{R} \mid a \le u \le b\}$$

the projection of u at t computes to

$$\Pi_{\mathcal{U}_{ad}}(u)(t) = \max\{a, \min\{b, u(t)\}\} = \begin{cases} a, & \text{if } u(t) < a, \\ u(t), & \text{if } a \le u(t) \le b, \\ b, & \text{if } u(t) > b. \end{cases}$$

In the sequel we assume that the projection onto $\mathcal{U}_{ad}$ is easy to compute like it is in the box-constrained case. Then, a conceptual projected gradient method reads as follows:

Algorithm 5.4 (Projected Gradient Method)

(0) Choose $u^0 \in \mathcal{U}_{ad}$, $\beta \in (0, 1)$, $\sigma \in (0, 1)$, $tol > 0$, and set $k := 0$.
(1) If $\|u^k - \Pi_{\mathcal{U}_{ad}}\left(u^k - \nabla J(u^k)\right)\|_\infty \le tol$, STOP.
(2) Set

$$\tilde{d}^k := -\nabla J(u^k),$$

compute

$$\tilde{u}^k := \Pi_{\mathcal{U}_{ad}}\left(u^k + \tilde{d}^k\right),$$

and set

$$d^k := \tilde{u}^k - u^k.$$

(4) Perform an Armijo line-search: Find smallest $j \in \{0, 1, 2, \ldots\}$ with

$$J(u^k + \beta^j d^k) \le J(u^k) + \sigma\beta^j J'(u^k)(d^k)$$

and set $\alpha_k := \beta^j$.
(5) Set $u^{k+1} := u^k + \alpha_k d^k$, $k := k + 1$, and go to (1).

Remark 5.2 An alternate version of the projected gradient method does not use the projection in step (2), but instead uses the projection directly in the line-search in step (4) as follows

$$J\left(\Pi_{\mathcal{U}_{ad}}\left(u^k + \beta^j d^k\right)\right) \le J(u^k) + \sigma\beta^j J'(u^k)(d^k),$$

compare [107, Sect. 3.7.1]. The new iterate in step (5) is then given by $u^{k+1} := \Pi_{\mathcal{U}_{ad}}\left(u^k + \alpha_k d^k\right)$.

It remains to answer one question: How does the gradient $\nabla J(u)$ look like?

In a Hilbert space U, this question is easy to answer, because the Fréchet derivative $J'(\tilde{u})$ of J at $\tilde{u}$ is a linear and continuous functional and by the Riesz theorem it possesses the representation $J'(\tilde{u})(u) = \langle \eta, u \rangle_U$ with some $\eta \in U$. Hence, the element η can be interpreted as the gradient of J at $\tilde{u}$. But Riesz' Theorem does not hold in general Banach spaces. However, we will make use of a formal Lagrange technique to obtain a similar representation in our setting. To this end define the auxiliary functional (with $\ell_0 = 1$)

$$\tilde{J}(\tilde{u}) := J(\tilde{u}) + \langle \lambda_f(\cdot), f(\tilde{x}(\cdot), \tilde{y}(\cdot), \tilde{u}(\cdot)) - \tilde{x}'(\cdot) \rangle_{L^2}$$
$$+ \langle \lambda_g(\cdot), g_x'(\tilde{x}(\cdot)) f(\tilde{x}(\cdot), \tilde{y}(\cdot), \tilde{u}(\cdot)) \rangle_{L^2}$$
$$= \varphi(\tilde{x}(t_f)) + \int_I \mathcal{H}(\tilde{x}(t), \tilde{y}(t), \tilde{u}(t), \lambda_f(t), \lambda_g(t), \ell_0) - \lambda_f(t)^\top \tilde{x}'(t) dt,$$

where $\lambda_f \in W^{1,\infty}(I, \mathbb{R}^{n_x})$ and $\lambda_g \in L^\infty(I, \mathbb{R}^{n_y})$ are functions to be specified later and $\tilde{x} := x(\tilde{u})$, $\tilde{y} := y(\tilde{u})$ are Fréchet differentiable functions of u. Partial integration of the last term yields

$$\tilde{J}(\tilde{u}) = \varphi(\tilde{x}(t_f)) - \left[\lambda_f(t)^\top \tilde{x}(t)\right]_{t_0}^{t_f}$$
$$+ \int_I \mathcal{H}(\tilde{x}(t), \tilde{y}(t), \tilde{u}(t), \lambda_f(t), \lambda_g(t), \ell_0) + \lambda_f'(t)^\top \tilde{x}(t) dt.$$

Formal differentiation of $\tilde{J}$ at $\tilde{u}$ in the direction u together with the Fréchet derivatives $S^x := x'(\tilde{u})(u)$ and $S^y := y'(\tilde{u})(u)$ yields

$$\tilde{J}'(\tilde{u})(u) = \left(\varphi'(\tilde{x}(t_f)) - \lambda_f(t_f)^\top\right) S^x(t_f)$$
$$+ \int_I \left(\mathcal{H}_x'[t] + \lambda_f'(t)^\top\right) S^x(t) + \mathcal{H}_y'[t] S^y(t) + \mathcal{H}_u'[t] u(t) dt,$$

where we exploited that the initial value of x does not depend on u and $[t]$ is an abbreviation for $(\tilde{x}(t), \tilde{y}(t), \tilde{u}(t), \lambda_f(t), \lambda_g(t), \ell_0)$.

As the derivatives S^x and S^y are expensive to compute, λ_f and λ_g are chosen in such a way that the terms involving S^x and S^y vanish. This yields the initial value problem with the index-one adjoint DAE

$$\lambda'_f(t) = -\mathscr{H}'_x(\tilde{x}(t), \tilde{y}(t), \tilde{u}(t), \lambda_f(t), \lambda_g(t), \ell_0)^\top, \tag{5.7}$$

$$0 = \mathscr{H}'_y(\tilde{x}(t), \tilde{y}(t), \tilde{u}(t), \lambda_f(t), \lambda_g(t), \ell_0)^\top, \tag{5.8}$$

$$\lambda_f(t_f) = \varphi'(\tilde{x}(t_f))^\top, \tag{5.9}$$

and $\tilde{J}'(\tilde{u})(u)$ reduces to

$$\tilde{J}'(\tilde{u})(u) = \int_I \mathscr{H}'_u[t]u(t)dt.$$

The adjoint DAE (5.7)–(5.9) is a linear semi-explicit DAE of index one and thus it is well-defined and possesses a unique solution.

One can show with similar techniques as in [48, Theorem 8.1.6] that $\tilde{J}'(\tilde{u})(u)$ and $J'(\tilde{u})(u)$ in fact coincide and thus

$$J'(\tilde{u})(u) = \tilde{J}'(\tilde{u})(u) = \int_I \mathscr{H}'_u[t]u(t)dt. \tag{5.10}$$

The representation in (5.10) reminds of the Riesz representation of a continuous linear functional and thus we define the gradient as follows:

Definition 5.1 (Gradient of Reduced Functional) The gradient $\nabla J(\tilde{u})$ of the reduced objective functional J in (5.5) at $\tilde{u}$ is defined by

$$\nabla J(\tilde{u})(\cdot) := \nabla_u \mathscr{H}(\tilde{x}(\cdot), \tilde{y}(\cdot), \tilde{u}(\cdot), \lambda_f(\cdot), \lambda_g(\cdot), \ell_0),$$

where $\ell_0 = 1$, $\tilde{x} = x(\tilde{u})$, $\tilde{y} = y(\tilde{u})$, and λ_f and λ_g solve the adjoint DAE (5.7)–(5.9).

In fact it is easy to verify that the normed direction $d = -\nabla J(\tilde{u})/\|\nabla J(\tilde{u})\|_2$ minimizes the directional derivative $J'(\tilde{u})(u)$ w.r.t. all directions u with $\|u\|_2 = 1$. This property justifies the alternate name "steepest descent method" for the gradient method.

Example 5.5 Consider Example 4.5. Since u is not restricted, i.e., $\mathscr{U} = \mathbb{R}$, the projected gradient method reduces to the classic gradient method. The gradient method with tolerance $tol = 10^{-5}$, $\alpha_1 = \alpha_2 = \alpha_4 = 1$, $\alpha_3 = 0.5 \cdot 10^{-2}$, $N = 1000$ and initial guess $u^0 \equiv -3$ produces the following iterates:

A look at the iterates reveals a major drawback of the gradient method: The convergence rate is only linear and it is rather slow in this example. Moreover, the gradient method can be quite sensitive w.r.t. to scaling and tolerances. For instance, if we reduce the tolerance tol to 10^{-6}, then the line-search procedure terminates with a step-size close to zero.

k	α_k	$J(u^k)$	$\|u^k - \Pi_{\mathscr{U}}\left(u^k - \nu J'(u^k)\right)\|_\infty$ $\nu = 10^{-3}$	$J'(u^k)(d^k)$
0	0.00000000E+00	0.13300873E+01	0.31148251E−02	−0.67634180E+01
1	0.65610000E+00	0.76882228E+00	0.25684209E−02	−0.37878631E+01
2	0.65610000E+00	0.45335163E+00	0.17268254E−02	−0.21216493E+01
3	0.65610000E+00	0.27650174E+00	0.14715161E−02	−0.11885594E+01
4	0.65610000E+00	0.17683468E+00	0.99728958E−03	−0.66603951E+00
5	0.65610000E+00	0.12081438E+00	0.85221832E−03	−0.37332995E+00
...	...	...	...	...
195	0.81000000E+00	0.41473570E−01	0.10260497E−04	−0.18966824E−04
196	0.72900000E+00	0.41471467E−01	0.18804354E−04	−0.17294724E−04
197	0.72900000E+00	0.41469326E−01	0.10490706E−04	−0.15639555E−04
198	0.72900000E+00	0.41467372E−01	0.18067697E−04	−0.14314419E−04
199	0.81000000E+00	0.41466091E−01	0.98482036E−05	−0.17769042E−04

The following pictures show some intermediate iterates (thin lines) and the converged solution (thick lines) of the gradient method for the states, the adjoints, and the control. □

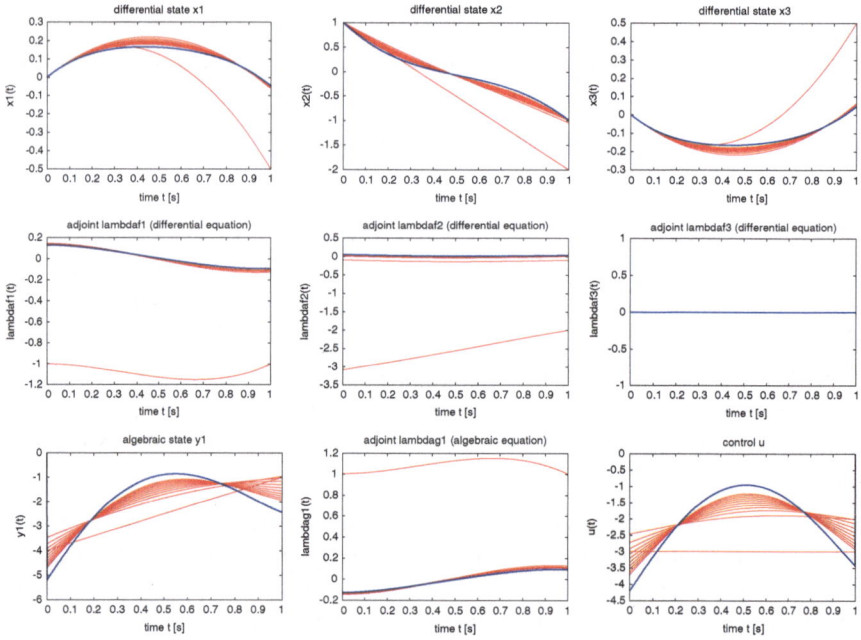

Example 5.6 Consider Example 4.6. The projected gradient method with tolerance $tol = 10^{-7}$, $\alpha_1 = \alpha_2 = \alpha_4 = 1$, $\alpha_3 = 0.5 \cdot 10^{-2}$, $N = 1000$ and initial guess

$u^0 \equiv -3$ produces the following iterates, which exhibit a slow linear convergence rate:

k	α_k	$J(u^k)$	$\left\| u^k - \Pi_{\mathscr{U}} \left(u^k - \nu J'(u^k) \right) \right\|_\infty$ $\nu = 10^{-3}$	$J'(u^k)(d^k)$
0	0.00000000E+00	0.13300873E+01	0.31148251E−02	−0.51621641E+01
1	0.81000000E+00	0.57297403E+00	0.21639411E−02	−0.25254696E+01
2	0.72900000E+00	0.35775984E+00	0.15304789E−02	−0.16252587E+01
3	0.65610000E+00	0.22253942E+00	0.12832913E−02	−0.91034675E+00
4	0.65610000E+00	0.14643376E+00	0.89053644E−03	−0.51008708E+00
5	0.65610000E+00	0.10367291E+00	0.74329723E−03	−0.28584962E+00
...	...	...	...	...
471	0.10000000E+01	0.42028359E−01	0.10342884E−06	−0.13237960E−08
472	0.10000000E+01	0.42028358E−01	0.10233230E−06	−0.12956173E−08
473	0.10000000E+01	0.42028356E−01	0.10124469E−06	−0.12680419E−08
474	0.10000000E+01	0.42028355E−01	0.10017421E−06	−0.12410509E−08
475	0.10000000E+01	0.42028354E−01	0.99108933E−07	−0.12146379E−08

The following pictures show some intermediate iterates (thin lines) and the converged solution (thick lines) of the projected gradient method for the states, the adjoints, and the control. □

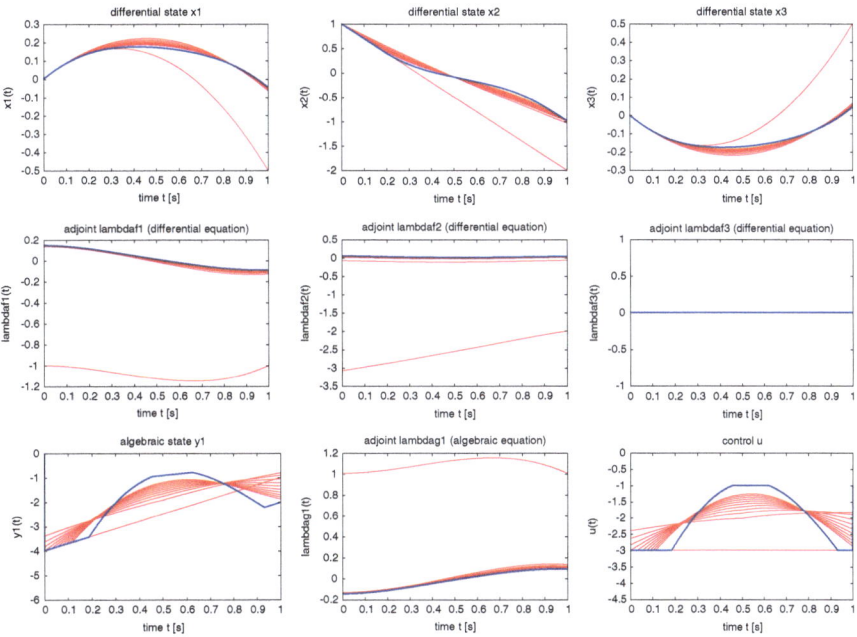

Example 5.7 Consider Example 4.7. The projected gradient method with tolerance $tol = 10^{-6}$, $\alpha_1 = \alpha_2 = \alpha_4 = 1$, $\alpha_3 = 0$, $N = 1000$ and initial guess $u^0 \equiv -3$ produces the following iterates with a slow linear convergence rate:

k	α_k	$J(u^k)$	$\left\| u^k - \Pi_{\mathcal{U}}\left(u^k - \nu J'(u^k)\right) \right\|_\infty$ $\nu = 10^{-3}$	$J'(u^k)(d^k)$
0	0.00000000E+00	0.12850873E+01	0.30848252E−02	−0.51021641E+01
1	0.81000000E+00	0.56345203E+00	0.21777410E−02	−0.25559953E+01
2	0.72900000E+00	0.33237684E+00	0.15172054E−02	−0.15910525E+01
3	0.65610000E+00	0.19663752E+00	0.12574239E−02	−0.87563698E+00
4	0.65610000E+00	0.12158166E+00	0.87048761E−03	−0.48210072E+00
5	0.65610000E+00	0.80129201E−01	0.71767043E−03	−0.26547456E+00
...	...	...	...	...
1000	0.10000000E+01	0.18185375E−01	0.10092448E−05	−0.18871142E−07
1001	0.10000000E+01	0.18185356E−01	0.10091735E−05	−0.18871079E−07
1002	0.10000000E+01	0.18185337E−01	0.10091040E−05	−0.18871008E−07
1003	0.10000000E+01	0.18185318E−01	0.10090356E−05	−0.18066969E−07
1004	0.10000000E+01	0.18185300E−01	0.99357979E−06	−0.17849778E−07

The following pictures show some intermediate iterates (thin lines) and the converged solution (thick lines) of the projected gradient method for the states, the adjoints, and the control. Note that the control is a bang–bang control.   □

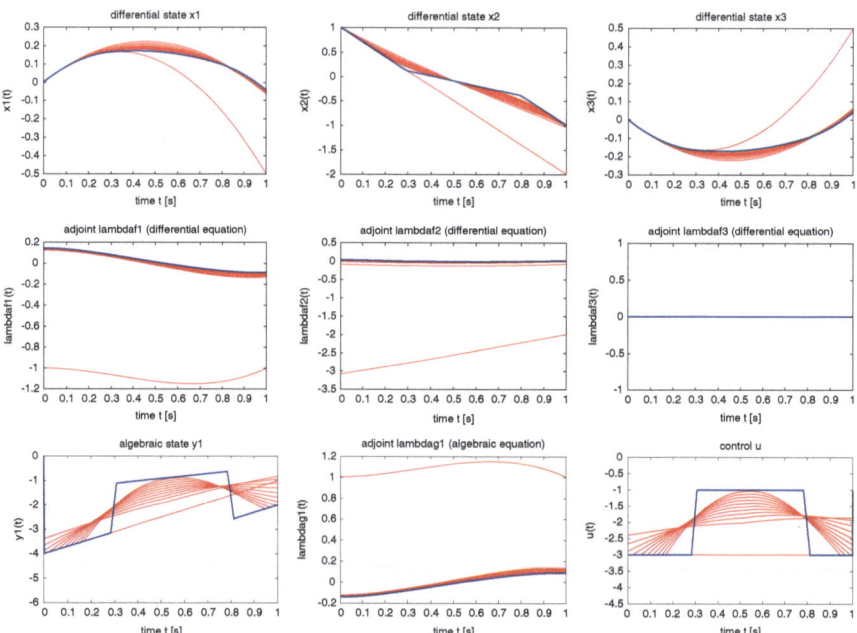

5.2 Lagrange–Newton Method

The Lagrange–Newton method and its extension to problems with inequality constraints, the SQP method, are analyzed in a Banach space setting in [1, 2]. The application of SQP methods to optimal control problems subject to ODEs can be found in [3, 4] and [85].

The basic idea of the Lagrange–Newton method is to solve the optimality system provided by the local minimum principle by Newton's method. Note that, although the global minimum principle provides stronger necessary conditions than the local one, the global minimum principle is difficult to exploit for numerical algorithms, because the optimality condition in (d) of Theorem 3.4 requires to find a global minimum of the Hamilton function on the set Ω. Since global minimization is expensive, we use the local minimum principle. To this end consider OCP-SE without control constraints and pure state constraints:

Problem 5.8 Minimize

$$\varphi(x(t_0), x(t_f)) + \int_I f_0(x(t), y(t), u(t)) \, dt$$

w.r.t. $x \in W^{1,\infty}(I, \mathbb{R}^{n_x})$, $y \in L^\infty(I, \mathbb{R}^{n_y})$, $u \in L^\infty(I, \mathbb{R}^{n_u})$ subject to the constraints

$$x'(t) = f(x(t), y(t), u(t)) \qquad \text{a.e. in } I,$$
$$0 = g(x(t)) \qquad \text{in } I,$$
$$0 = \psi(x(t_0), x(t_f)).$$

The local minimum principle in Theorem 3.3 (assuming $\ell_0 = 1$) yields the necessary conditions

$$T(x, y, u, \lambda_f, \lambda_g, \sigma, \zeta) := \begin{pmatrix} x'(\cdot) - f(x(\cdot), y(\cdot), u(\cdot)) \\ g(x(\cdot)) \\ \lambda_f'(\cdot) + \mathscr{H}_x'(x(\cdot), y(\cdot), u(\cdot), \lambda_f(\cdot), \lambda_g(\cdot), \ell_0)^\top \\ \mathscr{H}_y'(x(\cdot), y(\cdot), u(\cdot), \lambda_f(\cdot), \lambda_g(\cdot), \ell_0)^\top \\ \mathscr{H}_u'(x(\cdot), y(\cdot), u(\cdot), \lambda_f(\cdot), \lambda_g(\cdot), \ell_0)^\top \\ \psi(x(t_0), x(t_f)) \\ \lambda_f(t_0) + \kappa_{x_0}'(x(t_0), x(t_f), \sigma, \zeta)^\top \\ \lambda_f(t_f) - \kappa_{x_f}'(x(t_0), x(t_f), \sigma, \zeta)^\top \end{pmatrix} = 0,$$

$$(5.11)$$

where $\kappa(x_0, x_f, \sigma, \zeta) := \varphi(x_0, x_f) + \sigma^\top \psi(x_0, x_f) + \zeta^\top g(x_0)$.

Let the operator T be Fréchet-differentiable and let T' be surjective. Then Newton's method can be applied to the nonlinear equation $T(\eta) = 0$ with $\eta = (x, y, u, \lambda_f, \lambda_g, \sigma, \zeta)$. The resulting method is called Lagrange–Newton method and reads as follows:

Algorithm 5.9 (Lagrange–Newton Algorithm)

(0) Choose η^0, $tol > 0$, and set $k := 0$.
(1) If $\|T(\eta^k)\| \leq tol$, STOP.
(2) Solve the linear operator equation $T'(\eta^k)(d) = -T(\eta^k)$ for d and denote the solution by d^k.
(3) Set $\eta^{k+1} := \eta^k + d^k$, $k := k + 1$, and go to (1).

The standard convergence results on Newton's method apply under standard assumptions and globalized (or damped) versions can be considered. The linear operator equation in step (2) with $d = (x, y, u, \lambda_f, \lambda_g, \sigma, \zeta)$ corresponds to the linear DAE boundary value problem

$$
\begin{pmatrix} x' \\ \lambda'_f \\ 0 \\ 0 \\ 0 \end{pmatrix} + \begin{pmatrix} -f'_x & 0 & -f'_y & 0 & -f'_u \\ \mathcal{H}''_{xx} & \mathcal{H}''_{x\lambda_f} & \mathcal{H}''_{xy} & \mathcal{H}''_{x\lambda_g} & \mathcal{H}''_{xu} \\ g'_x & 0 & 0 & 0 & 0 \\ \mathcal{H}''_{yx} & \mathcal{H}''_{y\lambda_f} & \mathcal{H}''_{yy} & \mathcal{H}''_{y\lambda_g} & \mathcal{H}''_{yu} \\ \mathcal{H}''_{ux} & \mathcal{H}''_{u\lambda_f} & \mathcal{H}''_{uy} & \mathcal{H}''_{u\lambda_g} & \mathcal{H}''_{uu} \end{pmatrix} \begin{pmatrix} x \\ \lambda_f \\ y \\ \lambda_g \\ u \end{pmatrix}
$$

$$
= - \begin{pmatrix} (x^k)' - f \\ (\lambda^k_f)' + (\mathcal{H}'_x)^\top \\ g \\ (\mathcal{H}'_y)^\top \\ (\mathcal{H}'_u)^\top \end{pmatrix} \tag{5.12}
$$

with boundary conditions

$$
\psi'_{x_0} x(t_0) + \psi'_{x_f} x(t_f) = -\psi, \tag{5.13}
$$

$$
\lambda_f(t_0) + \kappa''_{x_0,x_0} x(t_0) + \kappa''_{x_0,x_f} x(t_f) + \kappa''_{x_0,\sigma} \sigma + \kappa''_{x_0,\zeta} \zeta = -\left(\lambda^k_f(t_0) + (\kappa'_{x_0})^\top\right), \tag{5.14}
$$

$$
\lambda_f(t_f) - \kappa''_{x_f,x_0} x(t_0) - \kappa''_{x_f,x_f} x(t_f) - \kappa''_{x_f,\sigma} \sigma = -\left(\lambda^k_f(t_f) - (\kappa'_{x_f})^\top\right), \tag{5.15}
$$

where all functions are evaluated at η^k. The components σ and ζ are treated via the trivial differential equations $\sigma' = 0$ and $\zeta' = 0$.

Since the boundary value problem (5.12)–(5.15) arose from a linearization of a nonlinear equation around some (arbitrary) iterate η^k, existence of a solution is not guaranteed. In particular it happens that the linearized boundary conditions are inconsistent. Moreover, the multiplier ζ might be redundant in the minimum principle and can be set to zero, e.g., if the initial value of $x(t_0)$ is fixed. This redundancy leads to rank deficiencies in the linear BVP. Regularization methods become necessary in these cases, compare [51]. For index one DAEs the situation is less involved, compare [48, Sect. 8.2].

A straightforward calculation shows that the index of the DAE in (5.12) is two, if the matrix

$$\begin{pmatrix} g'_x f'_y & 0 & g'_x f'_u \\ \mathscr{H}''_{yy} & \mathscr{H}''_{y\lambda_g} & \mathscr{H}''_{yu} \\ \mathscr{H}''_{uy} & \mathscr{H}''_{u\lambda_g} & \mathscr{H}''_{uu} \end{pmatrix}$$

is non-singular a.e. on I with an essentially bounded inverse on I.

The outlined Lagrange–Newton method can be applied to more general DAEs. Again, results for linear DAEs whether in leading term formulation or in descriptor form can be applied to the resulting boundary value problems in step (2) of the algorithm.

Example 5.10 Consider again Example 3.5 resp. Example 4.5. The Lagrange–Newton method with tolerance $tol = 10^{-15}$, $\alpha_1 = \alpha_2 = \alpha_4 = 1$, $\alpha_3 = 0.5 \cdot 10^{-2}$, $N = 1000$ (number of steps in the shooting method used for the numerical solution of BVP), and initial guess $x^0 \equiv 0$, $y^0 \equiv 0$, $u^0 \equiv -3$, $\lambda_f^0 \equiv 0$, $\lambda_g^0 \equiv 0$, $\sigma^0 \equiv 0$, $\zeta^0 \equiv 0$ converges in one step:

k	α_k	$\|T(\eta^k)\|_2^2$	$\|d^k\|_2^2$
0	0.000000E+00	0.140000E+02	0.271697E+04
1	0.100000E+01	0.172345E−26	0.438543E−09

This is not surprising since the optimal control problem is a linear-quadratic problem and the necessary optimality conditions form a linear operator equation.

The following pictures show the initial guess (constant lines) and the solution (thick lines) of the Lagrange–Newton method for the states, the adjoints, and the control.

□

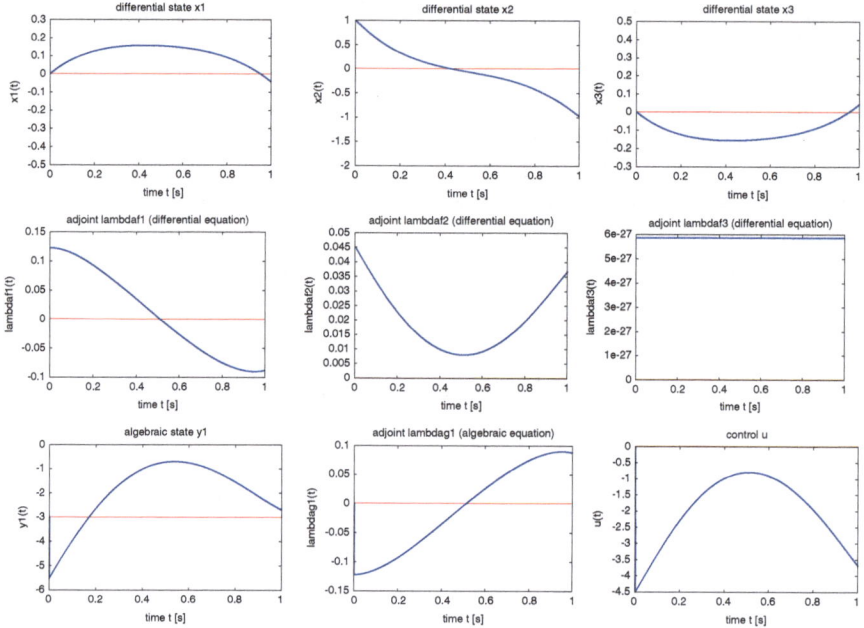

5.3 Treatment of Inequality Constraints

Inequality constraints can be handled conceptually by, e.g., SQP methods, see [3, 4, 85], interior-point methods, see [101, 110, 113], penalty methods, or semi-smooth Newton methods, see [31, 51, 62, 108, 109].

We focus on semi-smooth Newton methods and outline the main ideas. Semi-smooth Newton methods require the reformulation of the necessary optimality conditions in terms of a nonlinear equation. This can be achieved in several ways and we discuss two commonly used approaches.

Firstly, consider OCP-SE without pure state constraints but with control constraints $u(t) \in \mathcal{U}$, where $\mathcal{U}$ is a closed and convex set. In this case we have to deal with the variational inequality

$$\mathcal{H}_u'(\hat{x}(t), \hat{y}(t), \hat{u}(t), \lambda_f(t), \lambda_g(t), \ell_0)(u - \hat{u}(t)) \geq 0 \qquad \forall u \in \mathcal{U}$$

in (d) of Theorem 3.3. Likewise as in Sect. 5.1 this variational inequality can be reformulated as the equation

$$0 = \hat{u} - \Pi_{\mathcal{U}_{ad}}(\hat{u} - \nu \nabla_u \mathcal{H}) \quad (\nu > 0) \tag{5.16}$$

with

$$\mathcal{U}_{ad} := \{u \in L^\infty(I, \mathbb{R}^{n_u}) \mid u(t) \in \mathcal{U} \text{ a.e. in } I\}.$$

Herein, $\nabla_u \mathcal{H}$ is evaluated at $(\hat{x}, \hat{y}, \hat{u}, \lambda_f, \lambda_g, \ell_0)$ with $\ell_0 = 1$.

Replacing $\nabla_u \mathcal{H}$ in (5.11) by the right hand-side in (5.16) yields again a nonlinear equation similar to (5.11). Since the projection $\Pi_{\mathcal{U}_{ad}}$ is only Lipschitz continuous but not Fréchet differentiable, the nonlinear equation is non-smooth and modifications of Newton's method towards semi-smooth Newton methods are necessary. Herein, the Fréchet derivative $T'(\eta^k)$ has to be replaced by a suitably defined generalized Jacobian $\partial T(\eta^k)$, compare, e.g., [109] for details in the context of PDE constrained optimization.

Another transformation can be used, if the control constraints are expressed in terms of mixed control-state constraints of type $c(x(t), y(t), u(t)) \leq 0$. The resulting necessary optimality conditions then involve complementarity conditions of type

$$0 \leq \mu(t) \quad \perp \quad - c(x(t), y(t), u(t)) \geq 0.$$

These can be transformed to equivalent nonlinear and non-smooth equations by pointwise application of a suitable nonlinear complementarity function like the Fischer-Burmeister function $\phi(a, b) := \sqrt{a^2 + b^2} - a - b$. This Lipschitz-continuous function has the property that $\phi(a, b) = 0$ holds if and only if $0 \leq a \perp b \geq 0$. The resulting optimality system with the non-smooth equation

$$\phi(-c(x(t), y(t), u(t)), \mu(t)) = 0 \qquad \text{a.e. in } I$$

can be solved by a semi-smooth Newton method again, compare [31, 51]. We illustrate the performance of the semi-smooth Newton method, which we augmented by an Armijo line-search procedure, in the subsequent example.

Example 5.11 Consider again Example 5.6 resp. Example 4.6. The semi-smooth Newton method in [51] with tolerance $tol = 10^{-15}$, $\alpha_1 = \alpha_2 = \alpha_4 = 1$, $\alpha_3 = 0.5 \cdot 10^{-2}$, $N = 1000$ (number of steps in the shooting method used for the numerical solution of BVP), and initial guess $x^0 \equiv 0$, $y^0 \equiv 0$, $u^0 \equiv -3$, $\lambda_f^0 \equiv 0$, $\lambda_g^0 \equiv 0$, $\sigma^0 \equiv 0$, $\zeta^0 \equiv 0$ produces the following iterates, which exhibit a superlinear convergence rate:

The following pictures show some intermediate iterates (thin lines) and the converged solution (thick lines) of the semi-smooth Newton method for the states, the adjoints, the control, and the multipliers for the control constraints. $\square$

Pure state constraints in function space methods are often handled by relaxation or penalization, compare the virtual control concept in [32, 50, 66]. The idea of the virtual control concept is to introduce an artificial control $v \in L^\infty(I, \mathbb{R})$ (for a single state constraint) and to consider the relaxed state constraint

$$s(x(t)) - \gamma(\alpha)v(t) \leq 0$$

k	α_k	$\|F(z^k)\|_2^2$	$\|d\|_2^2$
0	0.000000E+00	0.180000E+02	0.136708E+05
1	0.100000E+01	0.343312E+01	0.880798E+04
2	0.100000E+01	0.910739E−01	0.965659E+05
3	0.100000E+01	0.108791E−02	0.385528E+04
4	0.523348E−01	0.107136E−02	0.281572E+04
5	0.471013E−01	0.104974E−02	0.433688E+04
…	…	…	…
19	0.100000E+01	0.214659E−03	0.113684E+02
20	0.100000E+01	0.989223E−08	0.106969E+01
21	0.100000E+01	0.532519E−10	0.425010E−02
22	0.100000E+01	0.502770E−13	0.108164E−03
23	0.100000E+01	0.720014E−19	0.470676E−06

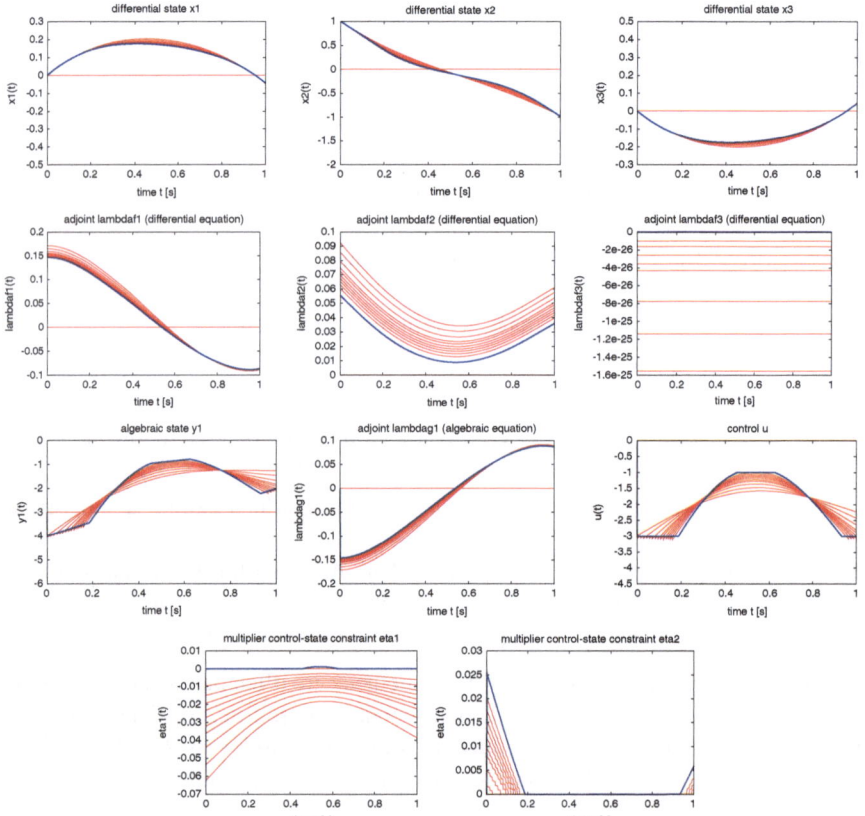

with a regularization parameter $\alpha > 0$ and some suitable positive function γ. In order to drive v to zero as $\alpha \to 0$, the term

$$\frac{\phi(\alpha)}{2} \int_I \|v(t)\|^2 dt$$

is added to the objective function, where ϕ is some suitable positive function. For any positive α the regularized problem has only mixed control-state constraints and no pure state constraints. It can be solved by the aforementioned methods. Under suitable conditions convergence of the regularized solutions can be shown. In fact, the virtual control concept is a penalty method since the optimal $\hat{v}_\alpha$ for $\alpha > 0$ satisfies

$$\hat{v}_\alpha(t) = \frac{1}{\gamma(\alpha)} \max\{0, s(\hat{x}_\alpha(t))\}$$

and instead of considering the relaxed state constraint one could add the penalty term

$$\frac{\phi(\alpha)}{2\gamma(\alpha)^2} \int_I \max\{0, s(x(t))\}^2 dt$$

to the objective function.

6   Conclusions and Future Directions

The paper aims to provide an overview on theoretical results and numerical techniques for optimal control problems subject to differential-algebraic equations. Naturally this is a vast area and not all topics could be covered in detail. Special attention was put on necessary and—whenever available—sufficient conditions of optimality for linear-quadratic and nonlinear optimal control problems and on numerical techniques. Although many theoretical and numerical issues in DAE optimal control have been addressed and solved today, there are several topics of interest with little attention so far. These topics, to the author's opinion, deserve more attention in the future. Amongst them are:

Sufficient conditions for nonlinear optimal control problems: Sufficient conditions of optimality are well-known for linear-quadratic DAE optimal control problems and for optimal control problems subject to explicit ODEs. However, sufficient conditions of optimality have not been obtained so far for optimal control problems subject to nonlinear DAEs with possibly higher index.

Sufficient conditions of optimality typically appear as well in the study of convergence properties of discretizations and parametric sensitivity analysis. Hence, they are of fundamental importance and should be investigated in more detail.

Parametric sensitivity analysis: Numerical techniques and theoretical properties with regard to the dependence of optimal solutions with respect to parameters are well established for finite dimensional optimization problems and hence on a discretization level, a parametric sensitivity analysis for DAE optimal control problems can be performed. The mathematically thorough function space counterpart for parametric DAE optimal control problems is missing so far. The task is to derive conditions that guarantee that the solution of a DAE optimal control problem is continuous, Lipschitz-continuous, or Fréchet-differentiable, respectively, with respect to some parameter entering the problem formulation. This analysis has to be carried out in a proper function space setting, which is suitable for the DAE optimal control problem.

Convergence properties of discretizations: Despite the fact that there are many sophisticated algorithms around with the ability to numerically solve discretized DAE optimal control problems, the convergence of discretizations of DAE optimal control problems has not been established so far. For semi-explicit DAEs of index one convergence proofs for ODE optimal control problems could be extended in a straightforward way. For higher index DAEs one is typically faced with a structural gap between the necessary conditions on discretization level, which usually do not involve index reduction, and on function space level, where some implicit index reduction takes place as explained in Sect. 3 and Remark 3.1.

Existence of optimal solutions: A fundamental question is the question of existence of an optimal solution for reasonable initial conditions. Apart from general existence results up to the author's knowledge there are no particular existence results for nonlinear (non-convex) DAE optimal control problems.

Optimal control of coupled DAE-PDE systems: Many practically important applications naturally lead to coupled systems of DAEs and partial differential equations, for instance if a truck, modeled by a mechanical multi-body system, carries some water tank, modeled by the Navier–Stokes equations. Such coupled systems have not been investigated systematically, neither from a numerical point of view nor from a theoretical point of view. Hence, the (optimal) control of coupled DAE-PDE systems provides a huge area for future research with regard to necessary and sufficient conditions, existence of solutions, and efficient structure exploiting numerical methods (both, discretization based and function space based). Extensions of the index concept towards PDAEs can be found in [26].

The above topics possess counterparts in ODE optimal control and for such problems most of the above issues have been solved to some satisfactory level, except for coupled ODE-PDE systems. It is expected that well-known ODE techniques and results regarding existence, sufficient conditions, and parametric sensitivity analysis can be extended in a more or less straightforward way to DAEs with a special structure like Hessenberg DAEs. More general DAEs, however, are likely to require new techniques. This is particularly true for the convergence analysis of higher index DAE optimal control problems owing to the gap between the necessary conditions for the discretized problem and those for the continuous problem.

7   Appendix: Auxiliary Results

Theorem 7.1 *Let $D : I \longrightarrow \mathbb{R}^{m \times n}$ be an essentially bounded matrix function in $L^{\infty}(I, \mathbb{R}^{m \times n})$. Then,*

$$W_D^{1,2}(I, \mathbb{R}^n) = \{x \in L^2(I, \mathbb{R}^n) \mid Dx \in W^{1,2}(I, \mathbb{R}^m)\}$$

is a Hilbert space with scalar product

$$\langle x, y \rangle_{W_D^{1,2}(I,\mathbb{R}^n)} = \langle x, y \rangle_{L^2(I,\mathbb{R}^n)} + \langle (Dx)', (Dy)' \rangle_{L^2(I,\mathbb{R}^m)}.$$

Proof Let $\{x_k\}$ be a Cauchy sequence in $W_D^{1,2}(I, \mathbb{R}^n)$. Then, $\{x_k\}$ is a Cauchy sequence in $L^2(I, \mathbb{R}^n)$ and $\{y_k'\}$ with $y_k := Dx_k$ is a Cauchy sequence in $L^2(I, \mathbb{R}^m)$. Since L^2 is complete, there exist limits $x \in L^2(I, \mathbb{R}^n)$ and $g \in L^2(I, \mathbb{R}^m)$ with $\lim_{k \to \infty} x_k = x$ and $\lim_{k \to \infty} y_k' = g$.

Moreover continuity of the mapping $x \mapsto Dx$ yields

$$\lim_{k \to \infty} y_k = \lim_{k \to \infty} Dx_k = D(\lim_{k \to \infty} x_k) = Dx =: y.$$

It remains to show that $y \in W^{1,2}(I, \mathbb{R}^m)$ and $y' = (Dx)' = g$. To this end for any $\phi \in C_0^{\infty}(I, \mathbb{R}^m)$ we find by partial integration

$$0 = \lim_{k \to \infty} \int_I (y_k'(t) - g(t))^\top \phi(t)\, dt$$

$$= - \lim_{k \to \infty} \int_I \left(y_k(t) - \int_{t_0}^t g(s)ds \right)^\top \phi'(t)\, dt$$

$$= - \int_I \left(y(t) - \int_{t_0}^t g(s)ds \right)^\top \phi'(t)\, dt.$$

Application of a variation lemma, e.g. [48, Lemma 3.1.9], yields a constant C with

$$y(t) = \int_{t_0}^t g(s)ds + C \qquad (a.e.\ in\ I).$$

Hence $y = Dx \in W^{1,2}(I, \mathbb{R}^m)$ and $y' = g$. Thus, $x \in W_D^{1,2}(I, \mathbb{R}^n)$ which shows the completeness of $W_D^{1,2}(I, \mathbb{R}^n)$.   □

Proof of Theorem 2.5:

Proof Let (x, u) be feasible for LQOCP and (2.11). Exploitation of the assumptions and neglecting the explicit time dependence for notational convenience yields

$$
\begin{aligned}
J(x, u) - J(\hat{x}, \hat{u}) &= \int_I \frac{1}{2} x^\top Q x - \frac{1}{2} u^\top R u - \frac{1}{2} \hat{x}^\top Q \hat{x} - \frac{1}{2} \hat{u}^\top R \hat{u} \, dt \\
&= \int_I \frac{1}{2} x^\top Q x + \frac{1}{2} u^\top R u - \frac{1}{2} \hat{x}^\top Q \hat{x} - \frac{1}{2} \hat{u}^\top R \hat{u} \\
&\quad + \lambda^\top \left(A(Dx)' + Bx + Pu - q \right) \\
&\quad - \lambda^\top \left(A(D\hat{x})' + B\hat{x} + P\hat{u} - q \right) \, dt \\
&= \int_I \frac{1}{2} x^\top Q x + \frac{1}{2} u^\top R u - \frac{1}{2} \hat{x}^\top Q \hat{x} - \frac{1}{2} \hat{u}^\top R \hat{u} \\
&\quad + \lambda^\top B(x - \hat{x}) + \lambda^\top P(u - \hat{u}) + (A^\top \lambda)^\top \left((Dx)' - (D\hat{x})' \right) \, dt \\
&= \int_I \frac{1}{2} x^\top Q x + \frac{1}{2} u^\top R u - \frac{1}{2} \hat{x}^\top Q \hat{x} - \frac{1}{2} \hat{u}^\top R \hat{u} \\
&\quad + \lambda^\top B(x - \hat{x}) + \lambda^\top P(u - \hat{u}) - \left((A^\top \lambda)' \right)^\top (Dx - D\hat{x}) \, dt \\
&= \int_I \frac{1}{2} x^\top Q x + \frac{1}{2} u^\top R u - \frac{1}{2} \hat{x}^\top Q \hat{x} - \frac{1}{2} \hat{u}^\top R \hat{u} \\
&\quad + \lambda^\top B(x - \hat{x}) + \lambda^\top P(u - \hat{u}) \\
&\quad - \hat{x}^\top Q(x - \hat{x}) - \lambda^\top B(x - \hat{x}) - \eta^\top S(x - \hat{x}) \, dt \\
&= \int_I \frac{1}{2} x^\top Q x + \frac{1}{2} u^\top R u - \frac{1}{2} \hat{x}^\top Q \hat{x} - \frac{1}{2} \hat{u}^\top R \hat{u} \\
&\quad - \hat{u}^\top R(u - \hat{u}) - \eta^\top F(u - \hat{u}) - \hat{x}^\top Q(x - \hat{x}) - \eta^\top S(x - \hat{x}) \, dt \\
&= \int_I \frac{1}{2} (x - \hat{x})^\top Q(x - \hat{x}) + \frac{1}{2} (u - \hat{u})^\top R(u - \hat{u}) \\
&\quad - \underbrace{\eta^\top}_{\geq 0} \underbrace{(Sx + Fu - r)}_{\leq 0} + \underbrace{\eta^\top (S\hat{x} + F\hat{u} - r)}_{=0} \, dt \\
&\geq 0,
\end{aligned}
$$

since Q and R are supposed to be uniformly positive semidefinite. □

References

1. Alt, W.: The Lagrange-Newton method for infinite dimensional optimization problems. Numer. Funct. Anal. Optim. **11**, 201–224 (1990)
2. Alt, W.: Sequential quadratic programming in banach spaces. In: Oettli, W., Pallaschke, D. (eds.) Advances in Optimization, pp. 281–301. Springer, Berlin (1991)
3. Alt, W., Malanowski, K.: The Lagrange-Newton method for nonlinear optimal control problems. Comput. Optim. Appl. **2**, 77–100 (1993)
4. Alt, W., Malanowski, K.: The Lagrange-Newton method for state constrained optimal control problems. Comput. Optim. Appl. **4**, 217–239 (1995)
5. Amodio, P., Mazzia, F.: Numerical solution of differential algebraic equations and computation of consistent initial/boundary conditions. J. Comput. Appl. Math. **87**, 135–146 (1997)
6. Ascher, U.M., Mattheij, R.M.M., Russell, R.D.: Numerical solution of boundary value problems for ordinary differential equations, In: Classics in Applied Mathematics, vol. 13. SIAM, Philadelphia (1995)
7. Ascher, U.M., Spiteri, R.J.: Collocation software for boundary value differential-algebraic equations. SIAM J. Sci. Comput. **15**(4), 938–952 (1994)
8. Backes, A.: Extremalbedingungen für optimierungs-probleme mit algebro-differentialgleichungen. PhD thesis, Mathematisch-naturwissenschaftliche fakultät, Humboldt-Universität Berlin (2006)
9. Balla, K., Kurina, G.A., März, R.: Index criteria for differential algebraic equations arising from linear-quadratic optimal control problems. J. Dyn. Control Syst. **12**(3), 289–311 (2006)
10. Balla, K., März, R.: A unified approach to linear differential algebraic equations and their adjoints. Z. Anal. Anwendungen **21**(3), 783–802 (2002)
11. Bauer, I.: Numerical methods for the solution of initial value problems and for the generation of first and second-order derivatives in optimisation problems of chemistry and process engineering. (Numerische verfahren zur lösung von anfangswertaufgaben und zur generierung von ersten und zweiten ableitungen mit anwendungen bei optimierungsaufgaben in chemie und verfahrenstechnik.). PhD thesis, Mathematisch-Naturwissenschaftliche Gesamtfakultät, iv, p. 169, Heidelberg (2000)
12. Bauer, I., Bock, H.G., Körkel, S., Schlöder, J.P.: Numerical methods for optimum experimental design in DAE systems. J. Comput. Appl. Math. **120**(1–2), 1–25 (2000)
13. Bertolazzi, E., Biral, F., Da Lio, M.: Symbolic-numeric efficient solution of optimal control problems for multibody systems. J. Comput. Appl. Math. **185**(2), 404–421 (2006)
14. Betts, J.T.: Practical methods for optimal control using nonlinear programming. In: Advances in Design and Control, vol. 3. SIAM, Philadelphia (2001)
15. Betts, J.T., Huffman, W.P.: Mesh refinement in direct transcription methods for optimal control. Optim. Control Appl. Meth. **19**, 1–21 (1998)
16. Biegler, L.T., Campbell, S.L., Mehrmann, V.(eds.): Control and optimization with differential-algebraic constraints. In: Advances in Design and Control, vol. 23. SIAM, Philadelphia (2012)
17. Biehn, N., Campbell, S.L., Jay, L., Westbrook, T.: Some comments on DAE theory for IRK methods and trajectory optimization. J. Comput. Appl. Math. **120**(1–2), 109–131 (2000)
18. Bock, H.G.: Randwertproblemmethoden zur parameteridentifizierung in systemen nichtlinearer differentialgleichungen. In: Bonner Mathematische Schriften, vol. 183. Fakultät der Universität, Bonn (1987)
19. Bock, H.G., Plitt, K.J.: A multiple shooting algorithm for direct solution of optimal control problems. Proceedings of the 9th IFAC Worldcongress, Budapest, Hungary (1984)
20. Brenan, K.E.: Differential-algebraic equations issues in the direct transcription of path constrained optimal control problems. Ann. Numer. Math. **1**, 247–263 (1994)
21. Brenan, K.E., Campbell, S.L., Petzold, L.R.: Numerical solution of initial-value problems in differential-algebraic equations. In: Classics In Applied Mathematics, vol. 14. SIAM, Philadelphia (1996)

22. Brown, P.N., Hindmarsh, A.C., Petzold, L.R.: Consistent initial condition calculation for differential-algebraic systems. SIAM J. Sci. Comput. **19**(5), 1495–1512 (1998)
23. Büskens, C.: Optimierungsmethoden und Sensitivitätsanalyse für optimale Steuerprozesse mit Steuer- und Zustandsbeschränkungen. PhD thesis, Fachbereich Mathematik, Westfälische Wilhems-Universität Münster, Munster (1998)
24. Büskens, C., Gerdts, M.: Real-time optimization of DAE systems. In: Grötschel, M., Krumke, S.O., Rambau, J. (eds.) Online Optimization of Large Scale Systems, pp. 117–128. Springer, Heidelberg (2001)
25. Campbell, S.L., Kunkel, P.: On the numerical treatment of linear-quadratic optimal control problems for general linear time-varying differential-algebraic equations. J. Comput. Appl. Math. **242**, 213–231 (2013)
26. Campbell, S.L., Marszalek, W.: The index of an infinite dimensional implicit system. Math. Comput. Model. Dyn. Syst. **5**(1), 18–42 (1999)
27. Campbell, S.L., März, R.: Direct transcription solution of high index optimal control problems and regular Euler-Lagrange equations. J. Comput. Appl. Math. **202**(2), 186–202 (2007)
28. Campbell, S.L., Kelley, C.T., Yeomans, K.D.: Consistent initial conditions for unstructured higher index DAEs: A computational study. In: Proceedings of the Computational Engineering in Systems Applications, pp. 416–421, France (1996)
29. Cao, Y., Li, S., Petzold, L.R., Serban, R.: Adjoint sensitivity analysis for differential-algebraic equations: The adjoint DAE system and its numerical solution. SIAM J. Sci. Comput. **24**(3), 1076–1089 (2003)
30. Caracotsios, M., Stewart, W.E.: Sensitivity analysis of initial-boundary-value problems with mixed PDEs and algebraic equations. Comput. chem. Eng. **19**(9), 1019–1030 (1985)
31. Chen, J., Gerdts, M.: Smoothing technique of nonsmooth Newton methods for control-state constrained optimal control problems. SIAM J. Numer. Anal. **50**(4), 1982–2011 (2012)
32. Cherednichenko, S., Krumbiegel, K., Rösch, A.: Error estimates for the Lavrientiev regularization of elliptic optimal control problems. Inverse Prob. **24**, 1–21 (2008)
33. de Pinho, M., Ilchmann, A.: Weak maximum principle for optimal control problems with mixed constraints. Nonlinear Anal. Theory Meth. Appl. **48**(8), 1179–1196 (2002)
34. de Pinho, M., Vinter, R.B.: Necessary conditions for optimal control problems involving nonlinear differential algebraic equations. J. Math. Anal. Appl. **212**, 493–516 (1997)
35. Deuflhard, P., Pesch, H.J., Rentrop, P.: A modified continuation method for the numerical solution of nonlinear two-point boundary value problems by shooting techniques. Numer. Math. **26**, 327–343 (1976)
36. Devdariani, E.N., Ledyaev, Yu. S.: Maximum principle for implicit control systems. Appl. Math. Optim. **40**, 79–103 (1999)
37. Diehl, M.: Real- time optimization for large scale nonlinear processes. PhD thesis, Naturwissenschaftlich-Mathematische Gesamtfakultät, Universität Heidelberg (2001)
38. Diehl, M., Bock, H.G., Schlöder, J.P.: A real-time iteration scheme for nonlinear optimization in optimal feedback control. SIAM J. Control Optim. **43**(5), 1714–1736 (2005)
39. Engelsone, A., Campbell, S.L., Betts, J.T.: Direct transcription solution of higher-index optimal control problems and the virtual index. Appl. Numer. Math. **57**(3), 281–296 (2007)
40. Fabien, B.C.: dsoa: The implementation of a dynamic system optimization algorithm. Optim. Control Appl. Meth. **31**(3), 231–247 (2010)
41. Feehery, W.F., Tolsma, J.E., Barton, P.I.: Efficient sensitivity analysis of large-scale differential-algebraic systems. Appl. Numer. Math. **25**, 41–54 (1997)
42. Führer, C., Leimkuhler, B.J.: Numerical solution of differential-algebraic equations for constraint mechanical motion. Numer. Math. **59**, 55–69 (1991)
43. Gear, C.W., Leimkuhler, B., Gupta, G.K.: Automatic integration of Euler-Lagrange equations with constraints. J. Comput. Appl. Math. **12/13**, 77–90 (1985)
44. Gerdts, M.: Direct shooting method for the numerical solution of higher index DAE optimal control problems. J. Optim. Theory Appl. **117**(2), 267–294 (2003)

45. Gerdts, M.: A variable time transformation method for mixed-integer optimal control problems. Optim. Control Appl. Meth. **27**(3), 169–182 (2006)
46. Gerdts, M.: Local minimum principle for optimal control problems subject to index-two differential-algebraic equations. J. Optim. Theory Appl. **130**(3), 443–462 (2006)
47. Gerdts, M.: Representation of the lagrange multipliers for optimal control problems subject to differential-algebraic equations of index two. J. Optim. Theory Appl. **130**(2), 231–251 (2006)
48. Gerdts, M.: Optimal Control of ODEs and DAEs. Walter de Gruyter, Berlin (2012)
49. Gerdts, M., Büskens, C.: Consisstent initialization of sensitivity matrices for a class of parametric DAE systems. BIT Numer. Math. **42**(4), 796–813 (2002)
50. Gerdts, M., Hüpping, B.: Virtual control regularization of state constrained linear quadratic optimal control problems. Comput. Optim. Appl. **51**(2), 867–882 (2012)
51. Gerdts, M., Kunkel, M.: A globally convergent semi-smooth Newton method for control-state constrained DAE optimal control problems. Comput. Optim. Appl. **48**(3), 601–633 (2011)
52. Gerdts, M., Sager, S.: Mixed-integer DAE optimal control problems: necessary conditions and bounds. In: Control and Optimization with Differential-Algebraic Constraints. Advances in Design and Control, vol. 23, pp. 189–212. SIAM, Philadelphia (2012)
53. Goh, C.J., Teo, K.L.: Control parametrization: A unified approach to optimal control problems with general constraints. Automatica **24**, 3–18 (1988)
54. Gopal, V., Biegler, L.T.: A successive linear programming approach for initialization and reinitialization after discontinuities of differential-algebraic equations. SIAM J. Sci. Comput. **20**(2), 447–467 (1998)
55. Griewank, A.: Evaluating Derivatives. Principles and Techniques of Algorithmic Differentiation. In: Frontiers in Applied Mathematics, vol. 19. SIAM, Philadelphia (2000)
56. Griewank, A., Walther, A.: Evaluating Derivatives. Principles and Techniques of Algorithmic Differentiation. 2nd edn. SIAM, Philadelphia (2008)
57. Gritsis, D.M., Pantelides, C.C., Sargent, R.W.H.: Optimal control of systems described by index two differential-algebraic equations. SIAM J. Sci. Comput. **16**(6), 1349–1366 (1995)
58. Hairer, E., Wanner, G.: Solving ordinary differential equations II: Stiff and differential-algebraic problems. 2nd edn. In: Computational Mathematics, vol. 14. Springer, New York (1996)
59. Hairer, E., Lubich, Ch., Roche, M.: The numerical solution of differential-algebraic systems by Runge-Kutta methods. In: Lecture Notes in Mathematics, vol. 1409. Springer, New York (1989)
60. Hargraves, C.R., Paris, S.W.: Direct trajectory optimization using nonlinear programming and collocation. J. Guid. Control Dyn. **10**, 338–342 (1987)
61. Hinze, M., Pinnau, R., Ulbrich, M., Ulbrich, S.: Optimization with PDE Constraints. In: Mathematical Modelling: Theory and Applications, vol. 23, p. 270. Springer, Dordrecht (2009)
62. Ito, K., Kunisch, K.: Lagrange Multiplier Approach to Variational Problems and Applications. In: Advances in Design and Control, vol. 15, p. 341. SIAM, Philadelphia (2008)
63. Jeon, M.: Parallel optimal control with multiple shooting, constraints aggregation and adjoint methods. J. Appl. Math. Comput. **19**(1–2), 215–229 (2005)
64. Kameswaran, S., Biegler, L.T.: Advantages of nonlinear-programming-based methodologies for inequality path-constrained optimal control problems-A numerical study. SIAM J. Sci. Comput. **30**(2), 957–981 (2008)
65. Kitzhofer, G., Koch, O., Pulverer, G., Simon, Ch. Weinmüller, E.B.: The new MATLAB code bvpsuite for the solution of singular implicit BVPs. J. Numer. Anal, Ind. Appl. Math. **5**(1–2), 113–134 (2010)
66. Krumbiegel, K., Rösch, A.: On the regularization error of state constrained Neumann control problems. Control Cybern. **37**(2), 369–392 (2008)
67. Kunkel, P., Mehrmann, V.: The linear quadratic optimal control problem for linear descriptor systems with variable coefficients. Math. Control Signal Syst. **10**(3), 247–264 (1997)

68. Kunkel, P., Mehrmann, V.:   Differential-Algebraic Equations. Analysis and Numerical Solution., p. 377. European Mathematical Society, Zürich (2006)
69. Kunkel, P., Mehrmann, V.: Optimal control for unstructured nonlinear differential-algebraic equations of arbitrary index. Math. Control Signal Syst. **20**(3), 227–269 (2008)
70. Kunkel, P., Mehrmann, V.: Formal adjoints of linear DAE operators and their role in optimal control. Electron. J. Linear Algebra **22**, 672–693 (2011)
71. Kunkel, P., Mehrmann, V., Stöver, R.:   Symmetric collocation for unstructered nonlinear differential-algebraic equations of arbitrary index. Numer. Math. **98**(2), 277–304 (2004)
72. Kunkel, P., Mehrmann, V., Stöver, R.:   Multiple shooting for unstructured nonlinear differential-algebraic equations of arbitrary index. SIAM J. Numer. Anal. **42**(6), 2277–2297 (2005)
73. Kunkel, P., Stöver, R.:   Symmetric collocation methods for linear differential-algebraic boundary value problems. Numer. Math. **91**(3), 475–501 (2002)
74. P. Kunkel, von dem Hagen, O.:   Numerical solution of infinite-horizon optimal-control problems. Comput. Econ. **16**(3), 189–205 (2000)
75. Kurcyusz, S.: On the existence and nonexistence of lagrange multipliers in banach spaces. J. Optim. Theory Appl. **20**(1), 81–110 (1976)
76. Kurina, G.A., März, R.:   On linear-quadratic optimal control problems for time-varying descriptor systems. SIAM J. Control Optim. **42**(6), 2062–2077 (2004)
77. Kurina, G.A., März, R.:   Feedback solutions of optimal control problems with DAE constraints. SIAM J. Control Optim. **46**(4), 1277–1298 (2007)
78. Lamour, R.: A well-posed shooting method for transferable DAE's. Numer. Math. **59**(8), 815–830 (1991)
79. Lamour, R.: A shooting method for fully implicit index-2 differential algebraic equations. SIAM J. Sci. Comput. **18**(1), 94–114 (1997)
80. Lamour, R., März, R., Tischendorf, C.: Differential-Algebraic Equations: A Projector Based Analysis. In: Differential-Algebraic Equations Forum. Springer, Berlin (2013)
81. Lee, H.W.J., Teo, K.L., Rehbock, V., Jennings. L.S.:   Control parametrization enhancing technique for optimal discrete-valued control problems. Automatica **35**(8), 1401–1407 (1999)
82. Leimkuhler, B., Petzold, L.R., Gear, C.W.:   Approximation methods for the consistent initialization of differential-algebraic equations.   SIAM J. Numer. Anal. **28**(1), 205–226 (1991)
83. Lempio, F.:   Tangentialmannigfaltigkeiten und infinite Optimierung. Habilitationsschrift, Universität Hamburg (1972)
84. Leyendecker, S., Ober-Blöbaum, S., Marsden, J.E., Ortiz, M.: Discrete mechanics and optimal control for constrained systems. Optim. Control Appl. Meth. **31**(6), 505–528 (2010)
85. Machielsen, K.C.P.: Numerical solution of optimal control problems with state constraints by sequential quadratic programming in function space. CWI Tract, vol. 53. Centrum voor Wiskunde en Informatica, Amsterdam (1988)
86. Maly, T., Petzold, L.R.:   Numerical methods and software for sensitivity analysis of differential-algebraic systems. Appl. Numer. Math. **20**(1), 57–79 (1996)
87. Mehrmann, V.:   The autonomous linear quadratic control problem. theory and numerical solution. In: Lecture Notes in Control and Information Sciences, vol 163. Springer, Berlin (1991)
88. Morison, K.R., Sargent, R.W.H.:   Optimization of multistage processes described by differential-algebraic equations. In: Numerical Aanalysis, Proceedings of the 4th IIMAS Workshop, Guanajuato/Mex. 1984. Lecture Notes in Mathematics, 1230, pp. 86–102. Springer, Berlin (1986)
89. Pantelides, C.C.: The consistent initialization of differential-algebraic systems. SIAM J. Sci. Stat. Comput. **9**(2), 213–231 (1988)
90. Pantelides, C.C., Sargent, R.W.H., Vassiliadis, V.S.: Optimal control of multistage systems described by high-index differential-algebraic equations. In: Bulirsch, R. (ed.) Computational Optimal Control. International Series of Numerical Mathematics, vol. 115, pp. 177–191. Birkhäuser, Basel (1994)

91. Pytlak, R.: Runge-Kutta based procedure for the optimal control of differential-algebraic equations. J. Optim. Theory Appl. **97**(3), 675–705 (1998)
92. Pytlak, R.: Numerical Methods for Optimal Control Problems with State Constraints. In: Lecture Notes in Mathematics, vol. 1707. Springer, New York (1999)
93. Pytlak, R.: Optimal control of differential-algebraic equations of higher index. I: First-order approximations. J. Optim. Theory Appl. **134**(1), 61–75 (2007)
94. Pytlak, R.: Optimal control of differential-algebraic equations of higher index. II: Necessary optimality conditions. J. Optim. Theory Appl. **134**(1), 77–90 (2007)
95. Pytlak, R.: Numerical procedure for optimal control of higher index DAEs. Discrete Contin. Dyn. Syst. **29**(2), 647–670 (2011)
96. Reich, S.: On an existence and uniqueness theory for nonlinear differential-algebraic equations. Circ. Syst. Signal Process. **10**(3), 343–359 (1991)
97. Roubicek, T., Valásek, M.: Optimal control of causal differential-algebraic systems. J. Math. Anal Appl. **269**(2), 616–641 (2002)
98. Sager, S.: Numerical methods for mixed-integer optimal control problems. PhD thesis, Naturwissenschaftlich-Mathematische Gesamtfakultät (Diss.), p. 219, Heidelberg (2006)
99. Sager, S., Bock, H.G., Reinelt, G.: Direct methods with maximal lower bound for mixed-integer optimal control problems. Math. Program.(A) **118**(1), 109–149 (2009)
100. Schanzer, G.F., Callies, R.: Multiple constrained rivaling actuators in the optimal control of miniaturized manipulators. Multibody Syst. Dyn. **19**(1–2), 21–43 (2008)
101. Schiela, A.: An interior point method in function space for the efficient solution of state constrained optimal control problems. Math. Program. **138**(1–2 (A)), 83–114 (2013)
102. Schulz, V.H.: Reduced SQP methods for large-scale optimal control problems in DAE with application to path planning problems for satellite mounted robots. PhD thesis, Interdisziplinäres Zentrum für Wissenschaftliches Rechnen, Universität Heidelberg (1996)
103. Schulz, V.H., Bock, H.G., Steinbach, M.C.: Exploiting invariants in the numerical solution of multipoint boundary value problems for DAE. SIAM J. Sci. Comput. **19**(2), 440–467 (1998)
104. Stöver, R.: Collocation methods for solving linear differential-algebraic boundary value problems. Numer. Math. **88**(4), 771–795 (2001)
105. Teo, K.L., Jennings, L.S., Lee, H.W.J., Rehbock, V.: The control parametrization enhancing transform for constrained optimal control problems. J. Aust. Math. Soc. B **40**, 314–335 (1999)
106. Trenn, S.: Solution concepts for linear DAEs: a survey. In: Ilchmann, A., Reis, T. (eds.) Surveys in Differential-Algebraic Equations I. Differential-Algebraic Equations Forum, pp. 137–172. Springer, Berlin (2013)
107. Tröltzsch, F.: Optimal control of partial differential equations. Theory, methods and applications. American Mathematical Society (AMS), Providence (2010)
108. Ulbrich, M.: Semismooth newton methods for operator equations in function spaces. SIAM J. Optim. **13**(3), 805–841 (2003)
109. Ulbrich, M.: Nonsmooth newton-like methods for variational inequalities and constrained optimization problems in function spaces. Habilitation, Technical University of Munich, Munich (2002)
110. Ulbrich, M., Ulbrich, S.: Primal-dual interior-point methods for PDE-constrained optimization. Math. Program. **117**(1–2 (B)), 435–485 (2009)
111. Vassiliades, V.C., Sargent, R.W.H., Pantelides, C.C.: Solution of a class of multistage dynamic optimization problems. 2. Problems with path constraints. Ind. Eng. Chem. Res. **33**, 2123–2133 (1994)
112. von Stryk, O.: Numerische Lösung optimaler Steuerungsprobleme: Diskretisierung, Parameteroptimierung und Berechnung der adjungierten Variablen. In: VDI Fortschrittberichte Reihe 8: Meß-, Steuerungs- und Regeleungstechnik, vol. 441. VDI, Düsseldorf (1994)
113. Weiser, M.: Interior point methods in function space. SIAM J. Control Optim. **44**(5), 1766–1786 (2005)

Differential-Algebraic Equations from a Functional-Analytic Viewpoint: A Survey

Roswitha März

Abstract The purpose of this paper is to provide an overview on the state of the art concerning functional-analytic properties associated with differential-algebraic equations (DAEs). We summarize the relevant literature and develop a basic theory of linear and nonlinear differential-algebraic operators. In particular, we consider Fredholm properties, normal solvability, generalized inverses, least-squares solutions, splittings of regular linear differential-algebraic operators, bounded outer inverses, local solvability of equations with regular nonlinear differential-algebraic operators, Newton–Kantorovich iterations, and regularizations of the ill-posed problems arising from higher-index operators.

Keywords Closure • Differential-algebraic operator • Fredholm operator • Generalized inverse • Ill-posedness • Least-squares solution • Linearization • Newton–Kantorovich iteration • Nonclosed-image operator • Normal solvability • Outer inverse • Regular differential-algebraic operator • Regularization • Tractability index • Well-posedness

Mathematics Subject Classification (2010) 34A09, 34L30, 47A05, 47A52, 47A53, 47E05, 47J06, 65L80, 34-02, 46-02, 47-02

Contents

R. März (✉)
Institut für Mathematik, Humboldt-Universität zu Berlin, Berlin, Germany
e-mail: maerz@mathematik.hu-berlin.de

© Springer International Publishing Switzerland 2015   163
A. Ilchmann, T. Reis (eds.), *Surveys in Differential-Algebraic Equations II*,
Differential-Algebraic Equations Forum, DOI 10.1007/978-3-319-11050-9_4

1 Introduction

Functional analysis is a child of the twentieth century. It provides us with a new language that allows us to formulate apparently different topics in a unique way [83, p. ix]. The topic of differential-algebraic equations (DAEs) is another, much younger child of the same century. To a large extent, DAEs are merely recognized as special ordinary differential equations (ODEs), a fact which dominates the hitherto existing analysis of DAEs. In contrast, the functional analysis of DAEs remains still in its initial stage so far, even though there have been various quite early approaches such as

- applying operator settings and aspects of functional-analytic discretization theory, e.g., [21,56,57],
- treating higher-index DAEs consistently as ill-posed operator equations, e.g., [36, 38,43,44],
- work concerning abstract differential equations which are not solved for the derivative in the context of optimization problems, cf. [52],
- diverse attempts in system and control theory to incorporate various types of solutions by appropriate function spaces, cf. [71,77],
- particular results on degenerate differential equations in function spaces, e.g., [27], and on DAEs within the context of partial differential-algebraic equations (PDAEs) and abstract DAEs, cf. [55, Chapter 12].

Nonetheless, as yet, an adequate sophisticated functional-analytic characterization of DAEs has not been accomplished. The purpose of the present paper is to provide an overview on basic functional-analytic properties of operators associated with DAEs and, furthermore, to summarize the relevant literature to the best of the author's knowledge.

We investigate operators associated with linear and nonlinear DAEs in standard form

$$E(t)x'(t) + F(t)x(t) = q(t), \tag{1.1}$$

$$\mathfrak{f}(x'(t), x(t), t) = 0, \tag{1.2}$$

and linear and nonlinear DAEs showing a properly involved derivative

$$A(t)(Dx)'(t) + B(t)x(t) = q(t), \tag{1.3}$$

$$f((Dx)'(t), x(t), t) = 0. \tag{1.4}$$

Such a DAE comprises m unknown functions and k equations. We follow the practice of control theory (e.g., [4,46]) and allow for rectangular DAEs with $k \neq m$ in Sect. 3. In contrast, in sensitivity analysis and simulation framework one is basically interested in uniquely solvable problems, that is here, in regular DAEs completed by appropriate boundary conditions. Regularity requires $k = m$, which is therefore supposed in Sects. 2 and 4.

We represent the DAEs as operator equations—we call the associated operators *differential-algebraic operators*—and apply functional analytic tools for their characterization and further treatment. In particular, different possible settings are inspected, also in view of eventual applications and numerical computations.

In essence, here we focus on compact intervals $\mathcal{I} = [t_a, t_e]$, which allows us to apply Banach spaces equipped with maximum-norms. Modifications for open intervals would require Fréchet spaces and more technicalities.

DAEs describe dynamical phenomena subject to certain constraints. They can be regarded as somehow uniformly singular implicit ODEs. DAEs arise in various application fields, often in specifically structured versions, see e.g., [6, 46, 51, 55, 70, 72]. In each application area, if available, the special DAE structure is separately exploited for numerical treatment, often with great success, e.g., in circuit simulation and the simulation of multibody dynamics. Currently, because of the generally increasing role of simulation in science and technology, more and more applications are coming along, resulting in a long-term demand for more comprehensive, more general DAE analysis and numerical treatment.

In the present paper, we do not address differently structured types of DAEs, but we direct our attention to the general linear and nonlinear DAEs (1.1), (1.2), and (1.3), (1.4), respectively. These two pairs of DAEs accompany each other in some sense. From the viewpoint of applications, the versions (1.3), (1.4), with properly involved derivatives, can be seen as advanced models (e.g., [3, 55, 79]). From the viewpoint of functional analysis, the differential-algebraic operators corresponding to (1.3), (1.4) turn out to be the closures of the unbounded operators associated with the standard form DAEs (1.1), (1.2) in natural settings.

At the beginning of the twentieth century, J.S.Hadamard formalized the classical concept of well-posedness for abstract equations

$$\mathcal{K}x = p, \tag{1.5}$$

where $\mathcal{K}$ is a mapping from some topological space X into a topological space Z. Equation (1.5) is said to be *well-posed*, e.g., [34, p. 3], if

(a) for each $p \in Z$, there is a solution $x \in X$ of (1.5);
(b) this solution x is unique in X;
(c) the dependence of x upon p is continuous.

An Eq. (1.5) which is not well-posed in this sense is said to be *ill-posed*. A method for solving approximately an ill-posed problem is called a *regularization method*.

Obviously, to answer the question whether a given equation is well-posed, one has to consider not only the operator $\mathcal{K}$ but also the spaces X and Z including their topologies.

Here we prefer topological spaces whose topology is defined by a norm. This allows to measure global physical quantities, responsibilities, sensitivities.

In a well-posed problem, the operator $\mathcal{K}$ must be bijective and the inverse $\mathcal{K}^{-1}$ must be continuous. Often the so-called *well-posedness principle*, see [84, p. 180], is helpful: If $\mathcal{K}$ is a bounded linear operator mapping a Banach X into a Banach space Z, then the Eq. (1.5) is well-posed, exactly if $\mathcal{K}$ is bijective.

In our context, the operator $\mathcal{K}$ is composed from an operator K acting in normed spaces X and Y such that the equation $Kx = q$ represents a DAE and the Eq. (1.5), with $Z = Y \times \mathbb{R}^\delta$, contains the DAE and an additional finite-dimensional part representing boundary and initial conditions.

Having a linear operator K, for eventually obtaining a well-posed problem (1.5) we are specifically interested in a finite-dimensional nullspace $\ker K$, in boundedness or closedness of K and last but not least a closed range $\operatorname{im} K$. The closed range property plays its role in the theory of Fredholm operators; it is the crucial ingredient of *normal solvability*. The related operator equations can indeed be ill-posed, but they become well-posed in slightly modified settings.

If $\mathcal{K}$ is an injective bounded linear operator acting in Banach spaces X and Z, and if the set $\operatorname{im} \mathcal{K}$ fails to be closed in Z, then Eq. (1.5) is no longer well-posed, since items (a) and (c) do not hold. Those kind of problems are said to be *essentially ill-posed in the sense of A.N. Tikhonov*.

We emphasize that, in standard settings, equations given by regular higher-index differential-algebraic operators turn out to be essentially ill-posed in this sense, and their solutions basically show an ambivalent character; they behave smoothly with regard to appropriately stated initial values, but, as a function of perturbations on the right-hand side, they are no longer continuous. We observe this behavior even in the case of regular constant coefficient linear DAEs with Kronecker index greater than one (cf. Sect. 2.2 below). The higher the index, the stronger the discontinuity.

Only index-0 and index-1 operators yield well-posed problems in standard settings. The commonly used index-reduction procedures which modify the original DAE by differentiation and elimination steps into an index-1 or index-0 DAE can be seen as particular regularization techniques.

Index-0 and index-1 problems can be solved numerically nearly as safely as explicit ODEs. In contrast, concerning the direct treatment of equations with higher index differential-algebraic operators as they are given, there remains a big gap between the practical needs and the theory available at this stage. Great future efforts are necessary to close this gap, and also to develop appropriate functional-analytic tools.

Having in mind the great impact of functional analysis on the treatment of partial differential equations and integral equations, we hope that the hitherto existing tentative functional-analytic approaches to DAEs will be promoted and extended to powerful tools in the future. Here we expose basic functional-analytic properties of differential-algebraic operators.

The material of this paper is organized in four sections that are almost independent of each other.

Serving as an easy introduction to the topic, Sect. 2 deals with operators associated with regular matrix pencils and provides the constitutive characteristics of regular differential-algebraic operators in different settings. We expose such settings that are reasonable in view of applications.

In Sect. 3 we regard differential-algebraic operators associated with standard form DAEs (1.1) and (1.2) comprising m unknowns and k equations stated in continuous resp. integrable function spaces. These operators are unbounded. We point out that their closures are accompanied by DAEs (1.3) and (1.4) showing properly involved derivatives via factorizations of the leading term. Turning to the corresponding special graph norm spaces we arrive at bounded operators. We characterize normally solvable linear differential-algebraic operators and construct

bounded inner inverses and least-squares solutions (LSS). We further provide a
class of nonlinear differential-algebraic operators being Fréchet differentiable in a
coherent Banach space setting and yielding normally solvable linearizations. No
special knowledge about DAEs is required in this section.

In Sect. 4 we accordingly turn to DAEs (1.3) and (1.4) with properly involved
derivatives stated in their natural Banach spaces. We study basic properties of
regular linear and Fréchet differentiable nonlinear differential-algebraic operators.
In particular, for linear operators we provide a characteristic operator splitting,
and describe the nullspace as well as the image. In case of a nonclosed image
we offer bounded outer inverses. For nonlinear differential-algebraic operators we
introduce the notions of regularity, characteristic values, and tractability index in
a rigorous functional-analytic sense. On this background we consider possible
Fredholm properties, well-posedness of boundary value problems, local solvability,
and Newton–Kantorovich iterations. This section relies on projector based analysis
[55]. It is recommended to take a look at the respective summary in the appendix.

Both Sects. 3 and 4 contain an extra subsection with notes and references to relate
the existing literature to the material.

Section 5 collects known regularization methods from the viewpoint of
functional-analysis.

The notation agreements, the functional-analytic background, and a short summary of projector based DAE analysis can be found in the appendices.

2 Constitutive Characteristics of Regular
Differential-Algebraic Operators: Exemplified by Means
of Regular Matrix Pencils

We consider the linear operator $\overset{\circ}{T} : X \to Y$,

$$\overset{\circ}{T}x := Ex' + Fx, \ x \in \text{dom}\,\overset{\circ}{T} \subseteq X, \tag{2.1}$$

associated with the constant coefficient DAE

$$Ex'(t) + Fx(t) = q(t), \quad t \in \mathcal{I}, \tag{2.2}$$

and formed by the ordered pair $\{E, F\}$ of real $m \times m$ matrices E, F. Diverse
appropriate function spaces X and Y will be specified below, in particular we apply
$\mathcal{C}(\mathcal{I}, \mathbb{R}^m)$ and $L^2(\mathcal{I}, \mathbb{R}^m)$.

2.1 Finite-Dimensional Nullspaces

The solutions of the homogeneous equation

$$Ex'(t) + Fx(t) = 0, \quad t \in \mathcal{I}. \tag{2.3}$$

constitute the nullspace ker $\overset{\circ}{T}$ of the operator $\overset{\circ}{T}$.

If E is nonsingular, the homogeneous equation (2.3) represents a regular implicit ODE, whose fundamental solution system forms an m-dimensional subspace in $\mathcal{C}^\infty(\mathcal{I}, \mathbb{R}^m)$; and then the associated operator $\overset{\circ}{T}$ has an m-dimensional nullspace— supposing the setting ensures the inclusion $\mathcal{C}^\infty(\mathcal{I}, \mathbb{R}^m) \subseteq \operatorname{dom} \overset{\circ}{T}$. However, what happens if E is singular? Is there a class of equations, such that Eq. (2.3) has a finite-dimensional solution space? The answer is closely related to regularity of the matrix pair $\{E, F\}$.

Definition 2.1 Given the ordered pair $\{E, F\}$ of matrices $E, F \in \mathcal{L}(\mathbb{R}^m)$, the matrix pencil $\lambda E + F$ is said to be *regular* if the polynomial $\rho(\lambda) := \det(\lambda E + F)$ does not vanish identically. Otherwise the matrix pencil is said to be *singular*.

Both the ordered pair $\{E, F\}$ and the DAE (2.2) are said to be *regular* if the accompanying matrix pencil is regular, and otherwise *nonregular*.

The operator $\overset{\circ}{T}$ given by (2.1) is called a *regular differential-algebraic operator* if the pair $\{E, F\}$ is regular.

If E is a nonsingular matrix, then the pair $\{E, F\}$ is always regular and the polynomial $p(\lambda)$ has degree m. If the first matrix E of a regular pair is singular, then the polynomial degree is lower. To be more precise we quote some well-known facts due to Weierstraß and Kronecker, cf. [30, 49, 80].

For any regular pair $\{E, F\}$, $E, F \in \mathcal{L}(\mathbb{R}^m)$, there exist nonsingular matrices $L, K \in \mathcal{L}(\mathbb{R}^m)$ and integers $0 \le l \le m$, $0 \le \mu \le l$, such that

$$LEK = \begin{bmatrix} I & \\ & N \end{bmatrix} \begin{matrix} \}m-l \\ \}l \end{matrix} \quad , \quad LFK = \begin{bmatrix} W & \\ & I \end{bmatrix} \begin{matrix} \}m-l \\ \}l \end{matrix} \quad . \tag{2.4}$$

Thereby, N is absent if $l = 0$, and otherwise N is nilpotent of order $\mu \ge 1$, i.e., $N^\mu = 0$, $N^{\mu-1} \ne 0$. For $l = 0$ we set $\mu = 0$. The integers l and μ as well as the eigenstructure of the blocks N and W are uniquely determined by the pair $\{E, F\}$. If $l = m$ then the upper blocks are absent.

The real matrix N has the eigenvalue zero only and can be transformed into its Jordan form by means of a real similarity transformation. Therefore, the transformation matrices L and K can be chosen in such a way that N is in Jordan form.

Now it is evident that the relation

$$\text{degree} \det(\lambda E + F) = m - l \leq \text{rank } E \qquad (2.5)$$

is given for each regular pencil.

The special pair given by (2.4) is said to be the *Weierstraß–Kronecker form* of the original pair $\{E, F\}$.

Definition 2.2 The *Kronecker index* of a regular matrix pair $\{E, F\}$, $E, F \in \mathcal{L}(\mathbb{R}^m)$, E singular, is defined to be the nilpotency order μ in the Weierstraß–Kronecker form (2.4). One writes $\text{ind}\{E, F\} = \mu$.

If E is nonsingular, one states $\text{ind}\{E, F\} = \mu := 0$.

The *Kronecker index of a regular DAE* (2.2) is given by $\text{ind}\{E, F\} = \mu$.

The Weierstraß–Kronecker form of a regular pair $\{E, F\}$ provides a broad insight into the structure of the associated DAE (2.2). Scaling of (2.2) by L, transforming $x = K \begin{bmatrix} y \\ z \end{bmatrix}$, and letting $Lq =: \begin{bmatrix} p \\ r \end{bmatrix}$ leads to the equivalent decoupled system

$$y'(t) + Wy(t) = p(t), \quad t \in \mathcal{I}, \qquad (2.6)$$

$$Nz'(t) + z(t) = r(t), \quad t \in \mathcal{I}. \qquad (2.7)$$

The first equation (2.6) represents an explicit ODE. The second one appears for $l > 0$, and it has the only solution

$$z(t) = \sum_{j=0}^{\mu-1} (-1)^j N^j r^{(j)}(t), \qquad (2.8)$$

provided that r is smooth enough. This becomes clear by deriving

$$z^j(t) = r^j(t) - Nz^{j+1}(t), \quad t \in \mathcal{I}, \quad j = 0, \dots, \mu - 1,$$

from (2.7), and then successively

$$z(t) = r(t) - Nz'(t) = r(t) - N(r'(t) - Nz''(t)) = r(t) - Nr'(t) + N^2 z''(t)$$

$$= \dots = \sum_{j=0}^{\mu-1} (-1)^j N^j r^{(j)}(t) + (-1)^\mu \underbrace{N^\mu}_{=0} z^\mu(t).$$

In case of the homogeneous DAE (2.3), the functions $p(\cdot)$ and $r(\cdot)$ vanish identically, and so does the solution component $z(\cdot)$. The solutions of the explicit ODE (2.6), with $p(\cdot) = 0$, form an $(m - l)$-dimensional subspace in $\mathcal{C}^\infty(\mathcal{I}, \mathbb{R}^m)$.

We summarize what we already know about our operator $\overset{\circ}{T}$.

Proposition 2.1 *If the differential-algebraic operator $\overset{\circ}{T}$ (2.1) is given by a regular matrix pair $\{E, F\}$ and if the setting ensures the inclusion $C^\infty(\mathcal{I}, \mathbb{R}^m) \subseteq \operatorname{dom} \overset{\circ}{T}$, then it holds that*

$$\dim \ker \overset{\circ}{T} = m - l, \quad \ker \overset{\circ}{T} \subseteq C^\infty(\mathcal{I}, \mathbb{R}^m). \tag{2.9}$$

By means of an appropriate initial condition at a point $t_a \in \mathcal{I}$,

$$Cx(t_a) = d \in \mathbb{R}^{m-l}, \tag{2.10}$$

with $C \in \mathcal{L}(\mathbb{R}^m, \mathbb{R}^{m-l})$, $\operatorname{rank} C = m - l$, $\ker C = \ker([I\ 0]K^{-1})$, we obtain the composed operator $\overset{\circ}{\mathcal{T}} : X \to Y \times \mathbb{R}^{m-l}$,

$$\overset{\circ}{\mathcal{T}}x := (\overset{\circ}{T}x, Cx(t_a)), \quad x \in \operatorname{dom} \overset{\circ}{T}, \tag{2.11}$$

which is then injective. We emphasize that, in contrast to regular ODEs, for describing an appropriate initial condition (2.10), the special structure of the DAE has to be attentively considered. The operator equation $\overset{\circ}{\mathcal{T}}x = (q, d)$ reflects the initial value problem (IVP) (2.2), (2.10).

As distinguished from the situation in Proposition 2.1, if the pair $\{E, F\}$ is nonregular, then $\ker \overset{\circ}{T}$ no longer has finite dimension, e.g., [55, Theorem 1.6]. We substantiate this fact by the following simple instance.

Example 2.2 (ker $\overset{\circ}{T}$ Fails to be Finite-Dimensional) The matrix pair $\{E, F\}$ associated with the operator

$$\overset{\circ}{T}x := \underbrace{\begin{bmatrix} 1 & 0 \\ 0 & 0 \end{bmatrix}}_{E} x' + \underbrace{\begin{bmatrix} 0 & -1 \\ 0 & 0 \end{bmatrix}}_{F} x, \quad x \in \operatorname{dom} \overset{\circ}{T},$$

is obviously nonregular. The nullspace $\ker \overset{\circ}{T} = \{x \in \operatorname{dom} \overset{\circ}{T} : x_1' = x_2\}$ depends on the choice of X; however, with $C^\infty(\mathcal{I}, \mathbb{R}^2) \subseteq \operatorname{dom} \overset{\circ}{T} \subseteq X$, the inclusion

$$\left\{ \begin{bmatrix} z \\ z' \end{bmatrix} : z \in C^\infty(\mathcal{I}, \mathbb{R}) \right\} \subseteq \ker \overset{\circ}{T}$$

attests that $\ker \overset{\circ}{T}$ no longer has finite dimension. □

2.2   Ill-Posed Behavior in Higher-Index Cases

In the more interesting case if $\mu \geq 1$, the solution expression (2.8) elucidates the dependence of the solution x on derivatives of the right-hand side q when indicated. The higher the index μ, the more differentiations are involved. Solely in the index-1 case we do have $N = 0$, hence $z(\cdot) = r(\cdot)$, and no derivatives are involved.

Since numerical differentiations in these circumstances may cause considerable trouble, it is very important to know the index μ as well as details of the structure responsible for the higher index when modeling and simulating in practice with DAEs. Not surprisingly, the typical solution behavior of so-called *ill-posed* problems can be observed in higher index DAEs: small perturbations of the right-hand side may cause enormous and somewhat discontinuous changes in the solution. We demonstrate this by the next example (cf. [55, Section 1.1]).

Example 2.3 (Ill-Posed Behavior) The IVP for the regular DAE

$$
\underbrace{\begin{bmatrix} 1 & 0 & 0 & 0 & 0 \\ 0 & 0 & 1 & 0 & 0 \\ 0 & 0 & 0 & 1 & 0 \\ 0 & 0 & 0 & 0 & 1 \\ 0 & 0 & 0 & 0 & 0 \end{bmatrix}}_{E} x'(t) + \underbrace{\begin{bmatrix} -\alpha & -1 & 0 & 0 & 0 \\ 0 & 1 & 0 & 0 & 0 \\ 0 & 0 & 1 & 0 & 0 \\ 0 & 0 & 0 & 1 & 0 \\ 0 & 0 & 0 & 0 & 1 \end{bmatrix}}_{F} x(t) = q(t), \quad t \in \mathcal{I} = [0,1],
$$

completed by the initial condition

$$
Cx(0) = \begin{bmatrix} 1 & 0 & 0 & 0 & 0 \end{bmatrix} x(0) = d, \tag{2.12}
$$

is uniquely solvable for each sufficiently smooth function q and each arbitrary $d \in \mathbb{R}$. The homogeneous IVP, with $q = 0$ and $d = 0$, has the identically zero solution only. The particular solution corresponding to the initial value $d \in \mathbb{R}$ and the excitation

$$
q_k(t) = \begin{bmatrix} 0 \\ 0 \\ 0 \\ 0 \\ \gamma_k(t) \end{bmatrix}, \quad \gamma_k(t) = \varepsilon \frac{1}{k} \sin kt, \quad k \in \mathbb{N}, \ \varepsilon \neq 0 \ \text{small}, \tag{2.13}
$$

reads as follows:

$$
x_{k,1}(t) = e^{\alpha t} d + \varepsilon \int_0^t k^2 e^{\alpha(t-s)} \cos ks \, ds, \quad x_{k,2}(t) = \varepsilon k^2 \cos kt,
$$

$$
x_{k,3}(t) = -\varepsilon k \sin kt, \quad x_{k,4}(t) = -\varepsilon \cos kt, \quad x_{k,5}(t) = \varepsilon \frac{1}{k} \sin kt.
$$

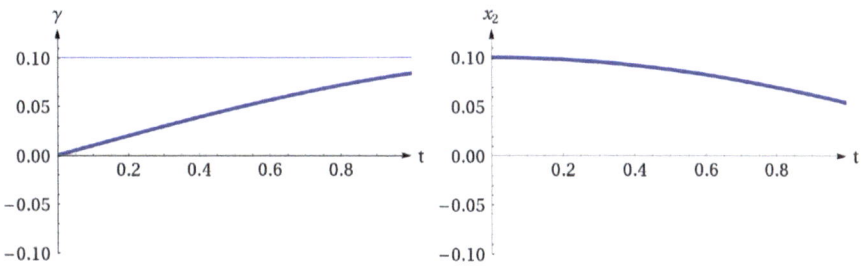

Fig. 1 γ_k and $x_{k,2}$ for $k = 1$

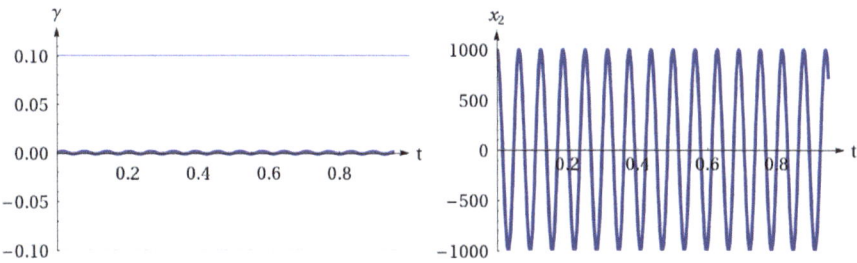

Fig. 2 γ_k and $x_{k,2}$ for $k = 100$

No doubt, the solution depends smoothly on d, but what about the dependence on the excitation q_k? To simplify matters we put $d = 0$. While the excitation q_k uniformly tends to zero for $k \to \infty$, the first three solution components grow unbounded, instead of also tending to zero. This is typical ill-posed behavior. The solution value at $t = 0$, that is,

$$x_{k,1}(0) = 0, \ x_{k,2}(0) = \varepsilon n^2, \ x_{k,3}(0) = 0, \ x_{k,4}(0) = -\varepsilon, \ x_{k,5}(0) = 0,$$

moves away from the origin with increasing k. For the perturbed system, the origin is no longer a consistent value at $t = 0$, as is the case for the unperturbed system. Figures 1 and 2 show the excitation γ_k and the response $x_{k,2}$ for $\varepsilon = 0.1$, $k = 1$, and $k = 100$.

Here the matrix pencil $\{E, F\}$ is regular with Kronecker index $\mu = 4$ and $l = 4$ (cf. (2.4)). The associated operator $\overset{\circ}{T}$ has the nullspace

$$\ker \overset{\circ}{T} = \text{im} \begin{bmatrix} w(\cdot) \\ 0 \\ 0 \\ 0 \\ 0 \end{bmatrix}, \ w(t) := e^{\alpha t}, \ t \in \mathcal{I},$$

if we suppose the setting satisfies $\mathcal{C}^{\infty}(\mathcal{I}, \mathbb{R}^5) \subseteq \mathrm{dom}\,\overset{\circ}{T}$. The corresponding composed operator $\overset{\circ}{\mathcal{T}}$ is injective. This is in full accordance with Proposition 2.1.□

This little constant coefficient example is alarming, but it is relatively harmless. In more general DAEs, time-dependent subspaces and nonlinear relations may considerably amplify the bad behavior. For this reason one should be careful in view of numerical simulations. It may well happen that an integration code seemingly works, however it generates wrong results.

2.3 Standard Settings: Looking for Boundedness, Closedness, and Normal Solvability

We inspect diverse standard settings of the operators $\overset{\circ}{T}$ and $\overset{\circ}{\mathcal{T}}$ given by (2.1) and (2.11), respectively. We ask if the operators and suitable extensions are bounded, closed, and eventually continuously invertible. We suppose the interval $\mathcal{I}$ to be compact such that maximum norms can be applied, $\mathcal{I} := [t_a, t_e]$. We consider for X the Banach spaces $\mathcal{C}^1$, $\mathcal{C}$, and the Hilbert space L^2.

2.3.1   Space of Continuously Differentiable Functions $X = \mathcal{C}^1(\mathcal{I}, \mathbb{R}^m)$

Most authors favor $\mathcal{C}^1$-solutions when dealing with DAEs. This corresponds to the setting $\mathrm{dom}\,\overset{\circ}{T} = X = \mathcal{C}^1(\mathcal{I}, \mathbb{R}^m)$, $Y = \mathcal{C}(\mathcal{I}, \mathbb{R}^m)$. Applying usual norms, the operator $\overset{\circ}{T}$ becomes bounded.

The special case given by the simple index-1 matrix pair in Weierstraß–Kronecker form

$$\overset{\circ}{T}x = Ex' + Fx = \begin{bmatrix} I & 0 \\ 0 & 0 \end{bmatrix}\begin{bmatrix} x_1' \\ x_2' \end{bmatrix} + \begin{bmatrix} W & 0 \\ 0 & I \end{bmatrix}\begin{bmatrix} x_1 \\ x_2 \end{bmatrix} = \begin{bmatrix} x_1' + Wx_1 \\ x_2 \end{bmatrix}$$

foreshadows the drawback of this setting: the derivative-free equation $x_2 = q_2$ which represents a part of the equation $\overset{\circ}{T}x = q$ has no solution unless one additionally supposes that q_2 is continuously differentiable. There is no natural reason for this demand. Formally, this setting yields

$$\mathrm{im}\,\overset{\circ}{T} = \{q \in \mathcal{C}(\mathcal{I}, \mathbb{R}^m) : q_2 \in \mathcal{C}^1(\mathcal{I}, \mathbb{R}^l)\}$$

which is a proper nonclosed subset in $Y = \mathcal{C}(\mathcal{I}, \mathbb{R}^m)$ such that $\overset{\circ}{T}$ fails to be normally solvable. This motivates us to turn to spaces X being richer in content. It seems to be more reasonable to accept instead a solution component x_2 being merely continuous.

2.3.2 Space of Continuous Functions $X = \mathcal{C}(\mathcal{I}, \mathbb{R}^m)$

Now we apply the function spaces $X = Y = \mathcal{C}(\mathcal{I}, \mathbb{R}^m)$, with the maximum norm, and consider the operator

$$\overset{\circ}{T}x = Ex' + Fx, \quad x \in \text{dom}\,\overset{\circ}{T} := \mathcal{C}^1(\mathcal{I}, \mathbb{R}^m). \tag{2.14}$$

The definition domain $\text{dom}\,\overset{\circ}{T} = \mathcal{C}^1(\mathcal{I}, \mathbb{R}^m)$ is dense in X, i.e., $\overset{\circ}{T}$ is densely defined. However, except for the dull case $E = 0$, the operator $\overset{\circ}{T}$ is unbounded in this setting. To simplify matters, we verify this fact supposing the interval $\mathcal{I} = [0, 1]$. Since E is not the zero matrix, there is a $c \in \mathbb{R}^m$ such that $Ec \neq 0$ and $|c| = 1$. The functions defined by $x_k(t) := t^k c$, $t \in [0, 1]$, belong to the definition domain of $\overset{\circ}{T}$ and $\|x_k\|_\infty = 1$ for all $k \in \mathbb{N}$. Derive

$$\|\overset{\circ}{T}x_k\|_\infty = \|Ex_k' + Fx_k\|_\infty \geq \|Ex_k'\|_\infty - \|Fx_k\|_\infty = k|Ec| - |Fc|.$$

If k increases, then $\|\overset{\circ}{T}x_k\|_\infty$ grows unboundedly, which shows that, in the given setting, the operator $\overset{\circ}{T}$ is no longer bounded. However, we may obtain a closure of the operator $\overset{\circ}{T}$, that is, a closed extension T of $\overset{\circ}{T}$.

First of all, the operator $\overset{\circ}{T}$ is closable. Namely, if $x_k \in \text{dom}\,\overset{\circ}{T}$ for $k \in \mathbb{N}$ and $\|x_k\|_\infty \to 0$, $\|\overset{\circ}{T}x_k - y_*\|_\infty \to 0$, then it follows that $\|Ex_k\|_\infty \to 0$, $\|(Ex_k)' - y_*\|_\infty = \|Ex_k' - y_*\|_\infty = \|\overset{\circ}{T}x_k - y_* - Fx_k\|_\infty \to 0$, and hence $y_* = 0$.

Now we look for the closure of $\overset{\circ}{T}$ in the given setting. First, we factorize $E = A\tilde{D}$, with $A \in \mathcal{L}(\mathbb{R}^n, \mathbb{R}^m)$, $\tilde{D} \in \mathcal{L}(\mathbb{R}^m, \mathbb{R}^n)$, $n \leq m$, in such a way that

$$\text{im}\,E = \text{im}\,A$$

is valid. We write

$$\overset{\circ}{T}x = Ex' + Fx = A\tilde{D}x' + Fx = A(\tilde{D}x)' + Fx, \quad x \in \text{dom}\,\overset{\circ}{T} = \mathcal{C}^1(\mathcal{I}, \mathbb{R}^m).$$

In particular, one can simply choose $A = E$, $\tilde{D} = I$, $n = m$ and $A = EE^+$, $\tilde{D} = E$, $n = m$. It should be emphasized that *the operator $\overset{\circ}{T}$ itself is independent of the description* by a special factorization.

Consider an arbitrary sequence of members $x_k \in \text{dom}\,\overset{\circ}{T}$, with $\|x_k - x_*\|_\infty \to 0$ and $\|\overset{\circ}{T}x_k - y_*\|_\infty \to 0$, and limits $x_* \in X$, $y_* \in Y$. It holds that

$$A(\tilde{D}x_k)' = \overset{\circ}{T}x_k - Fx_k \to y_* - Fx_* \in \text{im}\,A,$$

By means of a generalized inverse A^- of A (see Appendix 6.1.2) we express

$$(A^-A\tilde{D}x_k)' = A^-A(\tilde{D}x_k)' = A^-(\overset{\circ}{T}x_k - Fx_k) \to A^-(y_* - Fx_*).$$

We also have $A^- A\tilde{D}x_k \to A^- A\tilde{D}x_*$. Then, $A^- A\tilde{D}x_*$ is continuously differentiable and the relation $(A^- A\tilde{D}x_*)' = A^-(y_* - Fx_*)$ is given. Now it follows that $A(A^- A\tilde{D}x_*)' = AA^-(y_* - Fx_*) = y_* - Fx_*$, thus $A(A^- A\tilde{D}x_*)' + Fx_* = y_*$. It comes out that the operator T defined by

$$T := A(Dx)' + Fx, \quad x \in \text{dom } T = \{w \in \mathcal{C}(\mathcal{I}, \mathbb{R}^m) : Dw \in \mathcal{C}^1(\mathcal{I}, \mathbb{R}^n)\}, \quad (2.15)$$

with $D := A^- A\tilde{D}$, is the closure of the operator $\overset{\circ}{T}$ given by (2.14).

Yet another look at the resulting factorization $E = AD$ shows: We have $D := A^- A\tilde{D}$, $E = A\tilde{D}$ and im E = im A. It follows that rank AD = rank A, thus rank $D \geq$ rank A. Since, on the other hand the inclusion im $D \subseteq$ im $A^- A$ is valid we conclude the relations

$$\text{im } D = \text{im } A^- A, \quad \text{rank } D = \text{rank } A = \text{rank } E,$$

and further the decomposition

$$\ker A \oplus \text{im } D = \mathbb{R}^n. \quad (2.16)$$

In turn, (2.16) implies $\ker D = \ker AD = \ker E$.

Thereby it does not matter at all how the matrix $\tilde{D}$ and the generalized inverse A^- have been chosen. In particular, all resulting function spaces

$$\mathcal{C}_D^1(\mathcal{I}, \mathbb{R}^m) := \{w \in \mathcal{C}(\mathcal{I}, \mathbb{R}^m) : Dw \in \mathcal{C}^1(\mathcal{I}, \mathbb{R}^n)\} \quad (2.17)$$

coincide with

$$\{w \in \mathcal{C}(\mathcal{I}, \mathbb{R}^m) : Ew \in \mathcal{C}^1(\mathcal{I}, \mathbb{R}^m)\},$$

see Lemma 6.9, and for every $x \in \text{dom } T$ it holds that

$$Tx = A(Dx)' + Fx = A(DE^+Ex)' + Fx = ADE^+(Ex)' + Fx = EE^+(Ex)' + Fx.$$

In other words, to obtain the closure T of the operator $\overset{\circ}{T}$ we are recommended to reformulate the operator by means of a factorization $E = AD$ with so-called *well-matched* factors $A \in \mathcal{L}(\mathbb{R}^n, \mathbb{R}^m)$ and $D \in \mathcal{L}(\mathbb{R}^m, \mathbb{R}^n)$, $n \leq m$, satisfying the decomposition (2.16). Such a factorization is said to be *proper*. We underline once again the fact that the special factorization does not at all matter for the operator T itself. It is merely a representation tool. However, it should be noticed that, for instance, when numerically integrating a DAE, a so-called full-rank factorization (with $n = \text{rank } E$) appears to be quite preferable, see [55].

The function space $\operatorname{dom} T = C_D^1(\mathcal{I}, \mathbb{R}^m)$ equipped with the respective graph-norm is a Banach space. For all possible factorizations satisfying the condition (2.16), the norms

$$\|x\|_{C_D^1} := \|x\|_\infty + \|(Dx)'\|_\infty, \quad x \in C_D^1(\mathcal{I}, \mathbb{R}^m),$$

are equivalent to each other and also to the graph norm, see Lemma 6.9.

Proposition 2.4 *Let a pair* $\{E, F\}$ *of matrices* $E, F \in \mathcal{L}(\mathbb{R}^m)$ *be given and* $E \neq 0$. *Then the following assertions are valid for the differential-algebraic operator* $\overset{\circ}{T} \in \mathcal{L}(X)$ *defined by (2.14) with* $X = C(\mathcal{I}, \mathbb{R}^m)$:

1. *The operator* $\overset{\circ}{T}$ *is unbounded but closable. The closure* T *of* $\overset{\circ}{T}$ *can be expressed by (2.15), where* $E = AD$ *is any factorization satisfying the decomposition (2.16).*
2. $\ker T$ *is closed in* X.
3. *The closure* T *maps the Banach space* $C_D^1(\mathcal{I}, \mathbb{R}^m)$ *continuously into the Banach space* X.
4. *If* $\{E, F\}$ *is a regular pair with Kronecker index* $\mu = 1$, *then the closure* T *is a Fredholm operator. It holds that* $\operatorname{im} T = X$ *and* $\operatorname{ind}_{fred}(T) = \alpha(T) = m - l$, *whereby* l *is the structural size described in (2.4).*
5. *If* $\{E, F\}$ *is a regular pair with Kronecker index* $\mu > 1$, *then* T *is densely solvable and the nullspace of* T *has finite dimension. The range* $\operatorname{im} T$ *is a nonclosed proper subset in* X *such that* T *is neither Fredholm nor normally solvable.*
6. *If* $\{E, F\}$ *is a regular pair with Kronecker index* $\mu = 0$, *that is,* E *is nonsingular, then* $\overset{\circ}{T}$ *is already closed,* $T = \overset{\circ}{T}$, $\operatorname{im} T = X$, $\dim \ker T = m$, *and* T *is a Fredholm operator with* $\operatorname{ind}_{fred}(T) = m$.

Proof The statements (1) and (3) are already verified and (2) retrieves a general property of closed operators. Assertion (6) reflects well-known facts of the ODE theory. It remains to consider items (4) and (5). Applying the Weierstraß–Kronecker form (2.4) we factorize $E = AD$, $A := E$, $D := E^- E$ and express

$$E = L^{-1} \begin{bmatrix} I & 0 \\ 0 & N \end{bmatrix} K^{-1}, \quad E^- := K \begin{bmatrix} I & 0 \\ 0 & N^+ \end{bmatrix} L, \quad E^- E = K \begin{bmatrix} I & 0 \\ 0 & N^+ N \end{bmatrix} K^{-1},$$

and introduce $K^{-1} x =: \begin{bmatrix} y \\ z \end{bmatrix}$ such that

$$\operatorname{dom} T = \{x \in C(\mathcal{I}, \mathbb{R}^m) : K \begin{bmatrix} I & 0 \\ 0 & N^+ N \end{bmatrix} K^{-1} x \in C^1(\mathcal{I}, \mathbb{R}^m)\}$$

$$= \{x \in C(\mathcal{I}, \mathbb{R}^m) : \begin{bmatrix} I & 0 \\ 0 & N^+ N \end{bmatrix} K^{-1} x \in C^1(\mathcal{I}, \mathbb{R}^m)\}$$

$$= \{x \in C(\mathcal{I}, \mathbb{R}^m) : y \in C^1(\mathcal{I}, \mathbb{R}^{m-l}), N^+ N z \in C^1(\mathcal{I}, \mathbb{R}^l)\}.$$

The equation $Tx = q$ can be traced back to the decoupled system (cf. (2.6), (2.7))

$$y'(t) + Wy(t) = p(t), \quad N(N^+Nz)'(t) + z(t) = r(t). \tag{2.18}$$

If the index μ equals 1 then N is the zero matrix of size l, thus

$$\text{dom}\, T = \{x \in \mathcal{C}(\mathcal{I}, \mathbb{R}^m) : y \in \mathcal{C}^1(\mathcal{I}, \mathbb{R}^{m-l})\}.$$

In this case, for each arbitrary continuous right hand sides p and r, there exist a continuously differentiable solution y of the first equation and a continuous solution z of the second one, which is now trivial. Altogether, for each continuous q there exists an element $x \in \text{dom}\, T$ such that $Tx = q$. This confirms assertion (4).

If $\mu \geq 2$, then N is a nontrivial nilpotent matrix. The first equation in (2.18) has again a continuously differentiable solution y for each continuous p. However, the demand $N^+Nz \in \mathcal{C}^1$ implies necessarily $Nz \in \mathcal{C}^1$, thus $N^{\mu-1}z \in \mathcal{C}^1$. Multiplying the second equation by $N^{\mu-1}$ yields $N^{\mu-1}z = N^{\mu-1}r$. In other words, for solvability of the second equation it is necessary that $N^{\mu-1}r$ is continuously differentiable. In consequence, the range $\text{im}\, T$ contains only those continuous functions showing certain smoother components. In any case, at least one component has to be continuously differentiable. Such sets are not closed in the continuous function space. Regarding the inclusion $\mathcal{C}^\infty(\mathcal{I}, \mathbb{R}^m) \subseteq \text{im}\, T$ we are done with assertion (5).

$\square$

Example 2.5 (Continuation 1 of Example 2.3) A natural factorization is here

$$E = \begin{bmatrix} 1 & 0 & 0 & 0 & 0 \\ 0 & 0 & 1 & 0 & 0 \\ 0 & 0 & 0 & 1 & 0 \\ 0 & 0 & 0 & 0 & 1 \\ 0 & 0 & 0 & 0 & 0 \end{bmatrix} = \begin{bmatrix} 1 & 0 & 0 & 0 \\ 0 & 1 & 0 & 0 \\ 0 & 0 & 1 & 0 \\ 0 & 0 & 0 & 1 \\ 0 & 0 & 0 & 0 \end{bmatrix} \begin{bmatrix} 1 & 0 & 0 & 0 & 0 \\ 0 & 0 & 1 & 0 & 0 \\ 0 & 0 & 0 & 1 & 0 \\ 0 & 0 & 0 & 0 & 1 \end{bmatrix} =: AD,$$

and the closure T of the operator $\overset{\circ}{T}$ reads

$$Tx = A(Dx)' + Fx = \begin{bmatrix} x_1' - \alpha x_1 - x_2 \\ x_3' + x_2 \\ x_4' + x_3 \\ x_5' + x_4 \\ x_5 \end{bmatrix}, \quad x \in \text{dom}\, T = \mathcal{C}_D^1(\mathcal{I}, \mathbb{R}),$$

$$\mathcal{C}_D^1(\mathcal{I}, \mathbb{R}) = \{x \in \mathcal{C}(\mathcal{I}, \mathbb{R}^5) : x_1, x_3, x_4, x_5 \in \mathcal{C}^1(\mathcal{I}, \mathbb{R})\}.$$

The pair $\{E, F\}$ is regular with Kronecker index $\mu = 4$ so that Proposition 2.4(5) applies. In more detail it results that

$$\operatorname{im} T = \{q \in \mathcal{C}(\mathcal{I}, \mathbb{R}^5) : q_5, q_4 - q_5', q_3 - (q_4 - q_5')' \in \mathcal{C}^1(\mathcal{I}, \mathbb{R})\}.$$

The range $\operatorname{im} T$ is a proper nonclosed subset in $\mathcal{C}(\mathcal{I}, \mathbb{R}^5)$ which indicates essentially ill-posedness. The closure of the associated composed operator $\overset{\circ}{\mathcal{T}}$ is given by (cf. (2.12))

$$\mathcal{T}x = (Tx, x_1(0)), \quad x \in \operatorname{dom} T.$$

The operator $\mathcal{T}$ is injective, however, the inverse $\mathcal{T}^{-1}$ fails to be continuous. For instance, the sequence $(q_k, 0) \in \operatorname{im} T \times \mathbb{R}$ defined by (2.13) tends to zero for $k \to \infty$, that means $\|q_k\|_\infty \to 0$, whereas the sequence of the corresponding responses $x_k = \mathcal{T}^{-1}(q_k, 0)$ growths unboundedly,

$$\|x_k\|_\infty > \|x_{k,2}\|_\infty = \varepsilon k^2.$$

$\square$

2.3.3 Space of Integrable Functions $X = L^2(\mathcal{I}, \mathbb{R}^m)$

We set $X = Y = L^2(\mathcal{I}, \mathbb{R}^m)$ and apply the Hilbert space $L^2(\mathcal{I}, \mathbb{R}^m)$ with the usual scalar product and norm. Consider the operator

$$\overset{\circ}{T}x = Ex' + Fx, \quad x \in \operatorname{dom} \overset{\circ}{T} := \mathcal{C}^\infty(\mathcal{I}, \mathbb{R}^m). \tag{2.19}$$

The definition domain is dense in X, i.e., $\overset{\circ}{T}$ is densely defined. Except for the dull case $E = 0$, the operator $\overset{\circ}{T}$ is unbounded in this setting. To simplify matters, we show this fact supposing the interval $\mathcal{I} = [0, 1]$. Since E is not the zero matrix, there is a $c \in \mathbb{R}^m$ such that $Ec \neq 0$ and $|c| = 1$. The functions defined by $x_k(t) := \sqrt{2k+1}\, t^k c$, $t \in [0, 1]$, belong to the definition domain of $\overset{\circ}{T}$, and we have $\|x_k\|_{L^2} = 1$ for all $k \in \mathbb{N}$. Derive

$$\|\overset{\circ}{T}x_k\|_{L^2} = \|Ex_k' + Fx_k\|_{L^2} \geq \|Ex_k'\|_{L^2} - \|Fx_k\|_{L^2} = k\frac{\sqrt{2k+1}}{\sqrt{2k-1}}|Ec| - |Fc|.$$

If k increases, then $\|\overset{\circ}{T}x_k\|_{L^2}$ grows unboundedly, which shows that, also in this setting, the operator $\overset{\circ}{T}$ is unbounded.

After the idea of the previous subsection we introduce the function space

$$H_D^1(\mathcal{I}, \mathbb{R}^m) := \{w \in L^2(\mathcal{I}, \mathbb{R}^m) : Dw \in H^1(\mathcal{I}, \mathbb{R}^n)\}$$

and the operator T by

$$T := A(Dx)' + Fx, \ x \in \operatorname{dom} T = H_D^1(\mathcal{I}, \mathbb{R}^m), \tag{2.20}$$

whereby we apply a proper factorization $E = AD$ such that the decomposition (2.16) is valid, i.e., $\ker A \oplus \operatorname{im} D = \mathbb{R}^n$. Eventually, the operator T will be proved to be the closure of $\mathring{T}$. Again, the special choice of the factorization does not matter, neither for the function space $H_D^1(\mathcal{I}, \mathbb{R}^n)$ nor for the operator T, see Lemma 6.9. We follow the lines of [38] applying Hilbert space basics (see Appendix 6.1).

Lemma 2.6 *The operator $\mathring{T}$ given by (2.19) possesses an adjoint and a biadjoint. The adjoint and biadjoint operators $\mathring{T}^*$ and $\mathring{T}^{**}$ of $\mathring{T}$ are given on the domains*

$$\operatorname{dom} \mathring{T}^* \quad = \{w \in L^2(\mathcal{I}, \mathbb{R}^m) : A^*w \in H^1(\mathcal{I}, \mathbb{R}^n), A^*w(t_a) = 0, A^*w(t_e) = 0\},$$

$$\operatorname{dom} \mathring{T}^{**} \quad = \{w \in L^2(\mathcal{I}, \mathbb{R}^m) : Dw \in H^1(\mathcal{I}, \mathbb{R}^n)\},$$

respectively. These domains do not depend on the chosen proper factorization $E = AD$.

Proof For arbitrary $x \in C^\infty(\mathcal{I}, \mathbb{R}^m)$ and suitable functions y by partial integration we obtain

$$(\mathring{T}x, y) = \int_{t_a}^{t_e} \langle Ex'(t) + Fx(t), y(t)\rangle \mathrm{d}t = \int_{t_a}^{t_e} \langle A(Dx)'(t) + Fx(t), y(t)\rangle \mathrm{d}t$$

$$= \int_{t_a}^{t_e} \langle x(t), -D^*(A^*y)'(t) + F^*y(t)\rangle \mathrm{d}t$$

$$+ \langle Dx(t_e), A^*y(t_e)\rangle - \langle Dx(t_a), A^*y(t_a)\rangle.$$

Therefore, the adjoint operator is explicitly given on the set

$$\{w \in L^2(\mathcal{I}, \mathbb{R}^m) : A^*w \in H^1(\mathcal{I}, \mathbb{R}^n), A^*w(t_a) = 0, A^*w(t_e) = 0\} =: \operatorname{dom} \mathring{T}^*$$

by means of

$$\mathring{T}^*y = -D^*(A^*y)' + F^*y, \quad y \in \operatorname{dom} \mathring{T}^*.$$

In fact, we have $(\overset{\circ}{T}x, y) = (x, \overset{\circ}{T}^*y)$ for all $x \in \mathrm{dom}\,\overset{\circ}{T}$ and $y \in \mathrm{dom}\,\overset{\circ}{T}^*$. Since $\mathrm{dom}\,\overset{\circ}{T}^*$ is dense in $L^2(\mathcal{I}, \mathbb{R}^m)$, the adjoint of the operator $\overset{\circ}{T}^*$ exists. Compute

$$(\overset{\circ}{T}^*y, x) = \int_{t_a}^{t_e} \langle -D^*(A^*y)'(t) + F^*y(t), x(t)\rangle \mathrm{dt}$$

$$= \int_{t_a}^{t_e} \langle y(t), A(Dx)'(t) + Fx(t)\rangle \mathrm{dt}$$

$$=: (y, \overset{\circ}{T}^{**}x)$$

for all $y \in \mathrm{dom}\,\overset{\circ}{T}^*$ and all $x \in \{w \in L^2(\mathcal{I}, \mathbb{R}^m) : Dw \in H^1(\mathcal{I}, \mathbb{R}^n)\} =: \mathrm{dom}\,\overset{\circ}{T}^{**}$ which proves the first part of the assertion.

It remains to verify the invariance with respect to the factorization $E = AD$. Consider a further factorization $E = \bar{A}\bar{D}$, with $\bar{A} \in \mathcal{L}(\mathbb{R}^{\bar{n}}, \mathbb{R}^m)$, $\bar{D} \in \mathcal{L}(\mathbb{R}^m, \mathbb{R}^{\bar{n}})$, and $\ker \bar{A} \oplus \mathrm{im}\,\bar{D} = \mathbb{R}^{\bar{n}}$, $\bar{n} \le m$. Because $\ker D = \ker \bar{D}$ it holds that $D^+D = \bar{D}^+\bar{D}$. Therefore, $Dw \in H^1(\mathcal{I}, \mathbb{R}^n)$ implies $\bar{D}w = \bar{D}\bar{D}^+\bar{D}w = \bar{D}D^+Dw \in H^1(\mathcal{I}, \mathbb{R}^{\bar{n}})$, and hence $\mathrm{dom}\,\overset{\circ}{T}^{**}$ does not depend on the factorization.

Analogously, due to $\mathrm{im}\,A = \mathrm{im}\,\bar{A}$, $AA^+ = \bar{A}\bar{A}^+$, we derive from $A^*w \in H^1(\mathcal{I}, \mathbb{R}^n)$ that $\bar{A}^*w = \bar{A}^*\bar{A}^{*+}\bar{A}^*w = \bar{A}^*A^{*+}A^*w \in H^1(\mathcal{I}, \mathbb{R}^{\bar{n}})$.

Additionally, for $\tau = t_a, t_e$, we compute $\bar{A}^*w(\tau) = \bar{A}^*A^{*+}A^*w(\tau)$ and $A^*w(\tau) = A^*\bar{A}^{*+}\bar{A}^*w(\tau)$. This proves $\mathrm{dom}\,\overset{\circ}{T}^*$ to be independent of the special factorization. □

The next assertion is the counterpart of Proposition 2.4.

Proposition 2.7 *Let the pair $\{E, F\}$ of matrices $E, F \in \mathcal{L}(\mathbb{R}^m)$ be given and $E \ne 0$. Then the following assertions are valid for the differential-algebraic operator $\overset{\circ}{T} \in \mathcal{L}(X)$, defined by (2.19) with $X = L^2(\mathcal{I}, \mathbb{R}^m)$:*

1. *The operator $\overset{\circ}{T}$ is unbounded but closable. The closure T of $\overset{\circ}{T}$ can be expressed by (2.20), where $E = AD$ is any factorization satisfying the decomposition (2.16).*
2. *$\ker \hat{T}$ is closed in X.*
3. *The closure T maps the Hilbert space $H_D^1(\mathcal{I}, \mathbb{R}^m)$ continuously into the Hilbert space X.*
4. *If $\{E, F\}$ is a regular pair with Kronecker index $\mu = 1$, then the closure T is a Fredholm operator. It holds that $\mathrm{im}\,T = X$ and $\mathrm{ind}_{fred}(T) = \alpha(T) = m - l$, whereby l is the structural size described in (2.4).*
5. *If $\{E, F\}$ is a regular pair with Kronecker index $\mu > 1$, then T is densely solvable and the nullspace of T is finite-dimensional. The range $\mathrm{im}\,T$ is a nonclosed proper subset in X such that T is neither Fredholm nor normally solvable.*

6. *If* $\{E, F\}$ *is a regular pair with Kronecker index* $\mu = 0$, *that is,* E *is nonsingular,* *then* $\overset{\circ}{T}$ *is already closed,* $T = \overset{\circ}{T}$, $\operatorname{im} T = X$, $\dim \ker T = m$, *and* T *is a Fredholm operator with* $\operatorname{ind}_{fred}(T) = m$.

Proof Owing to the existence of the biadjoint $\overset{\circ}{T}^{**}$ one has $T = \overset{\circ}{T}^{**}$, and the assertion (1) is a consequence of Lemma 2.6. Assertion (2) reflects a general property of closed operators. Assertion (3) is due to the inequality

$$\|Tx\|_{L^2}^2 = \|A(Dx)' + Fx\|_{L^2}^2 \leq c \left(\|(Dx)'\|_{L^2}^2 + \|x\|_{L^2}^2\right) = c\|x\|_{H_D^1}^2,$$

which is valid for all $x \in H_D^1(\mathcal{I}, \mathbb{R}^m)$. The statements (4)–(6) can be proved along the lines of Proposition 2.4 by replacing the function spaces $\mathcal{C}$ and $\mathcal{C}^1$ by L^2 and H^1, correspondingly. $\square$

2.4 Peculiar Approaches

2.4.1 Enforcing Surjectivity by Image Space Adaption

Suppose that the pair $\{E, F\}$ is regular and the factorization $E = AD$ satisfies the condition (2.16). The operator

$$T \in \mathcal{L}(\mathcal{C}_D^1(\mathcal{I}, \mathbb{R}^m), \mathcal{C}(\mathcal{I}, \mathbb{R}^m)), \quad Tx := A(Dx)' + Fx, \ x \in \mathcal{C}_D^1(\mathcal{I}, \mathbb{R}^m),$$

is bounded. By Proposition 2.4, if $\operatorname{ind}\{E, F\} \geq 2$, the range $\operatorname{im} T$ is a proper, nonclosed subset in $\mathcal{C}(\mathcal{I}, \mathbb{R}^m)$. The resulting composed operator (cf. (2.11))

$$\mathcal{T} \in \mathcal{L}(\mathcal{C}_D^1(\mathcal{I}, \mathbb{R}^m), \mathcal{C}(\mathcal{I}, \mathbb{R}^m) \times \mathbb{R}^{m-l}),$$

is injective but inherits the nonclosed range. In the consequence, the equation

$$\mathcal{T}x = (q, d) \tag{2.21}$$

is essentially ill-posed in this setting. We refer once again to Examples 2.3 and 2.5 for the ill-posed feature of the solution which actually justifies that notion.

Applying the Weierstraß–Kronecker form or the projector based decoupling procedure described in [55, Chapter 1] one can specify in detail what the range of the operator T looks like. Being aware of the precise description of $\operatorname{im} T$ we can apply the new function space $Y_{new} := \operatorname{im} T$ as well as a suitable norm so that Y_{new} becomes a Banach space and $T \in \mathcal{L}(\mathcal{C}_D^1(\mathcal{I}, \mathbb{R}^m), Y_{new})$ remains bounded. Then, in this peculiar setting, the operator T is Fredholm and the associated composed operator $\mathcal{T}$ becomes a homeomorphism as a continuous bijection in Banach spaces. This way, in the new setting, the Eq. (2.21) becomes well-posed. This idea seems fine; however, it rather obscures the actual solution behavior and it is essentially factitious as our example demonstrates.

Example 2.8 (Continuation 2 of Example 2.3) The operator (see also Example 2.5)

$$Tx := \begin{bmatrix} x_1' - \alpha x_1 - x_2 \\ x_3' + x_2 \\ x_4' + x_3 \\ x_5' + x_4 \\ x_5 \end{bmatrix}, \quad x \in \mathcal{C}_D^1(\mathcal{I}, \mathbb{R}^5),$$

is bounded, where $\mathcal{C}_D^1(\mathcal{I}, \mathbb{R}^5) = \{x \in \mathcal{C}(\mathcal{I}, \mathbb{R}^5) : x_1, x_3, x_4, x_5 \in \mathcal{C}^1(\mathcal{I}, \mathbb{R})\}$. Its image

$$\text{im } T = \{q \in \mathcal{C}(\mathcal{I}, \mathbb{R}^5) : q_5, q_4 - q_5', q_3 - (q_4 - q_5')' \in \mathcal{C}^1(\mathcal{I}, \mathbb{R})\}$$

is a proper nonclosed subset in $\mathcal{C}(\mathcal{I}, \mathbb{R}^5)$. The resulting composed operator $\mathcal{T}$ has a discontinuous inverse. Set $Y_{new} = \text{im } T$ and define for each $q \in Y_{new}$ the norm

$$\|q\|_{Y_{new}} := \|q\|_\infty + \|q_5'\|_\infty + \|(q_4 - q_5')'\|_\infty + \|(q_3 - (q_4 - q_5')')'\|_\infty,$$

which yields a Banach space. In the new setting $T \in \mathcal{L}(\mathcal{C}_D^1(\mathcal{I}, \mathbb{R}^5), Y_{new})$, the operator T is surjective by construction and again bounded. Namely, it holds that

$$\|Tx\|_{Y_{new}} = \|Tx\|_\infty + \|x_5'\|_\infty + \|x_4\|_\infty + \|x_3\|_\infty$$

$$\leq c_{new}\|x\|_{\mathcal{C}_D^1}, \quad x \in \mathcal{C}_D^1(\mathcal{I}, \mathbb{R}^5).$$

The associated composed operator $\mathcal{T} \in \mathcal{L}(\mathcal{C}_D^1(\mathcal{I}, \mathbb{R}^5), Y_{new} \times \mathbb{R})$ actually becomes a homeomorphism. In the new setting, the continuity of $\mathcal{T}^{-1}$ is enforced by the stronger norm. Now, the sequence q_k given by (2.13) no longer converges to zero, instead, one has

$$\|q_k\|_{Y_{new}} > \|\gamma_k^{(3)}\|_\infty = \varepsilon k^2.$$

We emphasize that this problem manipulation does not change at all the actual bad solution behavior documented by Figs. 1 and 2.   □

The resulting operator $T \in \mathcal{L}(\mathcal{C}_D^1(\mathcal{I}, \mathbb{R}^m), Y_{new})$ is surjective and has a finite-dimensional nullspace (cf. Proposition 2.1), and hence, it is Fredholm, with $\text{ind}_{fred}(T) = \alpha(T) = m - l$.

The last result has quite limited importance: The information needed for this kind of manipulation is indispensable in general. Both the set Y_{new} and the norm $\|\cdot\|_{Y_{new}}$ strongly depend on the given matrix pair $\{E, F\}$; in higher index cases they are strongly individual ones for each pair.

As mentioned before, most authors favor C^1-solutions of DAEs. One can manipulate correspondingly the operator

$$T \in \mathcal{L}(C^1(\mathcal{I}, \mathbb{R}^m), C(\mathcal{I}, \mathbb{R}^m)), \quad Tx := Ex' + Fx, \ x \in X = C^1(\mathcal{I}, \mathbb{R}^m),$$

being of interest when insisting on C^1-solutions. Here we point out that all statements in [13, 14, 17] concerning the Fredholm index (there called Noetherian index) of differential-algebraic operators hold good for such a context only.

Example 2.9 (Continuation 3 of Example 2.3) For the operator (see also Examples 2.5, 2.8)

$$Tx := \begin{bmatrix} x_1' - \alpha x_1 - x_2 \\ x_3' + x_2 \\ x_4' + x_3 \\ x_5' + x_4 \\ x_5 \end{bmatrix}, \quad x \in C^1(\mathcal{I}, \mathbb{R}^5),$$

one immediately derives

$$Y_{new} = \operatorname{im} T =$$

$$\{q \in C(\mathcal{I}, \mathbb{R}^5) : q_5, q_4 - q_5', q_3 - (q_4 - q_5')', q_2 - q_2 - (q_3 - (q_4 - q_5')')' \in C^1(\mathcal{I}, \mathbb{R})\},$$

and

$$\|q\|_{Y_{new}} := \|q\|_\infty + \|q_5'\|_\infty + \|(q_4 - q_5')'\|_\infty$$
$$+ \|(q_3 - (q_4 - q_5')')'\|_\infty + \|(q_2 - (q_3 - (q_4 - q_5')')')'\|_\infty.$$

$\square$

We finish this subsubsection by revisiting once again the case of a regular index-1 pair $\{E.F\}$. There are different Banach space settings and bounded operators available. From Sect. 2.3.1 we know that then $T \in \mathcal{L}(C^1(\mathcal{I}, \mathbb{R}^m), C(\mathcal{I}, \mathbb{R}^m))$ fails to be normally solvable since $\operatorname{im} T$ is a nonclosed subset in $C(\mathcal{I}, \mathbb{R}^m)$.

The setting $T \in \mathcal{L}(C^1(\mathcal{I}, \mathbb{R}^m), Y_{new})$ yields a surjective operator, but needs special structural information also in case of regular index-1 pairs $\{E.F\}$. Namely, applying the corresponding Weierstraß–Kronecker form (2.4), with $N = 0$, we derive

$$Y_{new} = \operatorname{im} T = \{q \in C(\mathcal{I}, \mathbb{R}^m) : [0 \ I]Lq \in C^1(\mathcal{I}, \mathbb{R}^l)\}$$

and

$$\|q\|_{Y_{new}} = \|q\|_\infty + \|([0 \ I]Lq)'\|_\infty.$$

In comparison, in the setting $T \in \mathcal{L}(\mathcal{C}_D^1(\mathcal{I}, \mathbb{R}^m), \mathcal{C}(\mathcal{I}, \mathbb{R}^m))$, the operator T is also surjective. It seems to be an advantage of the latter setting that it uses original given information or easily available information concerning the coefficient E.

2.4.2 Topological Vector-Space $X = \mathcal{C}^\infty(\mathcal{I}, \mathbb{R}^m)$

The Weierstraß–Kronecker form (2.6), (2.7) tells us that, letting

$$X = Y = \mathcal{C}^\infty(\mathcal{I}, \mathbb{R}^m), \quad Tx = Ex' + Fx, \quad x \in \mathrm{dom}\, T = X,$$

the operator T induced by a regular matrix pair is surjective so that the equation $Tx = q$ is solvable in X for every $q \in Y$. Then the corresponding composed operator $\mathcal{T}$ acts bijectively between X and $Y \times \mathbb{R}^{m-l}$. This makes this topological vector-space setting quite popular, also for DAEs with real-analytic coefficients E and F.

The family of semi-norms

$$\eta_k(x) := \|x^{(k)}\|_\infty, \; x \in \mathcal{C}^\infty(\mathcal{I}, \mathbb{R}^m), \quad k = 0, 1, 2, \ldots,$$

generates (see [45, Section X.64]) a locally convex topology on $\mathcal{C}^\infty(\mathcal{I}, \mathbb{R}^m)$; and convergence $x_j \xrightarrow{j \to \infty} x_*$ actually means that $\|x_j^{(k)} - x_*^{(k)}\|_\infty \xrightarrow{j \to \infty} 0$ is valid for all $k = 0, 1, 2, \ldots$. Unfortunately, this vector-space setting does not offer valuable clues to the question whether the solution of the equation

$$\mathcal{T}x = (q, d).$$

depends continuously on the data q and d. Our alarming Example 2.3 suits this setting, however, now the problem is circumvented since the sequence (2.13) no longer converges to zero here.

2.4.3 A Too Lean Banach Space $X = \check{\mathcal{C}}(\mathcal{I}, \mathbb{R}^m) \subset \mathcal{C}^\infty(\mathcal{I}, \mathbb{R}^m)$

A bounded linear operator acting bijectively in Banach spaces has also a bounded inverse. To practice this well-known fact we refer briefly to the function space

$$\check{\mathcal{C}}(\mathcal{I}, \mathbb{R}^m) := \{x \in \mathcal{C}^\infty(\mathcal{I}, \mathbb{R}^m) : \sup_{j \geq 0} \|x^{(j)}\|_\infty < \infty\},$$

endowed with the norm

$$\|x\| := \sup_{j \geq 0} \|x^{(j)}\|_\infty, \quad x \in \check{\mathcal{C}}(\mathcal{I}, \mathbb{R}^m),$$

which is a Banach space. In the setting $\operatorname{dom} T = X = Y = \check{\mathcal{C}}(\mathcal{I}, \mathbb{R}^m)$, the composed operator $\mathcal{T}$ associated with a regular matrix pair becomes a bounded bijection between X and $Y \times \mathbb{R}^{m-\check{l}}$, with an appropriate $\check{l} \geq l$. This simulates continuous invertibility, which seems to be in contradiction to Example 2.3. However, the function space $\check{\mathcal{C}}(\mathcal{I}, \mathbb{R}^m)$ is much too lean in capacity. The basic inclusion $\mathcal{C}^\infty(\mathcal{I}, \mathbb{R}^m) \subseteq \operatorname{dom} T$ is no longer satisfied. In Example 2.3, to fit into this special setting, one has to assume $|\alpha| \leq 1$ and the sequence of excitations (2.13) considered there is not at all admissible. By this, the danger is factitious disregarded. The same would happen when applying equivalent weighted norms. Therefore, this special Banach space setting is much too restrictive, and hence inappropriate even for constant coefficient DAEs.

3   Normally Solvable Differential-Algebraic Operators

As is well known, if a densely defined, closed operator acting in Hilbert spaces has a closed image, then it is normally solvable, and there exist bounded inner inverses. In particular, the Moore–Penrose inverse is then bounded, and it makes good sense to seek LSS, see [62], also Appendix 6.1.2. For a normally solvable operator, the problem of calculating a pseudo solution becomes well-posed in Hadamard's sense. For this reason, if K is an operator with closed image, the equation $Kx = q$ is also said to be *well-posed in the sense of G. Fichera*.

We ask if the operator associated with the linear time-varying DAE

$$E(t)x'(t) + F(t)x(t) = q(t), \quad t \in \mathcal{I} = [t_a, t_e], \tag{3.1}$$

and the composed operator associated with the IVP

$$E(t)x'(t) + F(t)x(t) = q(t), \quad t \in \mathcal{I}, \quad Cx(t_a) = d, \tag{3.2}$$

have closed images. The matrix coefficients $E(t), F(t) \in \mathcal{L}(\mathbb{R}^m, \mathbb{R}^k)$ are continuous in t on the interval $\mathcal{I}$. The nullspace of E is a $\mathcal{C}^1$- subspace. As distinguished from the regular DAEs in Sect. 2, the DAE (3.1) comprises k equations but m unknown functions. These so-called *rectangular* DAEs play their role in optimization and system control, see [4].

Inspecting once again the settings discussed in Sect. 2, for the preimage space X we favor the two function spaces $\mathcal{C}(\mathcal{I}, \mathbb{R}^m)$, $L^2(\mathcal{I}, \mathbb{R}^m)$ being fully independent of DAE data, and, additionally, the two spaces $\mathcal{C}_D^1(\mathcal{I}, \mathbb{R}^m)$, $H_D^1(\mathcal{I}, \mathbb{R}^m)$ that incorporate certain problem data via a proper factorization $E = AD$.

Supposing a nonvanishing coefficient E, the differential-algebraic operator becomes unbounded, but densely defined and closable in the first two cases. We have to deal with the closure. The latter two function spaces already involve problem data, in essence, these spaces are the definition domains of the closures obtained in the first two cases, and endowed with the corresponding graph norms. In such a setting the differential-algebraic operator is bounded.

We establish practically useful sufficient criteria for normal solvability in terms of the original data, in Sect. 3.1 for closed operators and in Sect. 3.2 for bounded ones. Section 3.3 provides bounded generalized inverses and LSS.

In Sect. 3.4 we provide a large class of nonlinear DAEs

$$\mathfrak{f}(x'(t), x(t), t) = 0,$$

with m unknowns and k equations, yielding normally solvable linearizations in a natural Banach space setting. Again we establish the criteria of normal solvability in terms of the original data, and this serves as basis for Newton–Kantorovich iteration procedures etc.

3.1   Settings with Closed Differential-Algebraic Operators

We associate the linear DAE (3.1) with the operator

$$\overset{\circ}{T} \in \mathcal{L}(X, Y), \quad \overset{\circ}{T}x := Ex' + Fx, \ x \in \text{dom}\,\overset{\circ}{T} = \mathcal{C}^1(\mathcal{I}, \mathbb{R}^m) \subseteq X. \qquad (3.3)$$

We apply the function spaces $X = \mathcal{C}(\mathcal{I}, \mathbb{R}^m)$, $Y = \mathcal{C}(\mathcal{I}, \mathbb{R}^k)$ when seeking classical DAE solutions which satisfy the DAE at all $t \in \mathcal{I}$. Instead, the spaces $X = L^2(\mathcal{I}, \mathbb{R}^m)$, $Y = L^2(\mathcal{I}, \mathbb{R}^k)$ are applied, when we are interested in generalized solutions satisfying the DAE for almost all $t \in \mathcal{I}$.

We suppose that E does not vanish such that the DAE is nontrivial and $\overset{\circ}{T}$ is unbounded in both instances (cf. Sects. 2.3.2, 2.3.3).

Theorem 3.1 *If $E, F \in \mathcal{C}(\mathcal{I}, \mathcal{L}(\mathbb{R}^m, \mathbb{R}^k))$ and $\ker E$ is a $\mathcal{C}^1$-subspace, then the following statements are valid:*

1. *There exist factorizations $E = AD$ such that $A \in \mathcal{C}(\mathcal{I}, \mathcal{L}(\mathbb{R}^n, \mathbb{R}^k))$, $D \in \mathcal{C}^1(\mathcal{I}, \mathcal{L}(\mathbb{R}^m, \mathbb{R}^n))$, $\ker A$ and $\text{im}\,D$ are $\mathcal{C}^1$-subspaces, and the transversality condition*

$$\ker A(t) \oplus \text{im}\,D(t) = \mathbb{R}^n, \ t \in \mathcal{I}, \qquad (3.4)$$

   is valid. The border projector $R(t) \in \mathcal{L}(\mathbb{R}^n)$ onto $\text{im}\,D(t)$ along $\ker A(t)$ depends continuously differentiably on t.
   $B := F - AD'$ is continuous.

2. For $X = C(\mathcal{I}, \mathbb{R}^m)$, $Y = C(\mathcal{I}, \mathbb{R}^k)$, and each arbitrary factorization from (1), the operator

$$T \in \mathcal{L}(X, Y), \quad Tx := A(Dx)' + Bx, \ x \in \mathrm{dom}\, T = C_D^1(\mathcal{I}, \mathbb{R}^m) \subseteq X, \quad (3.5)$$

represents the closure of the operator $\overset{\circ}{T} \in \mathcal{L}(X, Y)$ given by (3.3).

3. For $X = L^2(\mathcal{I}, \mathbb{R}^m)$, $Y = L^2(\mathcal{I}, \mathbb{R}^k)$, and each arbitrary factorization from (1), the operator

$$T \in \mathcal{L}(X, Y), \quad Tx := A(Dx)' + Bx, \ x \in \mathrm{dom}\, T = H_D^1(\mathcal{I}, \mathbb{R}^m) \subseteq X, \quad (3.6)$$

represents the closure of the operator $\overset{\circ}{T} \in \mathcal{L}(X, Y)$ given by (3.3).

To shorten the wording, a factorization according to (1) is said to be a *proper factorization*.

Proof 1. Since $\ker E$ is a C^1-subspace, the orthoprojector function E^+E onto $(\ker E)^\perp$ is continuously differentiable. Put $A := E$ and $D := E^+E$. Then $E = EE^+E = AD$ and $R = E^+E$, which makes the statement evident.

2. The transversality condition (3.4) generalizes condition (2.16). The proof follows the arguments given in Sect. 2.3.2.

3. We proceed as in Lemma 2.6 and [38] to verify the existence of the biadjoint $\overset{\circ}{T}{}^{**}$. Then the closure equals the biadjoint, $T = \overset{\circ}{T}{}^{**}$.

For each arbitrary $x \in C^1(\mathcal{I}, \mathbb{R}^m)$ and suitable functions y we derive by partial integration that

$$
\begin{aligned}
(\overset{\circ}{T}x, y) &= \int_{t_a}^{t_e} \langle E(t)x'(t) + F(t)x(t), y(t) \rangle \mathrm{d}t \\
&= \int_{t_a}^{t_e} \langle A(t)(Dx)'(t) + B(t)x(t), y(t) \rangle \mathrm{d}t \\
&= \int_{t_a}^{t_e} \langle x(t), -D(t)^*(A^*y)'(t) + B(t)^*y(t) \rangle \mathrm{d}t \\
&\quad + \langle D(t_e)x(t_e), A(t_e)^*y(t_e) \rangle - \langle D(t_a)x(t_a), A(t_a)^*y(t_a) \rangle.
\end{aligned}
$$

Therefore, the adjoint operator $\overset{\circ}{T}{}^*$ is given on the set

$$\{w \in L^2(\mathcal{I}, \mathbb{R}^m) : A^*w \in H^1(\mathcal{I}, \mathbb{R}^n), A(t_a)^*w(t_a) = 0, A(t_e)^*w(t_e) = 0\} =: \mathrm{dom}\, \overset{\circ}{T}{}^*$$

by means of

$$\overset{\circ}{T}{}^*y = -D^*(A^*y)' + B^*y, \quad y \in \mathrm{dom}\, \overset{\circ}{T}{}^*.$$

In fact, we have $(\mathring{T}x, y) = (x, \mathring{T}^*y)$ for all $x \in \text{dom } \mathring{T}$ and $y \in \text{dom } \mathring{T}^*$. Since dom $\mathring{T}^*$ is dense in $L^2(\mathcal{I}, \mathbb{R}^m)$, in turn, the adjoint of the operator $\mathring{T}^*$ also exists. Compute further

$$(\mathring{T}^*y, x) = \int_{t_a}^{t_e} \langle -D^*(A^*y)'(t) + F^*y(t), x(t)\rangle \mathrm{d}t$$

$$= \int_{t_a}^{t_e} \langle y(t), A(Dx)'(t) + Fx(t)\rangle \mathrm{d}t$$

$$=: (y, \mathring{T}^{**}x)$$

for all $y \in \text{dom } \mathring{T}^*$ and $x \in \{w \in L^2(\mathcal{I}, \mathbb{R}^m) : Dw \in H^1(\mathcal{I}, \mathbb{R}^n)\} =: \text{dom } \mathring{T}^{**}$. $\square$

As a densely defined closed operator, T has a closed nullspace, however, now ker T is not necessarily finite-dimensional as is the case for regular differential-algebraic operators (e.g., Propositions 2.4, 2.7).

For time-invariant regular index-0 and index-1 pairs $\{E, F\}$, the operator T is surjective, thus im T is closed. For regular higher index pairs $\{E, F\}$ the closed-image property gets lost, see Propositions 2.4, 2.7. Next we characterize a further large class of linear time-varying DAEs (3.1) yielding closed images of the associated operators T, too.

Applying a proper factorization as described in Theorem 3.1, we introduce the following denotations to be used throughout Sect. 3:

$$G_0 := AD = E, \quad B = F - AD',$$

$$P_0 := G_0^+ G_0 = E^+ E, \quad Q_0 := I - P_0,$$

$$W_0 := I - G_0 G_0^+, \tag{3.7}$$

$$G_1 := G_0 + BQ_0 = E + (F - AD')Q_0 = E + FQ_0 + EQ_0' \tag{3.8}$$

$$P_1 := G_1^+ G_1, \quad Q_1 := I - P_1,$$

$$W_1 := I - G_1 G_1^+. \tag{3.9}$$

Since $E(t)$ has constant rank on the interval $\mathcal{I}$, rank $E(t) =: r$, the orthoprojector function W_0 is continuous. The orthoprojector functions P_0 and Q_0 are even continuously differentiable, since ker E is a $\mathcal{C}^1$-subspace. The matrix function G_1 is continuous. If $G_1(t)$ has constant rank, then W_1 is also continuous. We have further

$$\text{im } G_0 = \text{im } A = \text{im } E,$$

$$W_0 F = W_0 B,$$

$$\text{im } G_1 = \text{im } G_0 \oplus \text{im } W_0 B Q_0 = \text{im } E \oplus \text{im } W_0 F Q_0.$$

The matrix functions G_1 and W_0FQ_0 have constant rank at the same time. This fact will frequently be exploited later on for ensuring the continuity of W_1.

Let D^- denote the pointwise generalized inverse such that

$$DD^-D = D, \ D^-DD^- = D^-, \ D^-D = P_0, \ DD^- = R.$$

D^- is continuously differentiable (e.g. [55, Proposition A.17]). The factorization according to Theorem 3.1(1) can be chosen so that the sum (3.4) is orthogonal and it results that $D^- = D^+$.

The continuous projector-valued function $W_1 \in \mathcal{C}(\mathcal{I}, \mathcal{L}(\mathbb{R}^k))$ is uniformly bounded on the compact interval $\mathcal{I}$. The assignment

$$y \in \mathcal{C}(\mathcal{I}, \mathbb{R}^k) \to W_1(t)y(t) =: (W_1 y)(t), \ \text{for all } t \in \mathcal{I}, \tag{3.10}$$

defines a bounded projector acting on $\mathcal{C}(\mathcal{I}, \mathbb{R}^k)$, which we also denote by W_1, more precisely $W_1 \in \mathcal{L}_b(\mathcal{C}(\mathcal{I}, \mathbb{R}^k))$, and we write

$$\{y \in \mathcal{C}(\mathcal{I}, \mathbb{R}^k) : W_1 y = 0\} = \ker W_1 \subseteq \mathcal{C}(\mathcal{I}, \mathbb{R}^k). \tag{3.11}$$

Similarly, the assignment

$$y \in L^2(\mathcal{I}, \mathbb{R}^k) \to W_1(t)y(t) =: (W_1 y)(t), \ \text{for almost all } t \in \mathcal{I}, \tag{3.12}$$

defines a bounded projector acting on $L^2(\mathcal{I}, \mathbb{R}^k)$, which we denote by W_1, too, more precisely $W_1 \in \mathcal{L}_b(L^2(\mathcal{I}, \mathbb{R}^k))$, and we write

$$\{y \in L^2(\mathcal{I}, \mathbb{R}^k) : W_1 y = 0\} = \ker W_1 \subseteq L^2(\mathcal{I}, \mathbb{R}^k). \tag{3.13}$$

No confusion should arise from this within the respective context. We proceed analogously with further continuous projector-valued functions, for instance P_1.

Owing to the uniform boundedness of the continuous projector-valued functions, the resulting projectors acting on the function spaces are bounded, and hence, their nullspaces and images are closed.

The following theorem actually provides a first criterion of normal solvability.

Theorem 3.2 *Let $E, F \in \mathcal{C}(\mathcal{I}, \mathcal{L}(\mathbb{R}^m, \mathbb{R}^k))$ and $\ker E$ be a C^1-subspace. Let the matrix function $W_0FQ_0 = W_0BQ_0$ have constant rank on $\mathcal{I}$ and let the condition*

$$W_1FP_0 = 0 \tag{3.14}$$

be satisfied. Then the following statements are valid:

1. *The operator $T \in \mathcal{L}_c(\mathcal{C}(\mathcal{I}, \mathbb{R}^m), \mathcal{C}(\mathcal{I}, \mathbb{R}^k))$ from Theorem 3.1(2) has a closed image, namely $\operatorname{im} T = \ker W_1 \subseteq \mathcal{C}(\mathcal{I}, \mathbb{R}^k)$.*
2. *The operator $T \in \mathcal{L}_c(L^2(\mathcal{I}, \mathbb{R}^m), L^2(\mathcal{I}, \mathbb{R}^k))$ from Theorem 3.1(3) has a closed image, namely $\operatorname{im} T = \ker W_1 \subseteq L^2(\mathcal{I}, \mathbb{R}^k)$.*

Proof Since E and $W_0 F Q_0$ are continuous constant-rank matrix functions, so are G_1 and W_1, and the subspaces $\ker W_1 \subseteq \mathcal{C}(\mathcal{I}, \mathbb{R}^k)$ and $\ker W_1 \subseteq L^2(\mathcal{I}, \mathbb{R}^k)$ are closed in fact.

We derive for each arbitrary $x \in \operatorname{dom} T$:

$$Tx = A(Dx)' + Bx = G_1(D^-(Dx)' + Q_0 x) + B P_0 x$$
$$= G_1(D^-(Dx)' + Q_0 x) + G_1 G_1^+ B P_0 x + W_1 B P_0 x.$$

Condition (3.14) yields $W_1 B P_0 x = W_1 F P_0 x = 0$, which immediately implies $W_1(Tx) = 0$. It follows that $\operatorname{im} T \subseteq \ker W_1$. Next we verify that, for each arbitrary $q \in \ker W_1$, there exists at least one $x \in \operatorname{dom} T$ satisfying $Tx = q$.

The IVP

$$u' - R'u + D G_1^+ B D^- u = D G_1^+ q, \quad u(t_a) = 0$$

has a unique solution $u \in \mathcal{C}^1(\mathcal{I}, \mathbb{R}^n)$ and $u \in H^1(\mathcal{I}, \mathbb{R}^n)$, accordingly. It holds that $u = Ru$, thus $Ru' = -D G_1^+ B D^- u + D G_1^+ q$.

Put $x := D^- u - Q_0 G_1^+ B D^- u + Q_0 G_1^+ q$ such that $Dx = DD^- u = u$, thus $x \in \operatorname{dom} T$, $P_0 x = D^- u$, $Q_0 x = -Q_0 G_1^+ B D^- u + Q_0 G_1^+ q$, and further

$$Tx = A(Dx)' + Bx = G_1(D^-(Dx)' + Q_0 x) + B P_0 x$$
$$= G_1(D^-(Dx)' + Q_0 x) + G_1 G_1^+ B P_0 x$$
$$= G_1\{D^- u' + Q_0 x + G_1^+ B P_0 x\} = G_1\{D^-(u' - R'u) + Q_0 x + G_1^+ B P_0 x\}$$
$$= G_1\{D^-(-D G_1^+ B D^- u + D G_1^+ q) + Q_0 x + Q_0 G_1^+ B P_0 x + P_0 G_1^+ B P_0 x\}$$
$$= G_1\{-P_0 G_1^+ B P_0 x + P_0 G_1^+ q + Q_0 x + Q_0 G_1^+ B P_0 x + P_0 G_1^+ B P_0 x\}$$
$$= G_1\{P_0 G_1^+ q + Q_0 x + Q_0 G_1^+ B P_0 x\} = G_1\{P_0 G_1^+ q + Q_0 G_1^+ q\} = G_1 G_1^+ q = q.$$

In consequence, q belongs to $\operatorname{im} T$, and hence $\operatorname{im} T = \ker W_1$. $\qquad\square$

Though the operator T has here a closed image, it is not necessarily Fredholm. The nullspace may fail to be finite-dimensional. Furthermore, if $W_1 \neq 0$, then there is no finite codimension.

The DAE described in Theorem 3.2 is tractable with index 0, if $W_0 F Q_0 = 0$, and tractable with index 1 otherwise, see [55].

For $W_0 = 0, m = k$, the DAE is even regular with index 0. For $W_0 \neq 0, W_1 = 0$, and $m = k$, the DAE is regular with tractability index 1.

The matrix pair $\{E, F\}$ in Example 2.2 obviously satisfies the condition $W_0 F = 0$. The associated DAE is tractable with index 1. The same happens for [85, Example, p. 485]. In both instances, the differential-algebraic operators are normally solvable in fact. A large class of differential-algebraic operators which satisfy the conditions of Theorem 3.2 is associated with so-called *strangeness-free DAEs*, as we show by the following example.

Example 3.3 (Strangeness-Free DAEs) Many papers start supposing a so-called strangeness-free DAE. We recall the respective description of this class, see [50, Definition 2.4]: The DAE (3.1) is said to be strangeness-free, if there are pointwise orthogonal matrix functions $L \in \mathcal{C}(\mathcal{I}, \mathcal{L}(\mathbb{R}^k))$ and $K \in \mathcal{C}^1(\mathcal{I}, \mathcal{L}(\mathbb{R}^m))$ such that

$$LEK = \tilde{E} = \begin{bmatrix} \tilde{E}_{11} & 0 & 0 \\ 0 & 0 & 0 \\ 0 & 0 & 0 \end{bmatrix}, \quad LFK + LEK' = \tilde{F} = \begin{bmatrix} \tilde{F}_{11} & \tilde{F}_{12} & \tilde{F}_{13} \\ \tilde{F}_{21} & \tilde{F}_{22} & 0 \\ 0 & 0 & 0 \end{bmatrix},$$

with nonsingular blocks $\tilde{E}_{11}, \tilde{F}_{22}$, and all blocksizes are allowed to be zero.

For strangeness-free DAEs all conditions of Theorem 3.2 are satisfied so that the associated operator T is normally solvable. Namely, we have im $\tilde{E} = L$ im E, ker $E = K$ ker $\tilde{E}$. We set and compute

$$\tilde{Q}_0 = \tilde{W}_0 = \begin{bmatrix} 0 & 0 & 0 \\ 0 & I & 0 \\ 0 & 0 & I \end{bmatrix}, \quad \tilde{A} = \tilde{E}, \quad \tilde{D} = \tilde{P}_0, \quad \tilde{B} = \tilde{F},$$

$$\tilde{W}_0 \tilde{F} \tilde{Q}_0 = \begin{bmatrix} 0 & 0 & 0 \\ 0 & \tilde{F}_{22} & 0 \\ 0 & 0 & 0 \end{bmatrix}, \quad \tilde{G}_1 = \begin{bmatrix} \tilde{E}_{11} & \tilde{F}_{12} & \tilde{F}_{13} \\ 0 & \tilde{F}_{22} & 0 \\ 0 & 0 & 0 \end{bmatrix}, \quad \tilde{W}_1 = \begin{bmatrix} 0 & 0 & 0 \\ 0 & 0 & 0 \\ 0 & 0 & I \end{bmatrix},$$

thus $\tilde{W}_1 \tilde{F} = 0$. Then we set $P_0 = K \tilde{P}_0 K^{-1}$ and factorize $E = AD$, $A = E$, $D = P_0$. P_0 is the orthoprojector function onto ker E, and the decomposition ker $E \oplus$ im $P_0 = \mathbb{R}^m$ is valid.

$W_0 = L^{-1} \tilde{W}_0 L$ is the orthoprojector function with ker $W_0 =$ im E. Regarding that $\tilde{W}_0 LE = 0$ we derive

$$W_0 F Q_0 = L^{-1} \tilde{W}_0 LFK \tilde{Q}_0 K^{-1}$$
$$= L^{-1} \tilde{W}_0 (\tilde{F} - LEK') \tilde{Q}_0 K^{-1} = L^{-1} \tilde{W}_0 \tilde{F} \tilde{Q}_0 K^{-1},$$

thus rank $W_0 F Q_0 =$ rank $\tilde{W}_0 \tilde{F} \tilde{Q}_0 =$ rank $\tilde{F}_{22}$ is constant. Compute further

$$G_1 = E + B Q_0 = E + (F - E P_0') Q_0,$$

$$LG_1 K = LEK + L(F - EP_0')KK^{-1} Q_0 K = \tilde{E} + LFK \tilde{Q}_0 - LEP_0' K \tilde{Q}_0$$

$$= \tilde{E} + (\tilde{F} - LEK') \tilde{Q}_0 - LEP_0' K \tilde{Q}_0$$

$$= \tilde{E} + \tilde{F} \tilde{Q}_0 - LE(P_0 K' + P_0' K) \tilde{Q}_0 = \tilde{G}_1 - LE(P_0 K)' \tilde{Q}_0.$$

Next, $W_1 = L^{-1}\tilde{W}_1 L$ is the orthoprojector function with $\ker W_1 = \operatorname{im} G_1$. Regarding that $\tilde{W}_1 LE = 0$ we finally derive

$$W_1 FP_0 = L^{-1}\tilde{W}_1 LFK \tilde{P}_0 K^{-1}$$
$$= L^{-1}\tilde{W}_1 (\tilde{F} - LEK')\tilde{P}_0 K^{-1} = L^{-1}\tilde{W}_1 \tilde{F} \tilde{P}_0 K^{-1} = 0.$$

We mention that such a strangeness-free DAE has tractability index 0, if $\tilde{F}_{22}$ has size zero, and otherwise tractability index 1, see [55]. $\qquad\square$

Next we deal with the linear IVP

$$E(t)x'(t) + F(t)x(t) = q(t), \ t \in \mathcal{I}, \quad Cx(t_a) = d. \tag{3.15}$$

As before, the coefficients E and F are continuous, $\ker E$ is a $\mathcal{C}^1$-subspace, and E has constant rank $r > 0$. Let the matrix $C \in \mathcal{L}(\mathbb{R}^m, \mathbb{R}^r)$ have full row-rank r, and

$$\ker C = \ker E(t_a). \tag{3.16}$$

We consider the previous differential-algebraic operator $\overset{\circ}{T} \in \mathcal{L}(X, Y)$ corresponding to the DAE as well as its composed associate $\overset{\circ}{\mathcal{T}} \in \mathcal{L}(X, Y \times \mathbb{R}^r)$ defined by

$$\overset{\circ}{\mathcal{T}}x := (\overset{\circ}{T}x, Cx(t_a)), \ x \in \operatorname{dom}\overset{\circ}{\mathcal{T}} = \operatorname{dom}\overset{\circ}{T}.$$

Since the composed operator is also unbounded but closable, we immediately turn to its closure $\mathcal{T} \in \mathcal{L}_c(X, Y \times \mathbb{R}^r)$,

$$\mathcal{T}x := (Tx, Cx(t_a)), \ x \in \operatorname{dom}\mathcal{T} = \operatorname{dom}T.$$

We continue using the previous notations, e.g., (3.8), (3.9). Additionally, we introduce the matrix function $U \in \mathcal{C}^1(\mathcal{I}, \mathcal{L}(\mathbb{R}^r))$ to be the unique solution of the IVP

$$U' - R'U + DG_1^+ BD^- U = 0, \quad U(t_a) = I. \tag{3.17}$$

U is the fundamental solution matrix normalized at t_a of the ODE in (3.17). It satisfies the conditions

$$U(t)R(t_a) = R(t)U(t)R(t_a), \quad U(t)^{-1}R(t) = R(t_a)U(t)^{-1}R(t), \quad t \in \mathcal{I}.$$

Theorem 3.4 *Let E and F be continuous, $\ker E$ be a $\mathcal{C}^1$-subspace, and let the matrix function $W_0 FQ_0$ have constant rank. Let the matrix $C \in \mathcal{L}(\mathbb{R}^m, \mathbb{R}^r)$ have full row-rank $r = \operatorname{rank} E$ and satisfy the condition $\ker C = \ker E(t_a)$.*

Additionally, let the inclusion

$$\ker (E + FQ_0 + EQ_0') \subseteq \ker E \qquad (3.18)$$

be satisfied pointwise on $\mathcal{I}$. Then the following statements are true for the two choices concerning the function spaces

$$X = \mathcal{C}(\mathcal{I}, \mathbb{R}^m), \ Y = \mathcal{C}(\mathcal{I}, \mathbb{R}^k), \ X^1 = \mathcal{C}^1(\mathcal{I}, \mathbb{R}^m), \ X_D^1 = \mathcal{C}_D^1(\mathcal{I}, \mathbb{R}^m), \qquad (3.19)$$

and

$$X = L^2(\mathcal{I}, \mathbb{R}^m), \ Y = L^2(\mathcal{I}, \mathbb{R}^k), \ X^1 = H^1(\mathcal{I}, \mathbb{R}^m), \ X_D^1 = H_D^1(\mathcal{I}, \mathbb{R}^m), \qquad (3.20)$$

respectively, and for the closure $\mathcal{T} \in \mathcal{L}_c(X, Y \times \mathbb{R}^r)$, $\operatorname{dom} \mathcal{T} = X_D^1$, of the operator $\overset{\circ}{\mathcal{T}} \in \mathcal{L}(X, Y \times \mathbb{R}^r)$ associated with the IVP (3.15):

1. The operator $\mathcal{T}$ is normally solvable and

$$\operatorname{im} \mathcal{T} = \{(q, d) \in Y \times \mathbb{R}^r :$$

$$W_1 q = W_1 B D^- U(D(t_a) C^+ d + \int_{t_a} U(s)^{-1} D(s) G_1(s)^+ q(s) \mathrm{d}s)\}.$$

2. $\ker \mathcal{T} = \{x \in X : P_1 x = 0\} \subset \{x \in X : Dx = 0\} \subset X_D^1$.
3. If $E + FQ_0$ has full column-rank, then $\mathcal{T}$ is injective.
4. If $(q, d) \in \operatorname{im} \mathcal{T}$, then there exists a unique $x_* \in X_D^1$ satisfying $\mathcal{T} x_* = (q, d)$ and $Q_1 x_* = 0$, as well as the inequality

$$\|x_*\|_{L^2} < \|x\|_{L^2} \qquad (3.21)$$

for all other solutions x of the equation $\mathcal{T} x = (q, d)$,

Proof 1. Since E and $W_0 F Q_0$ are continuous constant-rank matrix functions, so are G_1 and W_1. The inclusion (3.18) means actually $\ker G_1 \subseteq \ker G_0$.
    For given $x \in X_D^1$ and $d := Cx(t_a)$, $q := Tx$ we derive

$$q = Tx = A(Dx)' + Bx = ADD^-(Dx)' + BQ_0 x + BP_0 x$$

$$= G_1(D^-(Dx)' + Q_0 x) + G_1 G_1^+ B P_0 x + W_1 B P_0 x$$

$$= G_1(D^-(Dx)' + Q_0 x + G_1^+ B P_0 x) + W_1 B P_0 x,$$

thus

$$G_1 G_1^+ q = G_1(D^-(Dx)' + Q_0 x + G_1^+ B P_0 x) \text{ and } W_1 q = W_1 B P_0 x. \qquad (3.22)$$

From the first part we obtain the relation

$$D^-(Dx) + Q_0x + G_1^+ BP_0x - G_1^+q = \xi, \tag{3.23}$$

whereby ξ is an arbitrary function belonging to $\ker P_1 \subseteq X$. Owing to the properties $\ker G_1 \subseteq \ker G_0$ and $\ker D = \ker G_0$ it follows that $D\xi = 0$, $\xi = Q_0\xi$, $\xi \in X_D^1$. Now (3.23) decomposes into the system

$$(Dx)' - R'Dx + DG_1^+ BP_0x - DG_1^+q = D\xi = 0,$$

$$Q_0x + Q_0G_1^+ BP_0x - Q_0G_1^+q = Q_0\xi = \xi.$$

From $d = Cx(t_a)$ we obtain $D(t_a)x(t_a) = D(t_a)C^+d$. This yields the representation

$$x = D^-Dx + Q_0x$$

$$= (I - Q_0G_1^+ B)D^-U\{D(t_a)C^+d + \int_{t_a} U(s)^{-1}D(s)G_1(s)^+q(s)ds\}$$

$$+ Q_0G_1^+q + \xi, \tag{3.24}$$

further

$$Dx = U\{D(t_a)C^+d + \int_{t_a} U(s)^{-1}D(s)G_1(s)^+q(s)ds\}. \tag{3.25}$$

Now it is evident that q must satisfy the consistency condition

$$W_1q = W_1BD^-U\{D(t_a)C^+d + \int_{t_a} U(s)^{-1}D(s)G_1(s)^+q(s)ds\}. \tag{3.26}$$

Conversely, the equation $\mathcal{T}x = (q, d)$ is solvable for each arbitrary $d \in \mathbb{R}^r$, $G_1G_1^+q \in Y$, and the corresponding W_1q defined by (3.26). Namely, for such d and q, we put

$$\tilde{x} := (I - Q_0G_1^+ B)D^-U\{D(t_a)C^+d + \int_{t_a} U(s)^{-1}D(s)G_1(s)^+q(s)ds\} + Q_0G_1^+q$$

and obtain

$$D\tilde{x} = DD^-U\{D(t_a)C^+d + \int_{t_a} U(s)^{-1}D(s)G_1(s)^+q(s)ds\}$$

$$= U\{D(t_a)C^+d + \int_{t_a} U(s)^{-1}D(s)G_1(s)^+q(s)ds\}.$$

It becomes clear that $\tilde{x}$ belongs to dom $T = X_D^1$ and

$$C\tilde{x}(t_a) = CD(t_a)^- D(t_a)\tilde{x}(t_a) = CD(t_a)^- D(t_a)C^+d = CC^+d = d.$$

Finally, one finds that $T\tilde{x} = q$. This proves that

$$\mathrm{im}\, T = \{(q,d) \in Y \times \mathbb{R}^r :$$

$$W_1 q = W_1 BD^- U(D(t_a)C^+d + \int_{t_a} U(s)^{-1}(DG_1^+ q)(s)\mathrm{d}s)\}$$

in fact. This set is obviously closed.

2. For $d = 0$ and $q = 0$, we obtain the general solution (cf. (3.24)) $x = \xi$, such that $\ker P_1 \supseteq \ker T$. Conversely, we compute for $x \in \ker P_1$, that is, for $x = (I - P_1)x = Q_0(I - P_1)x$:

$$Tx = A(Dx)' + Bx = G_1(D^-(Dx)' + Q_0 x) + BP_0 x = G_1 Q_0 x = G_1 x = 0.$$

3. Regarding the relation $\mathrm{rank}\, G_1 = \mathrm{rank}\,(E + FQ_0)$, this statement is a simple consequence of (2).
4. Since P_1, Q_1 are complementary orthoprojector functions we have

$$\|x\|_{L^2}^2 = \|P_1 x\|_{L^2}^2 + \|Q_1 x\|_{L^2}^2.$$

For all solutions x_*, the component $P_1 x_*$ is completely fixed by d and q. The only free component is $\xi_* = Q_1 \xi_*$. Letting

$$\xi_* = - Q_1(I - Q_0 G_1^+ B)D^- U\{D(t_a)C^+d + \int_{t_a} U(s)^{-1}D(s)G_1(s)^+ q(s)\mathrm{d}s\}$$

$$- Q_1 Q_0 G_1^+ q,$$

one arrives at the only solution with $Q_1 x_* = 0$.

$\square$

The DAEs captured by Theorem 3.4 comprise a number of equations k that can be less, equal to, or larger than the number of unknown functions m. The DAEs are tractable with index 0, if $W_0 F Q_0 = 0$, and otherwise tractable with index 1. The corollary below specifies the index-0 case.

Corollary 3.5 *Let E and F be continuous, $\ker E$ be a C^1-subspace. Let the matrix $C \in \mathcal{L}(\mathbb{R}^m, \mathbb{R}^r)$ have full row-rank $r = \mathrm{rank}\, E$ and satisfy $\ker C = \ker E(t_a)$. Additionally, let the condition*

$$FQ_0 + EQ_0' = 0 \tag{3.27}$$

be fulfilled. Then all statements of Theorem 3.4 are valid with W_0, P_0, G_0^+ instead of W_1, P_1, G_1^+.

Moreover, rank $E = \text{rank}\,(E + FQ_0)$.

Proof Condition (3.27) is a special case of the inclusion (3.18), which leads to $W_0FQ_0 = 0$, $G_1 = G_0$, $E + FQ_0 = E(I - P_0Q_0')$, and hence $W_1 = W_0$, $P_1 = P_0$, $\text{rank}\,(E + FQ_0) = \text{rank}\,E$.   □

Example 3.6 (Over-determined index-0 DAE) The IVP for the over-determined DAE (3.1) with $m = 2$, $k = 3$, and the coefficients

$$E = \begin{bmatrix} 1 & 0 \\ 0 & 1 \\ 0 & 0 \end{bmatrix}, \quad F = \begin{bmatrix} 0 & 0 \\ 1 & 0 \\ 0 & 1 \end{bmatrix}, \quad C = \begin{bmatrix} 1 & 0 \\ 0 & 1 \end{bmatrix},$$

leads to the closure $T = \overset{\circ}{T}$, with the trivial factorization $E = AD$, $A = E$, $D = I$, $n = 2$. The operator T is injective. The corresponding DAE is tractable with index 0. We have $R = I$, $D^- = I$, $B = F$ and

$$W_0 = \begin{bmatrix} 0 & 0 & 0 \\ 0 & 0 & 0 \\ 0 & 0 & 1 \end{bmatrix}, Q_0 = 0, \ P_0 = I, \ G_1 = G_0, \ W_1 = W_0, \ G_0^+ = \begin{bmatrix} 1 & 0 & 0 \\ 0 & 1 & 0 \end{bmatrix},$$

and, trivially, $W_0FQ_0 = 0$ so that Corollary 3.5 applies. Derive further

$$(DG_0^+BD^-)(t) = \begin{bmatrix} 0 & 0 \\ 1 & 0 \end{bmatrix}, \quad U(t) = \begin{bmatrix} 1 & 0 \\ -(t - t_a) & 1 \end{bmatrix}.$$

The consistency condition for q and d (cf. (3.26)) reads here

$$q_3 = -(t - t_a)(d_1 + \int_{t_a} q_1(s)\mathrm{d}s) + d_2 + \int_{t_a} ((s - t_a)q_1(s) + q_2(s))\mathrm{d}s.$$

□

The following proposition generalizes the previous results. The structural conditions (3.14) and (3.18) are special instances of the condition (3.28) below. This is obvious for (3.14). Since (3.18) implies the inclusion $\ker G_1 \subseteq \ker G_0 = \ker D$, we have also $P_0(I - P_1) = 0$. Also condition (3.31) below is given in these particular instances. The differential-algebraic operator in Example 3.8 below meets the assumptions of the next proposition, but it satisfies neither condition (3.14) nor condition (3.18).

Proposition 3.7 *Let E and F be continuous, $\ker E$ be a C^1-subspace, and let the matrix function W_0FQ_0 have constant rank. Let the matrix $C \in \mathcal{L}(\mathbb{R}^m, \mathbb{R}^r)$ have full row-rank $r = \text{rank}\,E$, and satisfy the condition $\ker C = \ker E(t_a)$.*

Additionally, let the condition

$$W_1 F P_0 (I - P_1) = 0 \tag{3.28}$$

be given. Then the following statements are true for the two choices concerning the function spaces

$$X = \mathcal{C}(\mathcal{I}, \mathbb{R}^m), \ Y = \mathcal{C}(\mathcal{I}, \mathbb{R}^k), \ X^1 = \mathcal{C}^1(\mathcal{I}, \mathbb{R}^m), \ X_D^1 = \mathcal{C}_D^1(\mathcal{I}, \mathbb{R}^m), \tag{3.29}$$

and

$$X = L^2(\mathcal{I}, \mathbb{R}^m), \ Y = L^2(\mathcal{I}, \mathbb{R}^k), \ X^1 = H^1(\mathcal{I}, \mathbb{R}^m), \ X_D^1 = H_D^1(\mathcal{I}, \mathbb{R}^m), \tag{3.30}$$

respectively, and for the closure $\mathcal{T} \in \mathcal{L}_c(X, Y \times \mathbb{R}^r)$, dom $\mathcal{T} = X_D^1$, of the operator $\mathring{\mathcal{T}} \in \mathcal{L}(X, Y \times \mathbb{R}^r)$ associated with the IVP (3.15):

1. The map $\mathcal{H} \in \mathcal{L}(X)$ defined by

$$\mathcal{H} p = Q_0 p + (I - Q_0 G_1^+ B) D^- U \int_{t_a} U(s)^{-1} D(s) p(s) \mathrm{d}s, \quad p \in X,$$

is bounded and

$$\operatorname{im} \mathcal{H} = \mathring{X}_D^1(\mathcal{I}, \mathbb{R}^m) := \{ x \in X_D^1(\mathcal{I}, \mathbb{R}^m) : D(t_a) x(t_a) = 0 \}.$$

2. $\|\mathcal{H} p\|_{X_D^1} \le C \|p\|_X$ for all $p \in X$.
3. If, additionally,

$$\xi \in \ker P_1 \subseteq X \text{ implies } W_1 B P_0 \mathcal{H} \xi = 0, \tag{3.31}$$

then $\ker \mathcal{T} = \mathcal{H}(\ker P_1)$ and $\mathcal{T}$ has a closed image, namely

$$\operatorname{im} \mathcal{T} = \{ (q, d) \in Y \times \mathbb{R}^r : W_1 q = W_1 B D^- (D \mathcal{H}(G_1^+ q) + U D(t_a) C^+ d) \}.$$

4. Supposed condition (3.31) is valid, for each arbitrary $(q, d) \in \operatorname{im} \mathcal{T}$, the solutions $x_* \in \operatorname{dom} \mathcal{T}$ of the equation $\mathcal{T} x = (q, d)$ have the form

$$x_* = (I - Q_0 G_1^+ B) D^- U D(t_a) C^+ d + \mathcal{H}(G_1^+ q + \xi),$$

with an arbitrary part $\xi \in \ker P_1$.

Proof 1. The boundedness of $\mathcal{H}$ is evident. We investigate the image of the map $\mathcal{H}$. $p \in X$ implies $\mathcal{H} p \in X$ and $D \mathcal{H} p = D D^- U \int_{t_a} U(s)^{-1} D(s) p(s) \mathrm{d}s \in X^1(\mathcal{I}, \mathbb{R}^n)$, and hence $\mathcal{H} p \in \mathring{X}_D^1(\mathcal{I}, \mathbb{R}^m)$.

Next we show that the equation $\mathcal{H}p = x$ is solvable for each arbitrary $x \in \mathring{X}_D^1(\mathcal{I}, \mathbb{R}^m)$. Let $x \in \mathring{X}_D^1(\mathcal{I}, \mathbb{R}^m)$ be given. We set $\tilde{p} = D^-D\tilde{p} + Q_0\tilde{p}$,

$$D\tilde{p} := (Dx)' - R'Dx + DG_1^+ BD^-Dx,$$

$$Q_0\tilde{p} := Q_0 x + Q_0 G_1^+ BD^- U \int_{t_a} U(s)^{-1} D(s)\tilde{p}(s)ds.$$

It follows that

$$Dx = U \int_{t_a} U(s)^{-1} D(s)\tilde{p}(s)ds,$$

$$\mathcal{H}\tilde{p} = Q_0\tilde{p} + (I - Q_0 G_1^+ B)D^- U \int_{t_a} U(s)^{-1} D(s)\tilde{p}(s)ds$$

$$= Q_0 x + D^- U \int_{t_a} U(s)^{-1} D(s)\tilde{p}(s)ds,$$

$$= Q_0 x + D^- Dx = x.$$

This proves that $\operatorname{im} \mathcal{H} = \mathring{X}_D^1(\mathcal{I}, \mathbb{R}^m)$.

2. Regarding the relation

$$(D\mathcal{H}p)' = (RU)' \int_{t_a} U(s)^{-1} D(s)p(s)ds + RUU^{-1}Dp, \quad p \in X,$$

the inequality follows immediately.

3. We decompose $Tx - q = 0$ into $G_1 G_1^+ (Tx - q) = 0$ and $W_1(Tx - q) = 0$, that is,

$$G_1(D^-(Dx)' + Q_0 x + G_1^+ BP_0 x - G_1^+ q) = 0, \quad W_1 BP_0 x - W_1 q = 0,$$

and, equivalently,

$$D^-(Dx)' + Q_0 x + G_1^+ BP_0 x - G_1^+ q = \xi \in \ker P_1, \quad W_1 q = W_1 BP_0 x. \tag{3.32}$$

For their part, the first equation decomposes into

$$DD^-(Dx)' + DG_1^+ BP_0 x - DG_1^+ q = D\xi,$$

$$Q_0 x + Q_0 G_1^+ BP_0 x - Q_0 G_1^+ q = Q_0 \xi.$$

We reformulate the last systems once more as

$$(Dx)' - R'Dx + DG_1^+ BD^- Dx = D(G_1^+ q + \xi), \qquad (3.33)$$

$$Q_0 x + Q_0 G_1^+ BD^- Dx = Q_0(G_1^+ q + \xi).$$

The initial condition $Cx(t_a) = d$ means $D(t_a)x(t_a) = D(t_a)C^+ d$. Together with (3.33), this yields the representations

$$Dx = UD(t_a)C^+ d + U \int_{t_a} U(s)^{-1} D(s)(G_1(s)^+ q(s) + \xi(s)) ds,$$

as well as

$$x = D^- Dx + Q_0 x = D^- Dx - Q_0 G_1^+ BD^- Dx + Q_0(G_1^+ q + \xi)$$

$$= (I - Q_0 G_1^+ B) D^- Dx + Q_0(G_1^+ q + \xi)$$

$$= \mathcal{H}(G_1^+ q + \xi) + (I - Q_0 G_1^+ B) D^- UD(t_a)C^+ d.$$

The second equation of (3.32) reformulates now as

$$W_1 q = W_1 BD^- Dx$$

$$= W_1 BD^- \{ UD(t_a)C^+ d + U \int_{t_a} U(s)^{-1} D(s)(G_1(s)^+ q(s) + \xi(s)) ds \}$$

$$= W_1 BD^- \{ UD(t_a)C^+ d + D\mathcal{H}(G_1^+ q + \xi) \}$$

$$= W_1 BD^- \{ UD(t_a)C^+ d + D\mathcal{H}(G_1^+ q) + D\mathcal{H}(\xi) \}.$$

Regarding condition (3.31) we find the relation

$$W_1 q = W_1 BD^- \{ UD(t_a)C^+ d + D\mathcal{H}(G_1^+ q) \}. \qquad (3.34)$$

Therefore, if $d \in \mathbb{R}^r$ and $q \in Y$ are given, and (3.34) is satisfied, then $\tilde{x} := \mathcal{H}(G_1^+ q + \xi) + (I - Q_0 G_1^+ B) D^- UD(t_a)C^+ d$, with arbitrary $\xi \in \ker P_1$ satisfies $T\tilde{x} = (q, d)$. This proves the statement.

4. If $x \in \operatorname{dom} T$ is given, $q := Tx$, $d := Cx(t_a)$, then condition (3.34) must be valid. The resulting set $\operatorname{im} \mathcal{T}$ is obviously closed in $Y \times \mathbb{R}^r$.

For $q = 0$, $d = 0$ we obtain the solution representation $x = \mathcal{H}\xi$. Therefore, the nullspace of $\mathcal{T}$ is formed by the functions $x = \mathcal{H}\xi$ with $\xi \in \ker P_1$.

$\square$

Condition (3.14) in Theorem 3.2 trivially implies both relations (3.28) and (3.31). In this case, $\operatorname{im} \mathcal{T}$ has a very simple structure, see Theorem 3.2.

Condition (3.18) Theorem 3.4 also implies both (3.28) and (3.31). Because of $D\xi = 0$ we have $\mathcal{H}\xi = Q_0\xi = \xi$, and hence $\ker \mathcal{T} = \ker P_1$.

In the following example, neither (3.14) nor (3.18) is satisfied, but (3.28) and (3.31) are valid. By this, Proposition 3.7 turns out to be a nontrivial generalization of both previous theorems.

Example 3.8 (Supplementary DAE for Proposition 3.7) The DAE with constant coefficients

$$E = \begin{bmatrix} 1 & 0 & 0 \\ 0 & 0 & 1 \\ 0 & 0 & 0 \end{bmatrix}, \quad F = \begin{bmatrix} 0 & 0 & 0 \\ 1 & 1 & 0 \\ 1 & 0 & 0 \end{bmatrix},$$

has (nonregular) tractability index 0, see [55]. We apply the factorization $E = AD$, $A = E$, $D = E^+E$ such that $D^- = D^+$. We have then

$$P_0 = \begin{bmatrix} 1 & 0 & 0 \\ 0 & 0 & 0 \\ 0 & 0 & 1 \end{bmatrix}, \quad G_1 = \begin{bmatrix} 1 & 0 & 0 \\ 0 & 1 & 1 \\ 0 & 0 & 0 \end{bmatrix}, \quad G_1^+ = \begin{bmatrix} 1 & 0 & 0 \\ 0 & \frac{1}{2} & 0 \\ 0 & \frac{1}{2} & 0 \end{bmatrix}, \quad P_1 = \begin{bmatrix} 1 & 0 & 0 \\ 0 & \frac{1}{2} & \frac{1}{2} \\ 0 & \frac{1}{2} & \frac{1}{2} \end{bmatrix},$$

further

$$W_1 = W_0 = \begin{bmatrix} 0 & 0 & 0 \\ 0 & 0 & 0 \\ 0 & 0 & 1 \end{bmatrix}, \quad W_1FP_0 = \begin{bmatrix} 0 & 0 & 0 \\ 0 & 0 & 0 \\ 1 & 0 & 0 \end{bmatrix} \neq 0, \quad W_1FP_0(I - P_1) = 0.$$

$\ker E$ does not include $\ker G_1$, and neither condition (3.14) nor (3.18) is satisfied, but condition (3.28) is given. Taking into account that

$$DG_1^+FD^- = \begin{bmatrix} 0 & 0 & 0 \\ 0 & 0 & 0 \\ \frac{1}{2} & 0 & 0 \end{bmatrix}, \quad U(t) = \begin{bmatrix} 1 & 0 & 0 \\ 0 & 1 & 0 \\ -\frac{1}{2}(t - t_a) & 0 & 1 \end{bmatrix},$$

we verify that $W_1FP_0\mathcal{H}\xi = 0$ for all functions $\xi \in \ker P_1$, thus condition (3.31) is valid in fact. $\square$

3.2 Settings with Bounded Differential-Algebraic Operators

The available theory of bounded operators in Banach spaces and Hilbert spaces is rich in content. Bounded operators are favorable in many situations; for instance, when aiming to investigate smooth nonlinear problems (cf. Sect. 4) and when looking for LSS (cf. Theorem 6.5).

Fixing a proper factorization $E = AD$ according to Theorem 3.1(3), we turn from the standard form DAE (3.1) to the DAE with properly stated leading term

$$A(t)(Dx)'(t) + B(t)x(t) = q(t), \quad t \in \mathcal{I} = [t_a, t_e], \tag{3.35}$$

whereby $A \in \mathcal{C}(\mathcal{I}, \mathcal{L}(\mathbb{R}^n, \mathbb{R}^k))$, $D \in \mathcal{C}^1(\mathcal{I}, \mathcal{L}(\mathbb{R}^m, \mathbb{R}^n))$, $B \in \mathcal{C}(\mathcal{I}, \mathcal{L}(\mathbb{R}^m, \mathbb{R}^k))$, and $\ker A$, $\operatorname{im} D$ are $\mathcal{C}^1$-subspaces satisfying the transversality condition (3.4).

The function spaces $\mathcal{C}_D^1(\mathcal{I}, \mathbb{R}^m)$ and $H_D^1(\mathcal{I}, \mathbb{R}^m)$ endowed with the norm $\|\cdot\|_{\mathcal{C}_D^1}$ and the inner product $(\cdot, \cdot)_{H_D^1}$ become a Banach space and a Hilbert space, respectively, see Lemma 6.9.

The differential-algebraic operator $T \in \mathcal{L}(\mathcal{C}_D^1(\mathcal{I}, \mathbb{R}^m), \mathcal{C}(\mathcal{I}, \mathbb{R}^m))$ defined by

$$Tx = A(Dx)' + Bx, \quad x \in \mathcal{C}_D^1(\mathcal{I}, \mathbb{R}^m), \tag{3.36}$$

is bounded. We continue using the denotations (3.7)–(3.9).

$$G_0 = AD, \ P_0 = G_0^+ G_0, \ Q_0 = I - P_0, \ W_0 = I - G_0 G_0^+,$$
$$G_1 = G_0 + BQ_0, \ W_1 = I - G_1 G_1^+.$$

The next theorem states sufficient conditions for T to be normally solvable in terms of the original data. It can be seen as natural counterpart of Theorem 3.2(1).

Theorem 3.9 *Assume $A \in \mathcal{C}(\mathcal{I}, \mathcal{L}(\mathbb{R}^n, \mathbb{R}^k))$, $D \in \mathcal{C}^1(\mathcal{I}, \mathcal{L}(\mathbb{R}^m, \mathbb{R}^n))$, $\ker A$ is a $\mathcal{C}^1$-subspace, and $B \in \mathcal{C}(\mathcal{I}, \mathcal{L}(\mathbb{R}^m, \mathbb{R}^k))$, and let the transversality condition (3.4) be given.*

Let the matrix function $W_0 B Q_0$ have constant rank on the interval $\mathcal{I}$ and let the condition

$$W_1 B P_0 = 0 \tag{3.37}$$

be satisfied. Then the operator T defined by (3.36) is bounded and normally solvable, and

$$\operatorname{im} T = \ker W_1 \subseteq \mathcal{C}(\mathcal{I}, \mathbb{R}^k). \tag{3.38}$$

Proof The arguments used in the proof of Theorem 3.2(1) apply again.   □

The particular case if G_1 has pointwise full row-rank is quite important for minimization with differential-algebraic constraints, see [61]. Then, it holds that $W_1 = 0$ and the corresponding differential-algebraic operator is surjective.

Strangeness-free DAEs yield a special class of bounded and normally solvable operators T, see Example 3.3.

Analogously, the operator $T \in \mathcal{L}(H_D^1(\mathcal{I}, \mathbb{R}^m), L^2(\mathcal{I}, \mathbb{R}^m))$ defined by

$$Tx = A(Dx)' + Bx, \quad x \in H_D^1(\mathcal{I}, \mathbb{R}^m), \tag{3.39}$$

is bounded. The first statement of the next theorem represents the counterpart of Theorem 3.2(2).

Theorem 3.10 *Let the assumptions of Theorem 3.9 be satisfied.*

1. Then the operator T defined by (3.39) is bounded and normally solvable, with

$$\text{im } T = \ker W_1 \subseteq L^2(\mathcal{I}, \mathbb{R}^k). \tag{3.40}$$

2. For each $q \in L^2(\mathcal{I}, \mathbb{R}^k)$,

$$\inf\{\|Tx - q\|_{H_D^1} : x \in H_D^1\} \tag{3.41}$$

is attained, and the orthogonal generalized (Moore–Penrose) inverse T^+ is bounded.

Proof The arguments used in the proof of Theorem 3.2(2) apply again, and verify part (1). Part (2) is then a consequence of Theorem 6.5. □

Naturally, also Theorem 3.4 and Proposition 3.7 can be adapted to settings with bounded differential-algebraic operators.

3.3 Least-Squares Solutions of IVPs and Bounded Generalized Inverses of the Composed Operator

Let X and Y be Hilbert spaces and $K \in \mathcal{L}(X, Y)$ be a bounded or closed and densely defined operator. Then the equation $Kx = y$ possesses a LSS $x_* \in \text{dom } K$ exactly if $y \in \text{im } K + (\text{im } K)^\perp$. Denote by $LSS_y \subseteq \text{dom } K$ the set of all LSS corresponding to y. Since the nullspace of K is closed, the set LSS_y is either empty or linear affine and closed. Therefore, for each $y \in \text{im } K + (\text{im } K)^\perp$, there exists a unique minimum-norm LSS or pseudo solution. So far so good! However, when computing LSS in practice, one is confronted with the question whether a small residuum $\|K\tilde{x} - y\|$ ensures that $\tilde{x}$ is actually closed to a LSS or even to the pseudo solution. Unfortunately, for differential-algebraic operators in standard settings, even if y is consistent and the pseudo solution x_* exists, a minimizing sequence $\{x_l\}$ such that $\delta_l = \|Kx_l - y\| \xrightarrow{l \to \infty} 0$ does not necessarily converge; instead, $\|x_l - x_*\|$ may grow unboundedly, see Example 2.3 and its continuation by Example 2.5. This essentially ill-posed behavior is caused by the nonclosedness of the image of the operator. *The calculation of a LSS and the pseudo solution is practically safe only for operators with closed image*, which means, for normally solvable operators. Otherwise regularization techniques should be applied, see Sect. 5.

We continue investigating the linear IVP (3.15), that is,

$$E(t)x'(t) + F(t)x(t) = q(t), \ t \in \mathcal{I}, \quad Cx(t_a) = d.$$

The coefficients E and F are continuous, $\ker E$ is a $\mathcal{C}^1$-subspace, and E has constant rank $r > 0$. The matrix $C \in \mathcal{L}(\mathbb{R}^m, \mathbb{R}^r)$ has full row-rank r, and condition (3.16) is valid, that is,

$$\ker C = \ker E(t_a).$$

We choose now the Hilbert spaces

$$X = L^2(\mathcal{I}, \mathbb{R}^m), \quad Y = L^2(\mathcal{I}, \mathbb{R}^k) \tag{3.42}$$

as pre-image and image spaces. The differential-algebraic operator $\overset{\circ}{T} \in \mathcal{L}(X, Y)$ as well as its composed associate $\overset{\circ}{\mathcal{T}} \in \mathcal{L}(X, Y \times \mathbb{R}^r)$ defined by

$$\overset{\circ}{T}x := Ex' + Fx, \ x \in \operatorname{dom} \overset{\circ}{T} = \mathcal{C}^1(\mathcal{I}, \mathbb{R}^m) \subseteq X,$$

$$\overset{\circ}{\mathcal{T}}x := (\overset{\circ}{T}x, Cx(t_a)), \ x \in \operatorname{dom} \overset{\circ}{\mathcal{T}} = \overset{\circ}{T},$$

is unbounded but closable. We immediately turn to their closures (cf. Sect. 3.1) given by $T \in \mathcal{L}_c(X, Y)$ and $\mathcal{T} \in \mathcal{L}_c(X, Y \times \mathbb{R}^r)$,

$$Tx := A(Dx)' + Bx, \ x \in \operatorname{dom} T = H_D^1(\mathcal{I}, \mathbb{R}^m) \subseteq X,$$

$$\mathcal{T}x := (Tx, Cx(t_a)), \ x \in \operatorname{dom} \mathcal{T} = \operatorname{dom} T,$$

whereby we apply a proper factorization $E = AD$ being consistent with Theorem 3.1. Owing to the property $\ker C = \ker E(t_a) = \ker D(t_a)$ the expression $Cx(t_a) = CD(t_a)^- D(t_a)x(t_a)$ makes sense for each arbitrary $x \in H_D^1(\mathcal{I}, \mathbb{R}^m)$.

We continue using the notations introduced in Sect. 3.1.

Definition 3.1 For given $q \in Y$, $d \in \mathbb{R}^r$, the function $x_* \in H_D^1(\mathcal{I}, \mathbb{R}^m)$ is called LSS of the IVP (3.15), if it represents a LSS of the operator equation $\mathcal{T}x = (q, d)$, that means

$$\|Tx_* - q\|_{L^2}^2 + |Cx_*(t_a) - d|^2 \tag{3.43}$$

$$= \inf\{\|Tx - q\|_{L^2}^2 + |Cx(t_a) - d|^2 : x \in H_D^1(\mathcal{I}, \mathbb{R}^m)\}.$$

A LSS x_* is called a pseudo solution of the IVP (3.15), if the inequality $\|x_*\|_{L^2} \leq \|\tilde{x}_*\|_{L^2}$ is valid for all further LSS $\tilde{x}_*$.

Proposition 3.7 ensures the normal solvability of the operator $\mathcal{T} \in \mathcal{L}_c(X, Y \times \mathbb{R}^r)$. In particular, since $\ker \mathcal{T}$ and $\operatorname{im} \mathcal{T}$ are closed, the orthogonal direct sum decompositions

$$X = \ker \mathcal{T} \oplus (\ker \mathcal{T})^{\perp}, \quad Y \times \mathbb{R}^r = \operatorname{im} \mathcal{T} \oplus (\operatorname{im} \mathcal{T})^{\perp}$$

are valid, and there exist symmetric bounded projectors $\mathfrak{P}$ and $\mathfrak{R}$ acting on X and $Y \times \mathbb{R}^r$, respectively, such that

$$\operatorname{im} \mathfrak{P} = (\ker \mathcal{T})^{\perp}, \quad \operatorname{im} \mathfrak{R} = \operatorname{im} \mathcal{T}.$$

For each arbitrary $(q, d) \in Y \times \mathbb{R}^r$, Proposition 3.7(4) provides the solutions x_* of the equation $\mathcal{T}x = \mathfrak{R}(q, d) =: (q_{\mathfrak{R}}, d_{\mathfrak{R}})$ as

$$x_* = (I - Q_0 G_1^+ B) D^- UD(t_a) C^+ d_{\mathfrak{R}} + \mathcal{H}(G_1^+ q_{\mathfrak{R}} + \xi), \tag{3.44}$$

whereby the component $\xi \in \ker P_1$ can be chosen arbitrarily. Since the inequality

$$\|\mathcal{T}x - (q, d)\|^2 = \|\mathcal{T}x - \mathfrak{R}(q, d)\|^2 + \|(I - \mathfrak{R})(q, d)\|^2$$
$$\geq \|(I - \mathfrak{R})(q, d)\|^2 = \|\mathcal{T}x_* - (q, d)\|^2$$

is valid for all $x \in \operatorname{dom} \mathcal{T}$, each of those x_* is a LSS of the given IVP (3.15).

By omitting the free component $\xi \in \ker P_1$, a *special bounded generalized inverse* $\mathcal{T}^-$ of $\mathcal{T}$ results, as the following theorem based on Proposition 3.7 shows.

Theorem 3.11 *Let E and F be continuous, $\ker E$ be a C^1-subspace, and let the matrix function $W_0 F Q_0$ have constant rank. Let the matrix $C \in \mathcal{L}(\mathbb{R}^m, \mathbb{R}^r)$ have full row-rank $r = \operatorname{rank} E$, and satisfy the condition $\ker C = \ker E(t_a)$.*
Additionally, assume that $W_1 F P_0 (I - P_1) = 0$ and

$$W_1 B P_0 \mathcal{H} \xi = 0 \text{ for } \xi \in \ker P_1 \subseteq X.$$

1. *Then the map $\mathcal{T}^- \in \mathcal{L}(Y \times \mathbb{R}^r, X)$ defined by*

$$\mathcal{T}^-(q, d) := (I - Q_0 G_1^+ B) D^- UD(t_a) C^+ d_{\mathfrak{R}} + \mathcal{H}(G_1^+ q_{\mathfrak{R}}),$$
$$\text{for } (q, d) \in Y \times \mathbb{R}^r, \quad (q_{\mathfrak{R}}, d_{\mathfrak{R}}) := \mathfrak{R}(q, d).$$

   is a bounded generalized inverse of the operator $\mathcal{T}$, and $\mathcal{T}\mathcal{T}^- = \mathfrak{R}$.
2. *If even $W_1 F P_0 = 0$, then it results that $\operatorname{im} \mathcal{T} = \ker W_1 \times \mathbb{R}^r$, $\mathfrak{R} = \operatorname{diag}(I - W_1, I)$, and*

$$\mathcal{T}^-(q, d) := (I - Q_0 G_1^+ B) D^- UD(t_a) C^+ d + \mathcal{H}(G_1^+ q),$$
$$\text{for } (q, d) \in Y \times \mathbb{R}^r.$$

Proof 1. The map $\mathcal{T}^-$ is bounded by construction.

For each arbitrary $(q, d) \in Y \times \mathbb{R}^r$, the value $x_{(q,d)} := \mathcal{T}^-(q, d)$ fulfills the relation $\mathcal{T}x_{(q,d)} = \mathfrak{R}(q, d)$, and hence $\mathcal{T}\mathcal{T}^-(q, d) = \mathfrak{R}(q, d)$ as well as $\mathcal{T}^-\mathcal{T}\mathcal{T}^-(q, d) = \mathcal{T}^-\mathfrak{R}(q, d) = \mathcal{T}^-(q, d)$.

For each arbitrary $x \in \operatorname{dom} T$ and $q_x := Tx$, $d_x := Cx(t_a)$, it holds that $(q_x, d_x) = \mathfrak{R}(q_x, d_x)$, thus $\mathcal{T}x = (q_x, d_x) = \mathcal{T}\mathcal{T}^-(q_x, d_x) = \mathcal{T}\mathcal{T}^-\mathcal{T}x$.

2. is a direct consequence of Theorem 3.2.

$\square$

The second statement applies in particular to the class of strangeness-free DAEs, see Example 3.3.

One can compute the value $\mathcal{T}^-(q, d)$ by solving first the standard IVP

$$u' = R'u - DG_1^+BD^-u + DG_1^+q_\mathfrak{R}, \quad u(t_a) = D(t_a)C^+d_\mathfrak{R}, \tag{3.45}$$

and letting

$$\mathcal{T}^-(q, d) = (I - Q_0G_1^+B)D^-u + Q_0G_1^+q_\mathfrak{R}. \tag{3.46}$$

Here, we mention that, for regular DAEs with tractability index 0 and 1, the operator $\mathcal{T}$ is invertible and $\mathfrak{R} = I$. Then, of course, $\mathcal{T}^-$ and $\mathcal{T}^{-1}$ coincide. In the index-0 case, it results that $x = u$, and the IVP (3.45) reads simply $u' = -E^{-1}Fu + E^{-1}q$, $u(t_a) = d$.

In general, $\mathcal{T}^-(q, d)$ is a special LSS, but it is not necessarily a pseudo solution. For reaching a pseudo solution, an extra minimization has to be carried out among the LSS. Nevertheless, the Theorem 3.4 allows a direct choice of the pseudo solution. Namely, owing to the assumptions of Theorem 3.4 the map $\mathcal{H}$ simplifies so that

$$\mathcal{H}\xi = Q_0\xi = \xi \quad \text{for } \xi \in \ker P_1,$$

and we have further

$$\ker \mathcal{T} = \ker P_1, \quad \mathfrak{P} = P_1.$$

Then the minimum norm LSS x_{**} is attained by applying (cf. Theorem 3.4)

$$\xi_* = -Q_1(I - Q_0G_1^+B)D^-U(D(t_a)C^+d_\mathfrak{R} - Q_1\mathcal{H}(G_1^+q_\mathfrak{R})$$

with $Q_1 := I - P_1$, yielding

$$x_{**} = P_1(I - Q_0G_1^+B)D^-UD(t_a)C^+d_\mathfrak{R} + P_1\mathcal{H}(G_1^+q_\mathfrak{R}).$$

Equivalently, we can compute the solution u of the IVP (3.45) and set

$$\xi_* = -Q_1(I - Q_0 G_1^+ B)D^- u - Q_1 Q_0 G_1^+ q_{\mathfrak{R}}.$$

Again, the resulting

$$x_{**} = P_1(I - Q_0 G_1^+ B)D^- u - P_1 Q_0 G_1^+ q_{\mathfrak{R}}$$

is the pseudo solution of the original IVP (3.15). Namely, Q_1, P_1 are complementary orthoprojectors and the part $P_1 x_*$ of each LSS x_* (cf. (3.44)) is completely independent of ξ. The choice $\xi = \xi_*$ leads to $Q_1 x_{**} = 0$, thus

$$\|x_{**}\|_{L^2}^2 = \|\mathfrak{P} x_{**}\|_{L^2}^2 \le (\|\mathfrak{P} x_{**}\|_{L^2}^2 + \|(I - \mathfrak{P})x_*\|_{L^2}^2) = \|x_*\|_{L^2}^2$$

is valid for all LSS x_*.

If, moreover, $G_1 = G_0$, then the DAE is tractable with index 0, and the pseudo solution is simply

$$x_{**} = D^- u.$$

Example 3.12 (Pseudo solution) The DAE with the coefficients

$$E(t) = \begin{bmatrix} -t & t^2 \\ -1 & t \end{bmatrix}, \quad F(t) = \begin{bmatrix} 1 & 0 \\ 0 & 1 \end{bmatrix},$$

is tractable with index 0. The associated closed differential-algebraic operator has the coefficients

$$A(t) = \begin{bmatrix} t \\ 1 \end{bmatrix}, D(t) = \begin{bmatrix} -1 & t \end{bmatrix}, B(t) = \begin{bmatrix} 1 & -t \\ 0 & 0 \end{bmatrix}, D(t)^- = \frac{1}{1+t^2} \begin{bmatrix} -1 \\ t \end{bmatrix}.$$

Put $t_a = 0$ and $C = [1\ 0]$. The related initial condition reads

$$Cx(0) = x_1(0) = d.$$

Compute $G_0 = AD = E$, $\ker B = \ker G_0$, thus $BQ_0 = 0$, $G_1 = G_0$. The DAE is tractable with index 0, and Corollary 3.5 applies.
Compute further

$$Q_0(t) = \frac{1}{1+t^2} \begin{bmatrix} t^2 & t \\ t & 1 \end{bmatrix}, W_0(t) = \frac{1}{1+t^2} \begin{bmatrix} 1 & -t \\ -t & t^2 \end{bmatrix}, G_1(t)^+ = \frac{1}{(1+t^2)^2} \begin{bmatrix} -t & -1 \\ t^2 & t \end{bmatrix},$$

$R = 1$, $P_1 = P_0 = W_0$, $DG_1^+ BD^- = -\frac{t}{1+t^2}$.
The solution of the IVP $U' + DG_1^+ BD^- U = 0, U(0) = 1$ is $U(t) = (1+t^2)^{\frac{1}{2}}$.

(q, d) belongs to im $\mathcal{T}$, exactly if the consistency condition

$$W_0 q = \frac{1}{(1 + t^2)^{\frac{1}{2}}} \begin{bmatrix} -1 \\ t \end{bmatrix} \{-d + \int_0^t \frac{1}{(1 + s^2)^{\frac{3}{2}}} (sq_1(s) + q_2(s)) \mathrm{d}s\}. \qquad (3.47)$$

is valid.

For consistent (q, d) we derive the pseudo solution

$$x_{**} = D^- u = \frac{1}{(1 + t^2)^{\frac{1}{2}}} \begin{bmatrix} -1 \\ t \end{bmatrix} \{-d + \int_0^t \frac{1}{(1 + s^2)^{\frac{3}{2}}} (sq_1(s) + q_2(s)) \mathrm{d}s\}.$$

Regarding the consistency condition (3.47) we find the further expression

$$x_{**}(t) = W_0(t)q(t) = \frac{1}{(1 + t^2)} \begin{bmatrix} 1 \\ -t \end{bmatrix} (q_1(t) - tq_2(t)).$$

In contrast, in [50], this DAE has strangeness index 1, and hence, it is first reduced to strangeness-free form via the derivative array approach (which needs derivatives also of q) and not till then the least-squares calculus is applied. Thereby, operator images are not at all considered. The function

$$x_{\circledast}(t) = \frac{1}{(1 + t^2)} \begin{bmatrix} 1 \\ -t \end{bmatrix} (q_1(t) - tq_2(t) - x_{01} + tx_{02}) + x_0$$

is offered as the pseudo solution of the IVP

$$E(t)x'(t) + F(t)x(t) = q(t), \quad x(0) = x_0.$$

Observe that $x_{\circledast}$ formally coincides with x_{**} for $x_0 = 0$. $\qquad\qquad\square$

3.4 Nonlinear Differential-Algebraic Operators with Normally Solvable Linearizations

There are rich resources in the functional-analytic literature concerning implicit function theorems, the Kantorovich-type analysis for Newton-like iteration methods, and related topics for operator equations and least-squares problems in Banach spaces, e.g., [19, 24, 63, 64]. Usually those investigations rely on Fréchet-differentiable operators. A Fréchet-derivative is linear and bounded by definition (cf. Appendix 6.1.3).

The standard form DAE (1.2), that is,

$$\mathfrak{f}(x'(t), x(t), t) = 0 \qquad\qquad (3.48)$$

is described by the continuous function $\mathfrak{f} : \mathbb{R}^m \times \mathcal{D}_\mathfrak{f} \times \mathcal{I}_\mathfrak{f} \to \mathbb{R}^k$ which has continuous partial derivatives $\mathfrak{f}_{x^1}$ and $\mathfrak{f}_x$. The set $\mathcal{D}_\mathfrak{f} \times \mathcal{I}_\mathfrak{f} \subseteq \mathbb{R}^m \times \mathbb{R}$ is open. Suppose the nullspace $\ker \mathfrak{f}_{x^1}$ to be a $\mathcal{C}^1$-subspace. We associate with the DAE (3.48) the nonlinear differential-algebraic operator

$$\mathring{F} : \operatorname{dom} \mathring{F} \subseteq X \to Y,$$

$$\operatorname{dom} \mathring{F} := \{x \in \mathcal{C}^1(\mathcal{I}, \mathbb{R}^m) : x(t) \in \mathcal{D}, \ t \in \mathcal{I}\},$$

$$(\mathring{F}x)(t) := \mathfrak{f}(x'(t), x(t), t), \quad t \in \mathcal{I}, \quad x \in \operatorname{dom} \mathring{F}, \tag{3.49}$$

whereby $\mathcal{D} \subseteq \mathcal{D}_\mathfrak{f}$ is an open, connected set and $\mathcal{I} = [t_a, t_e] \subset \mathcal{I}_\mathfrak{f}$ is a compact interval.

We put $Y = \mathcal{C}(\mathcal{I}, \mathbb{R}^k)$ and look for an appropriate Banach space X to serve as the pre-image space. The operator equation $\mathring{F}x = 0$ represents the DAE (3.48) in the classical sense.

As pointed out in Sect. 2.3.1, it is not reasonable to choose $\mathcal{C}^1(\mathcal{I}, \mathbb{R}^m)$ as the pre-image space. We momentarily attempt $X = \mathcal{C}(\mathcal{I}, \mathbb{R}^m)$. Then, for each fixed $x_* \in \operatorname{dom} \mathring{F}$ and arbitrary $x \in \mathcal{C}^1(\mathcal{I}, \mathbb{R}^m)$, the directional derivative

$$\lim_{\alpha \to 0} \frac{1}{\alpha}(\mathring{F}(x_* + \alpha x) - \mathring{F}x_*) = E_* x' + F_* x =: \mathring{F}'(x_*)x,$$

is well-defined and linear, with

$$E_*(t) := \mathfrak{f}_{x^1}(x'_*(t), x_*(t), t), \ F_*(t) := \mathfrak{f}_x(x'_*(t), x_*(t), t), \ t \in \mathcal{I}.$$

The linear operator equation $\mathring{F}'(x_*)x = q$ represents the linearized DAE

$$E_* x' + F_* x = q. \tag{3.50}$$

In this setting, the operator $\mathring{F}'(x_*)$ is closely defined, but, unfortunately, it is unbounded. By means of the proper factorization (cf. Theorem 3.1) $E_* = A_* D_*$, $A_* := E_*$, $D_* := E_*^+ E_*$, we obtain the closure of $\mathring{F}'(x_*)$,

$$F'(x_*)x = A_*(D_* x)' + B_* x, \quad x \in \operatorname{dom} F'(x_*),$$

$$\operatorname{dom} F'(x_*) = \{x \in \mathcal{C}(\mathcal{I}, \mathbb{R}^m) : D_* x \in \mathcal{C}^1(\mathcal{I}, \mathbb{R}^m)\},$$

with $B_* := F_* - A_* D'_*$. The closure $F'(x_*)$ has a definition domain individually for each x_*. This configuration does not meet the usual requirements of the functional-analytic procedures.

The following *structural restriction* of the DAE makes the situation much more comfortable: it ensures a common domain for the derivatives $F'(x_*)$ and allows then to turn to bounded derivatives in a Banach space setting. Fortunately, most applications meet this structural restriction.

Let the *nullspace of the leading Jacobian* $\mathfrak{f}_{x^1}(x^1, x, t)$ *be independent of* x^1 *and* x, such that

$$\ker \mathfrak{f}_{x^1}(x^1, x, t) =: N_0(t), \quad (x^1, x, t) \in \mathbb{R}^m \times \mathcal{D} \times \mathcal{I}. \tag{3.51}$$

Let the matrix $D(t) \in \mathcal{L}(\mathbb{R}^m)$ represent the orthoprojector along $N_0(t)$, $t \in \mathcal{I}$. The matrix function D is continuously differentiable since N_0 is a $\mathcal{C}^1$-subspace. The identity

$$\mathfrak{f}(x^1, x, t) \equiv \mathfrak{f}(D(t)x^1, x, t) \tag{3.52}$$

follows from

$$\mathfrak{f}(x^1, x, t) - \mathfrak{f}(D(t)x^1, x, t) = \int_a^1 \mathfrak{f}_{x^1}(sx^1 + (1-s)D(t)x^1, x, t)(I - D(t))x^1 ds = 0.$$

It results that $D_* = E_*^+ E_* = D$ uniformly for all $x_* \in \operatorname{dom} \overset{\circ}{F}$. Moreover, we have

$$(\overset{\circ}{F}x)(t) = \mathfrak{f}(x'(t), x(t), t) = \mathfrak{f}(D(t)x'(t), x(t), t)$$

$$= \mathfrak{f}((Dx)'(t) - D'(t)x(t), x(t), t), \quad t \in \mathcal{I}, \quad x \in \operatorname{dom} \overset{\circ}{F},$$

which suggests to turn to the extension F of $\overset{\circ}{F}$,

$$(Fx)(t) := \mathfrak{f}((Dx)'(t) - D'(t)x(t), x(t), t), \quad t \in \mathcal{I}, \quad x \in \operatorname{dom} F. \tag{3.53}$$

$$\operatorname{dom} F := \{x \in \mathcal{C}_D^1(\mathcal{I}, \mathbb{R}^m) : x(t) \in \mathcal{D}, \ t \in \mathcal{I}\}.$$

Motivated also by the experience in the previous sections, we *apply now the advanced setting*

$$F : \operatorname{dom} F \subseteq \mathcal{C}_D^1(\mathcal{I}, \mathbb{R}^m) \to \mathcal{C}(\mathcal{I}, \mathbb{R}^k) \tag{3.54}$$

for the extended nonlinear differential-algebraic operator F determined by (3.53). In this advanced setting, F is Fréchet differentiable,

$$F'(x_*)x = A_*(Dx)' + B_*x, \quad x \in \mathcal{C}_D^1(\mathcal{I}, \mathbb{R}^m), \quad x_* \in \operatorname{dom} F,$$

$$A_*(t) := \mathfrak{f}_{x^1}((Dx_*)'(t) - D'(t)x_*(t), x_*(t), t),$$

$$B_*(t) := \mathfrak{f}_x((Dx_*)'(t) - D'(t)x_*(t), x_*(t), t) - A_*(t)D'(t), \quad t \in \mathcal{I}.$$

The operator equation $Fx = 0$ represents the DAE (3.48) still in the classical sense. Together with the differential-algebraic operator F we investigate the composed operator

$$\mathcal{F} : \text{dom } F \subset C_D^1(\mathcal{I}, \mathbb{R}^m) \to C(\mathcal{I}, \mathbb{R}^k) \times \mathbb{R}^r,$$

$$\mathcal{F}x := (Fx, Cx(t_a)), \quad x \in \text{dom } F.$$

which represents the IVP. The $m \times r$ matrix C will be specified later on.

Next we define the matrix functions

$$G_0(x^1, x, t) := \mathfrak{f}_{x^1}(x^1, x, t) = \mathfrak{f}_{x^1}(x^1, x, t)D(t),$$

$$B_0(x^1, x, t) := \mathfrak{f}_x(x^1, x, t) - \mathfrak{f}_{x^1}(x^1, x, t)D'(t),$$

$$G_1(x^1, x, t) := G_0(x^1, x, t) + B_0(x^1, x, t)Q_0(t),$$

pointwise for $x^1 \in \mathbb{R}^m, x \in \mathcal{D}, t \in \mathcal{I}$ and introduce the projector-valued functions

$$P_0 := D, \quad Q_0 := I - P_0, \quad W_0 := I - G_0 G_0^+, \quad W_1 := I - G_1 G_1^+.$$

For a given reference function $x_* \in \text{dom } F$ we abbreviate

$$G_{*0}(t) := G_0((Dx_*)'(t) - D'(t)x_*(t), x_*(t), t) = A_*(t)D(t),$$

$$G_{*1}(t) := G_1((Dx_*)'(t) - D'(t)x_*(t), x_*(t), t), \quad t \in \mathcal{I},$$

and so on.

Now we are prepared to characterize nonlinear differential-algebraic operators via their derivatives. For instance, the following statement is a nonlinear counterpart of Theorem 3.9.

Theorem 3.13 *Let the function $\mathfrak{f}$ be continuous, with continuous partial derivatives $\mathfrak{f}_{x^1}$ and $\mathfrak{f}_x$. Let the nullspace of $\mathfrak{f}_{x^1}$ be a C^1-subspace who varies with t only, and let the matrix function $W_0 B_0 Q_0$ have constant rank.*

If, additionally,

$$W_1 B_0 P_0 = W_1 \mathfrak{f}_x P_0 = 0, \tag{3.55}$$

then the derivative $F'(x_*)$ of the differential-algebraic operator F is normally solvable for each arbitrary $x_* \in \text{dom } F$ and

$$\text{im } F'(x_*) = \{x \in C(\mathcal{I}, \mathbb{R}^m) : W_{*1}x = 0\}. \tag{3.56}$$

Proof For each arbitrary reference function $x_* \in \text{dom } F$ we obtain the linearization

$$F'(x_*)x = A_*(Dx)' + B_*x, \quad x \in C_D^1(\mathcal{I}, \mathbb{R}^m), \tag{3.57}$$

which satisfies all conditions of Theorem 3.9 (cf. also (3.7), (3.8), (3.9)). In detail, condition (3.55) implies relation (3.37), that is, for $t \in \mathcal{I}$,

$$W_{*1}(t) B_*(t) P_0(t) = (W_1 B_0)((D x_*)'(t) - D'(t) x_*(t), x_*(t), t) P_0(t) = 0.$$

□

Example 3.14 (Strangeness-Free Reduced DAEs) A particular instance of normally solvable nonlinear differential-algebraic operators is given by the so-called strangeness-free reduced DAEs in [51],

$$x_1'(t) + \mathcal{L}(x_1(t), x_2(t), x_3(t), t) = 0,$$
$$x_2(t) + \mathcal{R}(x_1(t), x_3(t), t) = 0,$$

which represent over-determined DAEs with free component x_3. One has simply

$$G_0 = \begin{bmatrix} I & 0 & 0 \\ 0 & 0 & 0 \end{bmatrix}, \quad W_0 B_0 Q_0 = \begin{bmatrix} 0 & 0 & 0 \\ 0 & I & * \end{bmatrix}, \quad G_1 = \begin{bmatrix} I & * & * \\ 0 & I & * \end{bmatrix}, \quad W_1 = 0.$$

□

Theorem 3.15 *Let the function $\mathfrak{f}$ be continuous, with continuous partial derivatives $\mathfrak{f}_{x^1}$ and $\mathfrak{f}_x$. Let the nullspace of $\mathfrak{f}_{x^1}$ be a C^1-subspace who varies with t only and let the matrix function $W_0 B_0 Q_0$ have constant rank.*

Let the matrix $C \in \mathcal{L}(\mathbb{R}^m, \mathbb{R}^r)$ have full row-rank $r := \operatorname{rank} D(t_a)$ and let $\ker C = \ker D(t_a)$. If, additionally,

$$\ker G_1 \subseteq \ker G_0. \tag{3.58}$$

then the derivative $\mathcal{F}'(x_*)$ of the composed operator $\mathcal{F}$ is normally solvable for each arbitrary $x_* \in \operatorname{dom} F$ and

$$\operatorname{im} \mathcal{F}'(x_*) = \{(q, d) \in \mathcal{C}(\mathcal{I}, \mathbb{R}^k) \times \mathbb{R}^r : \tag{3.59}$$

$$W_{*1} q = W_{*1} B_* D^+ U_* \left(D(t_a) C^+ d + \int_{t_a} U_*(s)^{-1} D(s) G_{*1}^+ q(s) ds \right) \},$$

whereby U_* is the fundamental solution matrix uniquely determined by

$$U' - P_0' U + D G_{*1}^+ B_* D^+ U = 0, \quad U(t_a) = I.$$

Further, it holds that

$$\ker \mathcal{F}'(x_*) = \{ x \in \mathcal{C}_D^1(\mathcal{I}, \mathbb{R}^m) : G_{*1} x = 0 \},$$

and $\mathcal{F}'(x_*)$ is injective, if the matrix function G_1 shows full column-rank.

Moreover, if $(q, d) \in \operatorname{im} \mathcal{F}'(x_*)$, *then there is a unique* $z_* \in \mathcal{C}_D^1(\mathcal{I}, \mathbb{R}^m)$ *such that*

$$\|z_*\|_{L^2} = \min\{\|z\|_{L^2} : z \in \mathcal{C}_D^1(\mathcal{I}, \mathbb{R}^m), \mathcal{F}'(x_*)z = (q, d)\}.$$

Proof For each arbitrary reference function $x_* \in \operatorname{dom} F$ we obtain the linearization

$$\mathcal{F}'(x_*)x = (F(x_*)x, Cx(t_a)) = (A_*(Dx)' + B_*x, Cx(t_a)), \quad x \in \mathcal{C}_D^1(\mathcal{I}, \mathbb{R}^m).$$

Theorem 3.4 can be adapted to a setting with bounded operators analogously as Theorem 3.9 has been obtained from Theorem 3.2. Our linearization satisfies all conditions of the adapted version of Theorem 3.2 which proves the statements here. Note that here the special choice of D leads to $R = P_0$ and $D^- = D^+$. $\qquad\Box$

Regular index-0 DAEs are characterized by an equal number of equations and unknowns, $m = k$, and a nonsingular matrix function G_0, so that $W_1 = W_0 = 0$. Regular index-1 DAEs are given, if $m = k$, the matrix function G_0 is singular, but has constant rank, and the matrix function G_1 is nonsingular (e.g. [32, 55]). Obviously, Theorem 3.15 applies in both instances.

General under-determined index-0 DAEs are characterized by $m > k$ and a matrix function G_0 that has full row-rank k. General under-determined index-1 DAEs are characterized by $m > k$, a matrix function G_0 that has constant rank smaller than k, and a matrix function G_1 with full row-rank k, see [55]. This leads also to $W_1 = 0$, and Theorem 3.15 applies again.

We emphasize once again that our criteria of normal solvability are given in terms of the original data. No transformation in a special reduced form is required.

3.5 Notes and References

Remark 3.1 In optimal control one often prefers the spaces of essentially bounded functions. Denote by $L^\infty(\mathcal{I}, \mathbb{R}^m)$ the space of essentially bounded functions and by $W^{1,\infty}(\mathcal{I}, \mathbb{R}^m)$ the space of essentially bounded functions with essentially bounded first derivatives, further

$$W_D^{1,\infty}(\mathcal{I}, \mathbb{R}^m) := \{x \in L^\infty(\mathcal{I}, \mathbb{R}^m) : Dx \in W^{1,\infty}(\mathcal{I}, \mathbb{R}^n)\}.$$

The differential-algebraic operator

$$T : \operatorname{dom} T := W_D^{1,\infty}(\mathcal{I}, \mathbb{R}^m) \subseteq X \to L^\infty(\mathcal{I}, \mathbb{R}^k)$$

can then be treated analogously, with $X = W_D^{1,\infty}(\mathcal{I}, \mathbb{R}^m)$ as a bounded operator and with $X = L^\infty(\mathcal{I}, \mathbb{R}^m)$ as a closed operator, accordingly.

Remark 3.2 To our knowledge, [38] was the very first paper providing the closure of an operator associated with a DAE, and formulating conditions ensuring normal solvability. The function space applied in [38] is $L^2(\mathcal{I}, \mathbb{R}^m)$. In contrast to the present paper, in [38] also the coefficients E and F are integrable functions. In [85] a constant leading coefficient E is supposed for obtaining the closure of the differential-algebraic operator acting in L^2-spaces, and, in essence, also F is constant when regarding normal solvability. In [85] different numbers of unknowns m and equations k are allowed, whereas $m = k$ is supposed in [38].

Theorem 3.4 substantially generalizes [85, Theorem 3] which is proved there for $X = L^2$ by means of a quite involved regularization procedure.

In essence, the DAEs described in Proposition 3.7 are tractable with index 0 and index 1. We conjecture that exactly the DAEs with tractability index 0 and 1 (cf. [55, Chapter 10]) yield normally solvable operators. Example 3.8 supports this idea.

Remark 3.3 LSS have been discussed since the beginning of DAE research. Already in the early contribution [12], linear DAEs are treated as least-squares problems, with function spaces H^1 and L^2, and by a gradient method. It is reported that satisfactory numerical results are obtained only for matrix pencils having a simple structure, that is, in the absence of higher order nilpotent blocks. This is in line with our theory.

An updated analysis is reported in [16]. For a certain integer κ, the cost functional

$$J_\kappa(x) := \sum_{j=0}^{\kappa} \|(Tx - q)^{(j)}\|_{L^2}^2$$

is to be minimized subject to the fixed initial condition $x(t_a) = x_a$. Special interest is accorded to $\kappa = 0$. A simple gradient descent method and discretization by polynomials are discussed.

Sobolev steepest descent methods are applied in [68] to achieve LSS.

Remark 3.4 For bounded operators $K_b(X, Y)$ acting in Hilbert spaces, possibly with nonclosed image im K, there is a reach spectrum of iteration methods relying on Gaussian symmetrization, i.e., on the normal equation (e.g.,[24, 78])

$$K^*Kx = K^*q, \tag{3.60}$$

and regularization, among others, Tikhonov's method. To the author's knowledge, those iteration procedures are not developed for DAE problems as yet.

Remark 3.5 Generalized inverses of a differential-algebraic operator associated with strangeness-free DAEs are constructed in [50, 75] by strongly exploiting the special reduced form (cf. Example 3.3). Theorems 3.2 and 3.11, in particular condition (3.14), apply to the class of strangeness-free DAEs. In essence, regardless of several inconsistencies in [50], the corresponding generalized inverses $\mathcal{T}^-$ reproduce those in [50] as special cases. The DAE

$$\begin{bmatrix} 1 & 0 & 0 \\ 0 & 0 & 0 \end{bmatrix} x'(t) + \begin{bmatrix} 1 - \frac{t}{2} & \frac{t}{2} & 1 \\ -1 & 1 & 0 \end{bmatrix} x(t) = q(t),$$

which serves as test example for numerical experiments in [75], fulfills all conditions of Theorem 3.2, with $W_1 = 0$.

Section 3 provides generalized inverses for a larger class of DAEs than that of strangeness-free ones. In contrast to [50], it is not required to perform a preliminary reduction procedure via a derivative array system. Note that this approach requires higher derivatives. Example 3.12 demonstrates that the pseudo solution obtained that way does not necessarily coincide with the pseudo solution in our context.

Remark 3.6 The structural condition (3.51) was introduced in [32]. Nonlinear DAEs in standard form (3.48) that satisfy this condition were associated with operator equations in Banach spaces, e.g., in [57, 58]. There, solvability results for regular index-1 DAEs were obtained via the classical abstract implicit function theorem, and numerical methods for IVPs and BVPs were treated as discretizations of operator equations. It was shown how to accomplish well-posed problems by appropriately chosen initial and boundary conditions. It was further pointed out, that, in natural settings as described in Sects. 3.2 and 3.4, DAEs with higher index yield necessarily operators with nonclosed images, and hence essentially ill-posed problems.

Remark 3.7 Relying on the structural condition (3.51) for the nonlinear DAE (3.48), well-posed BVPs in regular index-1 DAEs are established and treated by Newton–Kantorovich iterations in natural Banach spaces, e.g., in [59, 69], also Sect. 4.

Remark 3.8 Up to now it remains open whether the somewhat unwieldy condition (3.31) in Proposition 3.7 can be replaced by a more transparent condition saying that the DAE is tractable with index zero or one (in the sense of [55, Definition 10.2]). Then one could upgrade this proposition to the level of a theorem, and eventually incorporate Theorems 3.2 and 3.4 as particular instances.

In this context it is worthy of mention that the nonregular versions of the tractability index and the strangeness index may substantially differ from each other, in contrast to the known rule $\mu_{tractability} = \mu_{strangeness} + 1$ given in the regular case ([55, Theorem 2.79]). In particular, in Example 3.6 we have $\mu_{tractability} = 0$ and $\mu_{strangeness} = 2$.

Remark 3.9 A quite common approach to DAEs consists in repeatedly adding to the given DAE the differentiated version of its derivative-free equations. From the viewpoint of functional analysis, by this, one tries to assemble an appropriate normally solvable equation. This justifies to speak of a *regularization* and states a reason of so-called over-determined discretizations (e.g., [28]). To simplify matters we outline the procedure for the linear standard form DAE (3.1) only (cf. also Remark 4.11). One starts letting

$$E_{[0]} := E, \quad F_{[0]} := F, \quad q_{[0]} := q, \quad k_{[0]} := k,$$

and, for $i \geq 0$, as long as a full column-rank $E_{[i]}$ is reached, one chooses a matrix function $\mathfrak{W}_{[i]} : \mathcal{I} \to \mathcal{L}(\mathbb{R}^{k_{[i]}}, \mathbb{R}^{\tau_{[i]}})$, $\tau_{[i]} > 0$, such that $\mathfrak{W}_{[i]} E_{[i]} = 0$, differentiates the equation $\mathfrak{W}_{[i]} F_{[i]} x = \mathfrak{W}_{[i]} q_{[i]}$, and adds the differentiated form to the previous DAE, i.e., $k_{[i+1]} := k_{[i]} + \tau_{[i]}$,

$$\underbrace{\begin{bmatrix} E_{[i]} \\ \mathfrak{W}_{[i]} F_{[i]} \end{bmatrix}}_{=:E_{[i+1]}} x' + \underbrace{\begin{bmatrix} F_{[i]} \\ (\mathfrak{W}_{[i]} F_{[i]})' \end{bmatrix}}_{=:F_{[i+1]}} x = \underbrace{\begin{bmatrix} q_{[i]} \\ (\mathfrak{W}_{[i]} q_{[i]})' \end{bmatrix}}_{=:q_{[i+1]}}.$$

All matrix functions should be at least continuous and sufficiently smooth. The point is to choose $\mathfrak{W}_{[i]}$ so that in each level the rank of the leading coefficient increases, rank $E_{[i]} <$ rank $E_{[i+1]}$. $\mathfrak{W}_{[i]}$ is not necessarily a projector function. There is a considerable latitude for choosing $\mathfrak{W}_{[i]}$. Having reached a prolonged system with a full column-rank leading coefficient, owing to Theorem 3.4 we know the associated differential-algebraic operator to be normally solvable. Then, in particular, a LSS makes sense. The prolongation procedure towards a normally solvable problem seems to work at least for all regular and quasi-regular DAEs (cf. [55], also Appendix 6.3). To the authors knowledge, rigorous expositions of this topic are missing so far. Obviously, there is a close relation to the procedures extracting a so-called *completion* or *underlying ODE* (e.g.[10]) from a derivative array system and the *differentiation index* calculus.

4 Regular Differential-Algebraic Operators in Their Natural Banach Spaces

In this section we treat nonlinear DAEs as operator equations with Fréchet differentiable operators. Motivated by the expressions obtained for the closures of the unbounded operators in Theorem 3.1 and by the arguments in Sect. 3.4 we immediately turn to DAEs with properly stated leading terms. More precisely, we investigate the nonlinear DAE

$$f((Dx)'(t), x(t), t) = 0, \tag{4.1}$$

which exhibits the involved derivative by means of an extra matrix-valued function D, and which involves m unknown functions and $k = m$ equations. We characterize regular differential-algebraic operators and suitable initial and boundary conditions so that the linearized composed operator becomes injective. We assign the tractability index and characteristic values to each regular differential-algebraic operator. We describe the image of linear differential-algebraic operators and provide an operator splitting into two characteristic parts. For a higher-index linear operator, in its natural setting, bounded generalized inverses do not exist, but a nontrivial bounded outer inverse can be constructed. Therewith, local solvability statements for nonlinear DAEs are achieved. Further the setting of BVPs and Newton–Kantorovich-like iterations are analyzed and challenging open questions are specified as well.

4.1 Notations and Basic Assumptions

With regard to Eq. (4.1) we agree upon the following assumptions to be valid throughout this section:

The function $f : \mathbb{R}^n \times \mathcal{D}_f \times \mathcal{I}_f \longrightarrow \mathbb{R}^m$, $\mathcal{D}_f \times \mathcal{I}_f \subseteq \mathbb{R}^m \times \mathbb{R}$ open, is continuous and has continuous partial derivatives f_y and f_x with respect to the first two variables $y \in \mathbb{R}^n$, $x \in \mathcal{D}_f$. The partial Jacobian $f_y(y, x, t)$ is everywhere singular. The matrix function $D : \mathcal{I}_f \to \mathcal{L}(\mathbb{R}^m, \mathbb{R}^n)$ is continuously differentiable and $D(t)$ has constant rank r on the given interval $\mathcal{I}_f$. Let the transversality condition

$$\ker f_y(y, x, t) \oplus \operatorname{im} D(t) = \mathbb{R}^n, \quad (y, x, t) \in \mathbb{R}^n \times \mathcal{D}_f \times \mathcal{I}_f, \tag{4.2}$$

be valid and $\ker f_y$ be a $\mathcal{C}^1$-subspace. We say that the DAE (4.1) has a *properly stated leading term*, also a *properly involved derivative*.

Let $\mathcal{I} \subseteq \mathcal{I}_f$ be a compact interval, $\mathcal{I} =: [t_a, t_e]$, and $\mathcal{D}_F \subseteq \mathcal{D}_f$ be open. We associate with the DAE (4.1) the nonlinear operator

$$F : \operatorname{dom} F \subseteq \mathcal{C}_D^1(\mathcal{I}, \mathbb{R}^m) \to \mathcal{C}(\mathcal{I}, \mathbb{R}^m),$$

$$\operatorname{dom} F : = \{x \in C_D^1(\mathcal{I}, \mathbb{R}^m) : x(t) \in \mathcal{D}_F \text{ for all } t \in \mathcal{I}\},$$

$$(Fx)(t) : = f((Dx)'(t), x(t), t), \quad t \in \mathcal{I}, \quad x \in \operatorname{dom} F. \tag{4.3}$$

Since D is continuously differentiable, the inclusions

$$\mathcal{C}^\infty(\mathcal{I}, \mathbb{R}^m) \subseteq \mathcal{C}^\nu(\mathcal{I}, \mathbb{R}^m) \subseteq \mathcal{C}_D^1(\mathcal{I}, \mathbb{R}^m) \tag{4.4}$$

are valid for all $\nu \in \mathbb{N}$. Endowed with the norm

$$\|x\|_{\mathcal{C}_D^1} := \|x\|_\infty + \|(Dx)'\|_\infty, \quad x \in \mathcal{C}_D^1(\mathcal{I}, \mathbb{R}^m), \tag{4.5}$$

the function space $C_D^1(\mathcal{I}, \mathbb{R}^m)$ is a Banach space (see Lemma 6.9) and the DAE (4.1) is represented as the operator equation

$$Fx = 0. \tag{4.6}$$

The operator F is said to be a *nonlinear differential-algebraic*. The operator equation (4.6) reflects the classical view on a DAE: the solutions belong to $C_D^1(\mathcal{I}, \mathbb{R}^m)$ and satisfy the DAE pointwise for all $t \in \mathcal{I}$. The arguments in Sect. 2.3 enable us to speak of the *natural* Banach space setting.

For given $x_* \in \text{dom } F$ we denote

$$A_*(t) := f_y((Dx_*)'(t), x_*(t), t),$$
$$B_*(t) := f_x((Dx_*)'(t), x_*(t), t), \quad t \in \mathcal{I}.$$

The directional derivative

$$F'(x_*)x := \lim_{\tau \to 0} \frac{1}{\tau}(F(x_* + \tau x) - F(x_*)) = A_*(Dx)' + B_*x$$

is well defined for each arbitrary $x_* \in \text{dom } F$ and $x \in C_D^1(\mathcal{I}, \mathbb{R}^m)$. The resulting map

$$F'(x_*) : C_D^1(\mathcal{I}, \mathbb{R}^m) \to C(\mathcal{I}, \mathbb{R}^m) \tag{4.7}$$

is linear and bounded. Moreover, $F'(x_*)$ varies continuously with respect to x_*. This means that the operator F *is Fréchet differentiable* and the map $F'(x_*)$ defined by

$$F'(x_*)x = A_*(Dx)' + B_*x, \quad x \in C_D^1(\mathcal{I}, \mathbb{R}^m),$$

is the Fréchet derivative of F at x_*. The linear operator equation

$$F'(x_*)x = q$$

stands now for the *linearization* of the original DAE at x_*, that is, for the linear DAE

$$A_*(Dx)' + B_*x = q. \tag{4.8}$$

We complete the DAE (4.1) by the boundary condition

$$b(x(t_a), x(t_e)) = 0. \tag{4.9}$$

The continuously differentiable function $b : \mathbb{R}^m \times \mathbb{R}^m \to \mathbb{R}^{m-l}$ will be specified later. Often we apply the particular case of an initial condition

$$Cx(t_a) = 0, \tag{4.10}$$

by letting $b(x, \bar{x}) := Cx$. The composed operator

$$\mathcal{F} : \text{dom } F \subseteq \mathcal{C}_D^1(\mathcal{I}, \mathbb{R}^m) \to \mathcal{C}(\mathcal{I}, \mathbb{R}^m) \times \mathbb{R}^{m-l},$$

$$\mathcal{F}x := (Fx, \; b(x(t_a), x(t_e))), \quad x \in \text{dom } F, \tag{4.11}$$

is Fréchet differentiable since F is so. The equation $\mathcal{F}x = 0$ represents the BVP (4.1), (4.9), whereas the equation $\mathcal{F}x = (q, d)$ is the operator form of the perturbed BVP

$$f((D(t)x(t))', x(t), t) = q(t), \; t \in \mathcal{I}, \quad b(x(t_a), x(t_e)) = d. \tag{4.12}$$

4.2   Regular Linear Differential-Algebraic Operators

First we study the linear bounded operator $T \in \mathcal{L}(\mathcal{C}_D^1(\mathcal{I}, \mathbb{R}^m), \mathcal{C}(\mathcal{I}, \mathbb{R}^m))$ given by

$$Tx := A(Dx)' + Bx, \quad x \in \mathcal{C}_D^1(\mathcal{I}, \mathbb{R}^m), \tag{4.13}$$

with coefficients

$$A \in \mathcal{C}(\mathcal{I}, \mathcal{L}(\mathbb{R}^n, \mathbb{R}^m)), \; D \in \mathcal{C}^1(\mathcal{I}, \mathcal{L}(\mathbb{R}^m, \mathbb{R}^n)), \; B \in \mathcal{C}(\mathcal{I}, \mathcal{L}(\mathbb{R}^m, \mathbb{R}^m)),$$

in some detail. Let $\ker A$ and $\text{im } D$ be $\mathcal{C}^1$-subspaces and let the transversality condition

$$\ker A(t) \oplus \text{im } D(t) = \mathbb{R}^n, \quad t \in \mathcal{I}, \tag{4.14}$$

be satisfied. Denote by $R(t)$ the border projector matrix onto $\text{im } D(t)$ along $\ker A(t)$, $t \in \mathcal{I}$. The resulting function R is continuously differentiable.

Let $Q_0, P_0 \in \mathcal{C}(\mathcal{I}, \mathcal{L}(\mathbb{R}^m))$ be projector-valued functions such that $\text{im } Q_0(t) = \ker D(t)$ for $t \in \mathcal{I}$, and $P_0 = I - Q_0$.[1] Moreover, denote by $D^- \in \mathcal{C}(\mathcal{I}, \mathcal{L}(\mathbb{R}^n, \mathbb{R}^m))$ the pointwise generalized inverse of D which is uniquely determined by

$$DD^-D = D, \quad D^-DD^- = D^-, \quad DD^- = R, \quad D^-D = P_0.$$

---

[1] In contrast to Sect. 3, here we do not fix these projectors to be orthogonal.

Definition 4.1 For given coefficients A, D and B, and any level $\kappa \in \mathbb{N}$, the sequence $G_0, \ldots, G_\kappa$ is said to be an *admissible matrix function sequence*, if it is built pointwise for all $t \in \mathcal{I}$ by the rule:
set $G_0 := AD$, $B_0 := B$, $N_0 := \ker G_0$,
and for $i \geq 1$:

$$G_i := G_{i-1} + B_{i-1} Q_{i-1}, \tag{4.15}$$

$$N_i := \ker G_i, \quad \widehat{N}_i := (N_0 + \cdots + N_{i-1}) \cap N_i,$$

find a complement X_i such that $N_0 + \cdots + N_{i-1} = \widehat{N}_i \oplus X_i$,

choose a projector Q_i such that $\operatorname{im} Q_i = N_i$ and $X_i \subseteq \ker Q_i$,

set $P_i := I - Q_i$, $\Pi_i := \Pi_{i-1} P_i$,

$$B_i := B_{i-1} P_{i-1} - G_i D^- (D \Pi_i D^-)' D \Pi_{i-1}, \tag{4.16}$$

and, if additionally,

(a) the matrix function G_i has constant rank r_i, $i = 0, \ldots, \kappa$,
(b) the intersection $\widehat{N}_i$ has constant dimension $u_i := \dim \widehat{N}_i$,
(c) the product function Π_i is continuous and $D \Pi_i D^-$ is continuously differentiable, $i = 0, \ldots, \kappa$.

The projector functions $Q_0, \ldots, Q_\kappa$ associated with an admissible matrix function sequence are said to be *admissible* themselves.
    An admissible matrix function sequence G_0, ..., G_κ is said to be *regular admissible*, if

$$\widehat{N}_i = \{0\}, \quad u_i = 0, \quad \text{for all} \quad i = 1, \ldots, \kappa.$$

Then, also the projector functions $Q_0, \ldots, Q_\kappa$ are called *regular admissible*.
    The numbers $r_0 = \operatorname{rank} G_0, \ldots, r_\kappa = \operatorname{rank} G_\kappa$ and $u_1, \ldots, u_\kappa$ are called *characteristic values* of the DAE on the interval $\mathcal{I}$.

We refer to [55] for many useful properties of the admissible matrix function sequences. By construction, it holds that

$$r_0 \leq r_1 \leq \ldots \leq r_\kappa.$$

Now we are prepared to generalize the traditional notion of regular differential-algebraic operators given in Sect. 2.1 for time-invariant coefficients accordingly.

Definition 4.2 The differential-algebraic operator (4.13) is said to be

1. regular with tractability index 0, if $r_0 = m$;

2. regular with tractability index μ, if there is an admissible matrix function sequence such that

$$r_0 \leq r_1 \leq \ldots \leq r_{\mu-1} < r_\mu = m; \qquad (4.17)$$

3. regular, if it is regular with any index.

The numbers (4.17) and μ are said to be the characteristic values and the tractability index of the regular differential-algebraic operator T.

In case of constant coefficients A, D, B, the matrix function sequence simplifies to a sequence of matrices. In particular, the second term in the definition of B_i disappears. It is well known [33] that the pair $\{AD, B\}$ of $m \times m$ matrices AD, B is regular with Kronecker index μ (cf. Sect. 2.1) exactly if an admissible sequence of matrices starting with $G_0 = AD$, $B_0 := B$ yields

$$r_0 \leq \ldots \leq r_{\mu-1} < r_\mu = m. \qquad (4.18)$$

Thereby, neither the factorization nor the special choice of admissible projectors matters. The characteristic values describe the structure of the Weierstraß–Kronecker form (2.4): we have $l = \sum_{j=0}^{\mu-1}(m - r_j)$ and the nilpotent part N contains altogether $s = m - r_0$ Jordan blocks, among them $r_i - r_{i-1}$ Jordan blocks of order $i, i = 1, \ldots, \mu$.

Also in the general regular time-varying case, the ingredients of an admissible matrix function sequence allow a decoupling which is quite similar to the Weierstraß–Kronecker form (e.g., [55, Section 2.4], also Appendix 6.3). Thereby, the matrix function

$$\mathcal{K} := (I - \Pi_{\mu-1})G_\mu^{-1}B_{\mu-1}\Pi_{\mu-1} + \sum_{l=1}^{\mu-1}(I - \Pi_{l-1})(P_l - Q_l)(D\Pi_l D^-)'D\Pi_{\mu-1},$$

plays its role. The decoupling of the two basic parts yielding the inherent explicit ODE and the differentiation problems is *complete* as in the Weierstraß–Kronecker form, exactly if $\mathcal{K}$ vanishes identically. If the original data A, D, B show some additional smoothness, the admissible projector functions can be chosen in such a way that $\mathcal{K}$ disappears (e.g.[55, Section 2.4]). In this case, we speak of *completely decoupling projector functions* $Q_0, \ldots, Q_{\mu-1}$.

Definition 4.3 The differential-algebraic operator (4.13) is said to be fine, if it is regular and the coefficients A, D, B are as smooth as required for the existence of completely decoupling projectors $Q_0, \ldots, Q_{\mu-1}$.

Example 4.1 (Continuation 4 of Example 2.3) The following matrix sequence is admissible for the pair $\{E, F\} = \{AD, B\}$ from Example 2.3 which is regular with index 4:

$$
G_0 = \begin{bmatrix} 1 & 0 & 0 & 0 & 0 \\ 0 & 0 & 1 & 0 & 0 \\ 0 & 0 & 0 & 1 & 0 \\ 0 & 0 & 0 & 0 & 1 \\ 0 & 0 & 0 & 0 & 0 \end{bmatrix}, Q_0 = \begin{bmatrix} 0 & 0 & 0 & 0 & 0 \\ 0 & 1 & 0 & 0 & 0 \\ 0 & 0 & 0 & 0 & 0 \\ 0 & 0 & 0 & 0 & 0 \\ 0 & 0 & 0 & 0 & 0 \end{bmatrix}, B_0 = \begin{bmatrix} -\alpha & -1 & 0 & 0 & 0 \\ 0 & 1 & 0 & 0 & 0 \\ 0 & 0 & 1 & 0 & 0 \\ 0 & 0 & 0 & 1 & 0 \\ 0 & 0 & 0 & 0 & 1 \end{bmatrix},
$$

$$
G_1 = \begin{bmatrix} 1 & -1 & 0 & 0 & 0 \\ 0 & 1 & 1 & 0 & 0 \\ 0 & 0 & 0 & 1 & 0 \\ 0 & 0 & 0 & 0 & 1 \\ 0 & 0 & 0 & 0 & 0 \end{bmatrix}, Q_1 = \begin{bmatrix} 0 & 0 & -1 & 0 & 0 \\ 0 & 0 & -1 & 0 & 0 \\ 0 & 0 & 1 & 0 & 0 \\ 0 & 0 & 0 & 0 & 0 \\ 0 & 0 & 0 & 0 & 0 \end{bmatrix}, \Pi_0 Q_1 = \begin{bmatrix} 0 & 0 & -1 & 0 & 0 \\ 0 & 0 & 0 & 0 & 0 \\ 0 & 0 & 1 & 0 & 0 \\ 0 & 0 & 0 & 0 & 0 \\ 0 & 0 & 0 & 0 & 0 \end{bmatrix},
$$

$$
G_2 = \begin{bmatrix} 1 & -1 & \alpha & 0 & 0 \\ 0 & 1 & 1 & 0 & 0 \\ 0 & 0 & 1 & 1 & 0 \\ 0 & 0 & 0 & 0 & 1 \\ 0 & 0 & 0 & 0 & 0 \end{bmatrix}, Q_2 = \begin{bmatrix} 0 & 0 & 0 & 1+\alpha & 0 \\ 0 & 0 & 0 & 1 & 0 \\ 0 & 0 & 0 & -1 & 0 \\ 0 & 0 & 0 & 1 & 0 \\ 0 & 0 & 0 & 0 & 0 \end{bmatrix}, \Pi_1 Q_2 = \begin{bmatrix} 0 & 0 & 0 & \alpha & 0 \\ 0 & 0 & 0 & 0 & 0 \\ 0 & 0 & 0 & 0 & 0 \\ 0 & 0 & 0 & 1 & 0 \\ 0 & 0 & 0 & 0 & 0 \end{bmatrix},
$$

$$
G_3 = \begin{bmatrix} 1 & -1 & \alpha & -\alpha^2 & 0 \\ 0 & 1 & 1 & 0 & 0 \\ 0 & 0 & 1 & 1 & 0 \\ 0 & 0 & 0 & 1 & 1 \\ 0 & 0 & 0 & 0 & 0 \end{bmatrix}, Q_3 = \begin{bmatrix} 0 & 0 & 0 & 0 & -1-\alpha-\alpha^2 \\ 0 & 0 & 0 & 0 & -1 \\ 0 & 0 & 0 & 0 & 1 \\ 0 & 0 & 0 & 0 & -1 \\ 0 & 0 & 0 & 0 & 1 \end{bmatrix}, \Pi_2 Q_3 = \begin{bmatrix} 0 & 0 & 0 & 0 & -\alpha^2 \\ 0 & 0 & 0 & 0 & 0 \\ 0 & 0 & 0 & 0 & 0 \\ 0 & 0 & 0 & 0 & 0 \\ 0 & 0 & 0 & 0 & 1 \end{bmatrix},
$$

$$
G_4 = \begin{bmatrix} 1 & -1 & \alpha & -\alpha^2 & \alpha^3 \\ 0 & 1 & 1 & 0 & 0 \\ 0 & 0 & 1 & 1 & 0 \\ 0 & 0 & 0 & 1 & 1 \\ 0 & 0 & 0 & 0 & 1 \end{bmatrix}, \Pi_3 = \begin{bmatrix} 1 & 0 & 1 & -\alpha & \alpha^2 \\ 0 & 0 & 0 & 0 & 0 \\ 0 & 0 & 0 & 0 & 0 \\ 0 & 0 & 0 & 0 & 0 \\ 0 & 0 & 0 & 0 & 0 \end{bmatrix},
$$

and the characteristic values are $r_0 = r_1 = r_2 = r_3 = 4$, $r_4 = 5$, and $\mu = 4$. Additionally, it follows that

$$Q_3 G_4^{-1} B_0 \Pi_3 = 0, \qquad\qquad Q_2 P_3 G_4^{-1} B_0 \Pi_3 = 0,$$

$$Q_1 P_2 P_3 G_4^{-1} B_0 \Pi_3 = 0, \qquad\qquad Q_0 P_1 P_2 P_3 G_4^{-1} B_0 \Pi_3 = 0,$$

and

$$\Pi_3 G_4^{-1} B_0 \Pi_3 = -\alpha \Pi_3. \tag{4.19}$$

The projectors Q_0, Q_1, Q_2, Q_3 provide a complete decoupling of the given DAE $A(Dx)'(t) + Bx(t) = q(t)$. The projectors $Q_0, \Pi_0 Q_1, \Pi_1 Q_2$, and $\Pi_2 Q_3$ represent the variables x_2, x_3, x_4, and x_5, respectively. The projector Π_3 and the coefficient (4.19) determine the inherent regular ODE, namely (the zero rows are dropped)

$$(x_1 + x_3 - \alpha x_4 + \alpha^2 x_5)' - \alpha(x_1 + x_3 - \alpha x_4 + \alpha^2 x_5)$$
$$= q_1 + q_2 - \alpha q_3 + \alpha^2 q_4 - \alpha^3 q_5. \tag{4.20}$$

It is noteworthy that no derivatives of the excitation q encroach in this part.

Here Π_3 is the spectral projector of the pair $\{E, F\}$. The decoupling projector of the basic parts is

$$G_4 \Pi_3 G_4^{-1} = \begin{bmatrix} 1 & 1 & -\alpha & \alpha^2 & -\alpha^3 \\ 0 & 0 & 0 & 0 & 0 \\ 0 & 0 & 0 & 0 & 0 \\ 0 & 0 & 0 & 0 & 0 \\ 0 & 0 & 0 & 0 & 0 \end{bmatrix}$$

and the equation $G_4 \Pi_3 G_4^{-1} Tx = G_4 \Pi_3 G_4^{-1} q$ results in the ODE (4.20).   □

Parts of the following theorem can be seen as counterparts and generalizations of Proposition 2.4.

Theorem 4.2 *Let the linear differential-algebraic operator (4.13) be fine with tractability index $\mu \geq 1$ and characteristic values $r_0 \leq \cdots \leq r_{\mu-1} < r_\mu = m$. Let the projector valued functions $Q_0, \ldots, Q_{\mu-1} \in \mathcal{C}(\mathcal{I}, \mathcal{L}(\mathbb{R}^m))$ be associated with a complete decoupling. Then the following statements are true:*

1. The topological direct sum decomposition

$$\mathcal{C}_D^1(\mathcal{I}, \mathbb{R}^m)$$
$$= \{x \in \mathcal{C}_D^1(\mathcal{I}, \mathbb{R}^m) : \Pi_{\mu-1}x = 0\} \oplus \{x \in \mathcal{C}_D^1(\mathcal{I}, \mathbb{R}^m) : (I - \Pi_{\mu-1})x = 0\}$$
$$=: \ker \Pi_{\mu-1} \oplus \operatorname{im} \Pi_{\mu-1}$$

is valid.[2]

---

[2]$\ker \Pi_{\mu-1}$ and $\operatorname{im} \Pi_{\mu-1}$ are used in $\mathbb{R}^m$ and in $\mathcal{C}_D^1$, but no confusion should arise.

2. *The operator splits into the sum of two bounded operators*

$$T = \underbrace{G_\mu \Pi_{\mu-1} G_\mu^{-1} T}_{T_{dyn}} + \underbrace{(I - G_\mu \Pi_{\mu-1} G_\mu^{-1})T}_{T_{alg}} =: T_{dyn} + T_{alg},$$

in which the operator T_{dyn} is normally solvable, and

$$\operatorname{im} T_{dyn} = \operatorname{im} G_\mu \Pi_{\mu-1} G_\mu^{-1}, \quad \ker T_{alg} = \operatorname{im} \Pi_{\mu-1}.$$

3. $\ker T = \ker T_{dyn} \cap \ker T_{alg} = \ker T_{dyn} \cap \operatorname{im} \Pi_{\mu-1}$ *has the finite dimension* $\delta :=$ $\operatorname{rank} \Pi_{\mu-1} = m - \sum_{i=0}^{\mu-1}(m - r_i) = m - l.$
4. T *is normally solvable exactly if* T_{alg} *is so;*

$$\operatorname{im} T = \operatorname{im} T_{dyn} \dotplus \operatorname{im} T_{alg} = \operatorname{im} G_\mu \Pi_{\mu-1} G_\mu^{-1} \dotplus \operatorname{im} T_{alg}. \tag{4.21}$$

5. *If* $\mu = 1$, *then* T_{alg} *is normally solvable and* T *is surjective, thus Fredholm.*
6. *If* $\mu > 1$, *then* $\operatorname{im} T_{alg}$ *and* $\operatorname{im} T$ *are nonclosed proper subsets of* $\mathcal{C}(\mathcal{I}.\mathbb{R}^m))$, *and the equation* $Tx = q$ *is essentially ill-posed.*

Proof 1. Since $D\Pi_{\mu-1}(I - D^+ D) = 0$ and $D\Pi_{\mu-1}D^+ = D\Pi_{\mu-1}D^- DD^+$ is continuously differentiable, this statement is a consequence of Lemma 6.10.

2. The boundedness of both operators T_{dyn} and T_{alg} is evident. Rearranging several terms as described in [55, pp. 93–96] we express

$$T_{dyn}x = G_\mu \Pi_{\mu-1} G_\mu^{-1}(A(Dx)' + Bx)$$
$$= G_\mu \{D^- D\Pi_{\mu-1}D^-(D\Pi_{\mu-1}x)' + \Pi_{\mu-1}G_\mu^{-1}B_\mu D^- D\Pi_{\mu-1}x\}, \tag{4.22}$$

for $x \in \mathcal{C}_D^1(\mathcal{I}, \mathbb{R}^m)$, and

$$T_{alg}x = (I - G_\mu \Pi_{\mu-1} G_\mu^{-1})(A(Dx)' + Bx)$$
$$= G_\mu \{\sum_{l=0}^{\mu-1} Q_l x - \sum_{l=0}^{\mu-2}(I - \Pi_l)Q_{l+1}D^-(D\Pi_l Q_{l+1}x)' + \sum_{l=0}^{\mu-2} \mathcal{M}_{l+1}D\Pi_l Q_{l+1}x\}, \tag{4.23}$$

with coefficients, see [55, p. 95], also Appendix 6.3.2,

$$\mathcal{M}_j := \sum_{k=0}^{j-1}(I-\Pi_k)\{P_k D^-(D\Pi_k D^-)' - Q_{k+1}D^-(D\Pi_{k+1}D^-)'\}D\Pi_{j-1}Q_l D^-,$$

for $l = 1, \ldots, \mu - 1$. The additional coefficient $\mathcal{K}$ arising in [55, p. 95] here disappears owing to the complete decoupling, cf. Appendix 6.3.2.

Expression (4.23) shows that $T_{alg}x = 0$ if $x = \Pi_{\mu-1}x$. Conversely, if x satisfies the equation $T_{alg}x = 0$, then

$$\sum_{l=0}^{\mu-1} Q_l x - \sum_{l=0}^{\mu-2}(I - \Pi_l)Q_{l+1}D^-(D\Pi_l Q_{l+1}x)' + \sum_{l=0}^{\mu-2}\mathcal{M}_{l+1}D\Pi_l Q_{l+1}x = 0$$

follows. Multiplying by $\Pi_{\mu-2}Q_{\mu-1}$ yields $\Pi_{\mu-2}Q_{\mu-1}x = 0$, $Q_{\mu-1}x = 0$, and

$$\sum_{l=0}^{\mu-2} Q_l x - \sum_{l=0}^{\mu-3}(I - \Pi_l)Q_{l+1}D^-(D\Pi_l Q_{l+1}x)' + \sum_{l=0}^{\mu-3}\mathcal{M}_{l+1}D\Pi_l Q_{l+1}x = 0.$$

Multiplying successively by $\Pi_{\mu-3}Q_{\mu-2},\ldots\Pi_0 Q_1$, we obtain $Q_{\mu-2}x = 0,\ldots,Q_1 x = 0$, respectively, and finally $Q_0 x = 0$. Therefore, each element x of the nullspace of T_{alg} has the form $x = \Pi_{\mu-1}x$, such that actually $\ker T_{alg} = \operatorname{im} \Pi_{\mu-1}$.

Next we turn to the operator T_{dyn}. For each arbitrary $q \in \operatorname{im} G_\mu \Pi_{\mu-1}G_\mu^{-1}$, the equation $T_{dyn}x = q = G_\mu \Pi_{\mu-1}G_\mu^{-1}q$ is equivalent to

$$D^- D\Pi_{\mu-1}D^-(D\Pi_{\mu-1}x)' + \Pi_{\mu-1}G_\mu^{-1}B_\mu D^- D\Pi_{\mu-1}x = \Pi_{\mu-1}G_\mu^{-1}q,$$

and further to

$$D\Pi_{\mu-1}D^-(D\Pi_{\mu-1}x)' + D\Pi_{\mu-1}G_\mu^{-1}B_\mu D^- D\Pi_{\mu-1}x = D\Pi_{\mu-1}G_\mu^{-1}q,$$

The standard IVP

$$u' - (D\Pi_{\mu-1}D^-)'u + D\Pi_{\mu-1}G_\mu^{-1}B_\mu D^- u = D\Pi_{\mu-1}G_\mu^{-1}q, \quad u(t_a) = 0, \tag{4.24}$$

possesses the unique solution $u_* \in \mathcal{C}^1(\mathcal{I},\mathbb{R}^n)$, and the relation $u_* = D\Pi_{\mu-1}D^-u_*$ is given. The function $x_* := D^-u_* = \Pi_{\mu-1}D^-u_*$ belongs to $\mathcal{C}_D^1(\mathcal{I},\mathbb{R}^m)$ and satisfies $T_{dyn}x_* = G_\mu \Pi_{\mu-1}G_\mu^{-1}q = q$. This verifies the property $\operatorname{im} T_{dyn} = \operatorname{im} G_\mu \Pi_{\mu-1}G_\mu^{-1}$. This subspace is closed in $\mathcal{C}(\mathcal{I},\mathbb{R}^m)$ and T_{dyn} is normally solvable.

3. By construction, $Tx = 0$ means $T_{dyn}x = 0$ and $T_{alg}x = 0$, and equivalently,

$$D\Pi_{\mu-1}D^-(D\Pi_{\mu-1}x)' + D\Pi_{\mu-1}G_\mu^{-1}B_\mu D^- D\Pi_{\mu-1}x = 0, \quad x = \Pi_{\mu-1}x.$$

Therefore, if $x_* \in \ker T$, then $Dx_* = D\Pi_{\mu-1}x_*$ is a solution of the ODE from (4.24) with $q = 0$.

Denote by $U \in \mathcal{C}^1(\mathcal{I}, \mathcal{L}(\mathbb{R}^\delta))$ the fundamental solution matrix normalized at t_a of the ODE from (4.24). We have then

$$(I - D\Pi_{\mu-1}D^-)UD(t_a)\Pi_{\mu-1}(t_a) = 0.$$

Each function $x_* := D^- UD(t_a)\Pi_{\mu-1}(t_a)c$, $c \in \mathbb{R}^m$ belongs to $\mathcal{C}_D^1(\mathcal{I}, \mathbb{R}^m)$, since $Dx_* = DD^- UD(t_a)\Pi_{\mu-1}(t_a)c = DD^- D\Pi_{\mu-1}D^- UD(t_a)\Pi_{\mu-1}(t_a)c = UD(t_a)\Pi_{\mu-1}(t_a)c$ is continuously differentiable. Moreover, Dx_* satisfies the homogeneous ODE from (4.24), with $q = 0$, and it holds that $x_* = \Pi_{\mu-1}x_*$, and hence $x_* \in \ker T$. It follows that

$$\ker T = \{x = D^- U\eta : \eta \in \operatorname{im} D(t_a)\Pi_{\mu-1}(t_a)\},$$

and $\dim \ker T = \operatorname{rank} D(t_a)\Pi_{\mu-1}(t_a) = \Pi_{\mu-1}(t_a) = \delta$.

4. is a direct consequence of (2).

5. If $\mu = 1$ then (4.23) simplifies to

$$T_{alg}x = G_1\{Q_0 x\}.$$

For each $q = (I - G_1\Pi_0 G_1^{-1})q = G_1 Q_0 G_1^{-1}q$, the equation $T_{alg}x = q$ has the solutions $x = Q_0 G_1^{-1}q + h$, with a free component $h = P_0 h$. This proves that $\operatorname{im} T_{alg} = \ker G_1 \Pi_0 G_1^{-1}$ is closed and T_{alg} is normally solvable. Moreover, T is surjective, thus Fredholm.

6. Suppose $\mu \geq 2$. Applying once more the representation (4.23) we find for given $x \in \mathcal{C}_D^1(\mathcal{I}, \mathbb{R}^m)$ and $q := T_{alg}x$ the relation

$$q = G_\mu\{\sum_{l=0}^{\mu-1} Q_l x - \sum_{l=0}^{\mu-2}(I - \Pi_l)Q_{l+1}D^-(D\Pi_l Q_{l+1}x)' + \sum_{l=0}^{\mu-2}\mathcal{M}_{l+1}D\Pi_l Q_{l+1}x\},$$

which implies

$$D\Pi_{\mu-2}Q_{\mu-1}q = D\Pi_{\mu-2}Q_{\mu-1}T_{alg}x = D\Pi_{\mu-2}Q_{\mu-1}x.$$

The component $D\Pi_{\mu-2}Q_{\mu-1}x = D\Pi_{\mu-2}Q_{\mu-1}D^-Dx$ is continuously differentiable, hence $D\Pi_{\mu-2}Q_{\mu-1}q$ is also necessarily continuously differentiable. Therefore, $\operatorname{im} T_{alg}$ contains continuous functions having certain smoother components, which means that $\operatorname{im} T_{alg}$ is a nonclosed subset within the continuous function space.

□

Ill-posed problems are known to need so-called *regularizations*. Therefore, also regular higher-index DAEs need those regularizations, which sounds in a way confusing. Note once more that regularity in the traditional DAE theory is tied to regular matrix pairs and their generalizations.

Corollary 4.3 *Under the assumptions of Theorem 4.2, the solutions of the equation*

$$T_{dyn}x = G_\mu \Pi_{\mu-1} G_\mu^{-1} q \tag{4.25}$$

have the form $x = v + D^- u$, *whereby* v *is an arbitrary function from* $\ker \Pi_{\mu-1}$ *and* u *is a solution of the ODE*

$$u' - (D\Pi_{\mu-1}D^-)'u + D\Pi_{\mu-1}G_\mu^{-1} B_\mu D^- u = D\Pi_{\mu-1}G_\mu^{-1} q, \tag{4.26}$$

with $u(t_a) = D(t_a)\Pi_{\mu-1}(t_a)c$, $c \in \mathbb{R}^m$.

In the DAE analysis, the ODE (4.26) plays a central role, it is said to be the *inherent explicit regular ODE* (IERODE) of the DAE. The operator T_{dyn} represents the dynamical part of the DAE, which motivates the subscript *dyn*.

It may happen that the projector function $\Pi_{\mu-1}$ vanishes identically. Then T_{dyn} is the zero operator and $T = T_{alg}$ is injective, with an inverse being a differential operator again.

Example 4.4 ($T_{dyn} = 0$) The operator

$$Tx = \begin{bmatrix} 1 \\ 0 \end{bmatrix} ([1\ 0]\, x)' + \begin{bmatrix} 0 & 1 \\ 1 & 0 \end{bmatrix} x \tag{4.27}$$

is fine with tractability index 2. It leads to $\Pi_1 = 0$, thus $T = T_{alg}$. We observe that $\operatorname{im} T_{alg} = \{q \in \mathcal{C}(\mathcal{I}, \mathbb{R}^2) : q_2 \in \mathcal{C}^1(\mathcal{I}, \mathbb{R})\}$ is a proper subset in $\mathcal{C}(\mathcal{I}, \mathbb{R}^2)$. Moreover, the inverse

$$T_{alg}^{-1} q = \begin{bmatrix} q_2 \\ q_1 - q_2' \end{bmatrix}, \quad q \in \operatorname{im} T, \tag{4.28}$$

is again a differential-algebraic operator.   □

Theorem 4.2 is meaningful also for the composed operator

$$\mathcal{T} \in \mathcal{L}(\mathcal{C}_D^1(\mathcal{I}, \mathbb{R}^m), \mathcal{C}(\mathcal{I}, \mathbb{R}^m) \times \mathbb{R}^\delta),$$
$$\mathcal{T}x = (Tx, Cx(t_a)), \quad x \in \mathcal{C}_D^1(\mathcal{I}, \mathbb{R}^m).$$

The operator equation $\mathcal{T}x = (q, d)$ reflects the IVP

$$A(Dx)' + Bx = q, \quad Cx(t_a) = d. \tag{4.29}$$

The composed operator $\mathcal{T}$ is bounded together with T. $\mathcal{T}$ is injective, if T is regular and the matrix $C \in \mathcal{L}(\mathbb{R}^m, \mathbb{R}^\delta)$ is such that

$$\ker C = \ker \Pi_{\mu-1}, \quad \delta = \operatorname{rank} \Pi_{\mu-1}. \tag{4.30}$$

If $\mu = 1$, then T is even surjective. Then, as a bijective bounded operator acting in Banach spaces, it has a bounded inverse, and hence the IVP (4.29) is well-posed. In contrast, if $\mu \geq 2$, then the surjectivity gets lost and $\operatorname{im} T$ is a nonclosed subset in $\mathcal{C}(\mathcal{I}, \mathbb{R}^m) \times \mathbb{R}^\delta$. Then the IVP (4.29) is no longer well-posed, but essentially ill-posed.

Owing to Theorem 4.2, in all higher index cases, the composed operator T is bounded and injective, but the inverse T^{-1} is unbounded. Nevertheless, there are nontrivial bounded linear outer inverses as the next theorem states.

Theorem 4.5 *Let the linear differential-algebraic operator T be fine with tractability index $\mu \geq 2$. Let the projector-valued functions $Q_0 \ldots Q_{\mu-1} \in \mathcal{C}(\mathcal{I}, \mathcal{L}(\mathbb{R}^m))$ be associated with a complete decoupling. Let the matrix $C \in \mathcal{L}(\mathbb{R}^m, \mathbb{R}^\delta)$ satisfy condition (4.30). Then the following statements are valid:*

1. For each $y \in \mathcal{C}(\mathcal{I}, \mathbb{R}^m)$ and $d \in \mathbb{R}^\delta$, the IVP

$$Tx = G_\mu P_1 \cdots P_{\mu-1} G_\mu^{-1} y, \quad Cx(t_a) = d, \tag{4.31}$$

is uniquely solvable in $\mathcal{C}_D^1(\mathcal{I}, \mathbb{R}^m)$ and the solution satisfies the inequality

$$\|x\|_{\mathcal{C}_D^1} \leq c\left(|d| + \|G_\mu P_1 \cdots P_{\mu-1} G_\mu^{-1} y\|_\infty\right) \leq \tilde{c}\left(|d| + \|y\|_\infty\right). \tag{4.32}$$

2. The operator $T^- \in \mathcal{L}(\mathcal{C}(\mathcal{I}, \mathbb{R}^m) \times \mathbb{R}^\delta, \mathcal{C}_D^1(\mathcal{I}, \mathbb{R}^m))$ defined by

$$T^-(y, z) := \text{solution of } (4.31) \text{ for } (y, z) \in \operatorname{dom} T^- = \mathcal{C}(\mathcal{I}, \mathbb{R}^m) \times \mathbb{R}^\delta, \tag{4.33}$$

is a bounded outer inverse of the composed operator T. The topological direct sum decomposition $\ker T^- \oplus \operatorname{im} TT^- = \mathcal{C}(\mathcal{I}, \mathbb{R}^m) \times \mathbb{R}^\delta$ is valid with

$$\ker T^- = \{(y, z) \in \mathcal{C}(\mathcal{I}, \mathbb{R}^m) \times \mathbb{R}^\delta : G_\mu P_1 \cdots P_{\mu-1} G_\mu^{-1} y = 0, \; z = 0\},$$

$$\operatorname{im} TT^- = \{(y, z) \in \mathcal{C}(\mathcal{I}, \mathbb{R}^m) \times \mathbb{R}^\delta : (I - G_\mu P_1 \cdots P_{\mu-1} G_\mu^{-1}) y = 0\}.$$

Proof 1. Because of $\Pi_{\mu-1} P_1 \cdots P_{\mu-1} = \Pi_{\mu-1}$ and $P_1 \cdots P_{\mu-1} - \Pi_{\mu-1} = Q_0 P_1 \cdots P_{\mu-1}$, the equation $Tx = G_\mu P_1 \cdots P_{\mu-1} G_\mu^{-1} y$ decomposes into the system

$$T_{dyn}x = G_\mu \Pi_{\mu-1} G_\mu^{-1} y, \quad T_{alg}x = G_\mu Q_0 P_1 \cdots P_{\mu-1} G_\mu^{-1} y.$$

Equation $T_{alg}x = G_\mu Q_0 P_1 \cdots P_{\mu-1} G_\mu^{-1} y$ means in detail, see (4.23),

$$G_\mu \left\{ \sum_{l=0}^{\mu-1} Q_l x - \sum_{l=0}^{\mu-2} (I - \Pi_l) Q_{l+1} D^-(D \Pi_l Q_{l+1} x)' + \sum_{l=0}^{\mu-2} \mathcal{M}_{l+1} D \Pi_l Q_{l+1} x \right\}$$

$$= G_\mu Q_0 P_1 \cdots P_{\mu-1} G_\mu^{-1} y.$$

Multiplying successively by $\Pi_{\mu-2}Q_{\mu-1}G_\mu^{-1},\ldots,\Pi_0 Q_1 G_\mu^{-1}$, we obtain the relations $\Pi_{\mu-2}Q_{\mu-1}x = 0,\ldots,\Pi_0 Q_1 x = 0$. It results that $(I - \Pi_{\mu-1})x = Q_0 x = Q_0 P_1 \cdots P_{\mu-1} G_\mu^{-1} y$.

Regarding condition (4.30) we obtain

$$D(t_a)\Pi_{\mu-1}(t_a) = D(t_a)C^+ C \Pi_{\mu-1}(t_a) = D(t_a)C^+C.$$

The IVP $T_{dyn}x = G_\mu \Pi_{\mu-1} G_\mu^{-1} y$, $D(t_a)\Pi_{\mu-1}x(t_a) = D(t_a)C^+z$ has the solutions $x = D^- u + v$, where $v \in \ker \Pi_{\mu-1}$ is arbitrary, and u is the unique solution of the ODE (4.26) (with y replacing q) satisfying the initial condition $u(t_a) = D(t_a)C^+z$, see Corollary 4.3.

Summarizing we know that $v = Q_0 P_1 \cdots P_{\mu-1} G_\mu^{-1} y$ and

$$Cx(t_a) = CD(t_a)^- u(t_a) = CD(t_a)^- D(t_a)C^+z = CC^+z = z,$$

and hence,

$$x = D^- u + Q_0 P_1 \cdots P_{\mu-1} G_\mu^{-1} y \in \mathcal{C}_D^1(\mathcal{I},\mathbb{R}^m)$$

is the unique solution of the IVP (4.31). Because of $Dx = u$, the inequality (4.32) is evident.

2. The inequality (4.32) actually means that $\mathcal{T}^-$ is bounded. The relations $\mathcal{T}\mathcal{T}^-(y,z) = (G_\mu P_1 \cdots P_{\mu-1} G_\mu^{-1} y, z)$ and $\mathcal{T}^-\mathcal{T}\mathcal{T}^- = \mathcal{T}^-$ follow immediately, and hence $\mathcal{T}^-$ is an outer inverse. Furthermore, $G_\mu P_1 \cdots P_{\mu-1}G_\mu^{-1} y = 0$, $z = 0$ yield $\mathcal{T}^-(y,z) = 0$, and vice versa.

$\square$

4.3 Regular Nonlinear Differential-Algebraic Operators

As we have seen in Sect. 4.1, the nonlinear differential-algebraic operator (4.3) is Fréchet differentiable under natural assumptions.

Definition 4.4 The nonlinear differential-algebraic operator F defined by (4.3) is said to be regular, if the derivative $F'(x_*) \in \mathcal{L}(\mathcal{C}_D^1(\mathcal{I},\mathbb{R}^m), \mathcal{C}(\mathcal{I},\mathbb{R}^m))$ is a fine regular differential-algebraic operator at least for each arbitrary reference function $x_* \in \text{dom } F \cap \mathcal{C}^m(\mathcal{I},\mathbb{R}^m)$.

Owing to Corollary 6.13 the derivative $F'(x_*)$ of a regular operator must have characteristics $r_0,\ldots,r_\mu$ and tractability index μ being uniform for all reference functions x_*. *We assign these characteristic values and the tractability index also to the nonlinear operator F.*

Theorem 4.6 *Let the nonlinear differential-algebraic operator F defined by (4.3) be regular. Then F is Fredholm, if and only if it has tractability index $\mu \in \{0, 1\}$. When indicated, the Fredholm index equals $\operatorname{ind}_{fred} F = \operatorname{rank} D = r_0$.*

Proof If F is regular with index 0 and 1, the linearization $F'(x_*)$ is fine not only for $x_* \in \operatorname{dom} F \cap C^m(\mathcal{I}, \mathbb{R}^m)$ but for all $x_* \in \operatorname{dom} F$. Then the statement concerning $\mu = 1$ follows from Theorem 4.2. In the less interesting case $\mu = 0$ the matrix function D remains nonsingular, which makes the statement evident.

If F is regular with tractability index $\mu > 1$, applying Theorem 4.2 once again, we know that F fails to be Fredholm.   □

In higher-index cases, the linearization $F'(x_*)$ has a nonclosed image $\operatorname{im} F'(x_*)$ in $C(\mathcal{I}, \mathbb{R}^m)$ and, what makes matters worse, the nonclosed proper subset $\operatorname{im} F'(x_*)$ may actually depend on the reference function x_*. We demonstrate this fact by the next example.

Example 4.7 (Image Depends on the Reference Function) The operator F defined by $m = 4, n = 1$,

$$f(y, x, t) = \begin{bmatrix} 1 \\ 0 \\ 0 \\ 0 \end{bmatrix} y + \begin{bmatrix} x_4 - \gamma(t) \\ x_1 + x_2 x_3 \\ x_2 \\ x_3 \end{bmatrix}, \quad y \in \mathbb{R}, x \in \mathbb{R}^4, t \in \mathcal{I}, \quad D = \begin{bmatrix} 1 & 0 & 0 & 0 \end{bmatrix},$$

with $\gamma \in C(\mathcal{I}, \mathbb{R})$, $\operatorname{dom} F = C_D^1(\mathcal{I}, \mathbb{R}^4) = \{x \in C(\mathcal{I}, \mathbb{R}^4) : x_1 \in C^1(\mathcal{I}, \mathbb{R})\}$, is associated with the simple DAE taken from [74, p. 41]

$$x_1'(t) + x_4(t) - \gamma(t) = 0,$$
$$x_1(t) + x_2(t)x_3(t) = 0.$$
$$x_2(t) = 0,$$
$$x_3(t) = 0.$$

For any reference function $x_* \in \operatorname{dom} F$ the linearization reads

$$F'(x_*)x = \begin{bmatrix} 1 \\ 0 \\ 0 \\ 0 \end{bmatrix} ([1\ 0\ 0\ 0] x)' + \begin{bmatrix} 0 & 0 & 0 & 1 \\ 1 & x_{*3} & x_{*2} & 0 \\ 0 & 1 & 0 & 0 \\ 0 & 0 & 1 & 0 \end{bmatrix} x, \quad x \in C_D^1(\mathcal{I}, \mathbb{R}^4),$$

the linear operator $F'(x_*)$ is regular with tractability index 2, and its image is

$$\operatorname{im} F'(x_*) = \{q \in C(\mathcal{I}, \mathbb{R}^4) : q_2 - x_{*3}q_3 - x_{*2}q_4 \in C^1(\mathcal{I}, \mathbb{R})\}.$$

For instance, if we choose the reference functions

$$x_* = \begin{bmatrix} 0 \\ 0 \\ 0 \\ \gamma \end{bmatrix} \quad \text{and} \quad x_{**} = \begin{bmatrix} 0 \\ 0 \\ \epsilon \\ \gamma \end{bmatrix},$$

then we are confronted with the different sets

$$\operatorname{im} F'(x_*) = \{q \in \mathcal{C}(\mathcal{I}, \mathbb{R}^4) : q_2 \in \mathcal{C}^1(\mathcal{I}, \mathbb{R})\},$$
$$\operatorname{im} F'(x_{**}) = \{q \in \mathcal{C}(\mathcal{I}, \mathbb{R}^4) : q_2 - \epsilon q_3 \in \mathcal{C}^1(\mathcal{I}, \mathbb{R})\}.$$

$\square$

4.3.1   Local Solvability by Outer Inverses

We apply a generalized implicit function theorem from [19] which does not base upon a bounded inverse of the derivative. Instead, a suitably chosen *approximate outer inverse* is used (see Appendix 6.1.2, Theorem 6.8).

If F is regular with tractability index $\mu \in \{0, 1\}$, then the linear IVP

$$F'(x_*)x = y, \quad Cx(t_a) = 0, \tag{4.34}$$

with $C \in \mathcal{L}(\mathbb{R}^m, \mathbb{R}^{m-l})$, $l = m - r_0$, and $\ker C = \ker D(t_a)$, is uniquely solvable for each arbitrary $y \in \mathcal{C}(\mathcal{I}, \mathbb{R}^m)$. The operator

$$T_*^- \in \mathcal{L}(\mathcal{C}(\mathcal{I}, \mathbb{R}^m), \mathcal{C}_D^1(\mathcal{I}, \mathbb{R}^m)),$$
$$T_*^- y := \text{ solution of the IVP (4.34)}, \quad y \in \mathcal{C}(\mathcal{I}, \mathbb{R}^m),$$

is actually an injective bounded outer inverse of $F'(x_*)$. Namely, it holds that $\ker T_*^- = \{0\}$ and

$$F'(x_*)T_*^- y = y, \quad T_*^- F'(x_*)T_*^- y = T_*^- y, \quad y \in \mathcal{C}(\mathcal{I}, \mathbb{R}^m).$$

Theorem 4.8 *Let F be the differential-algebraic operator described in Sect. 4.1, $x_* \in \operatorname{dom} F$, and $F(x_*) = 0$. Let F be regular with tractability index $\mu \in \{0, 1\}$.*

Then, whenever $z_* \in \mathcal{C}_D^1(\mathcal{I}, \mathbb{R}^m)$ satisfies $\|z_*\|_{\mathcal{C}_D^1} = 1$ and $F(x_*)z_* = 0$, there exists a solution $x(s) = x_* + s z_* + o(s)$ to $Fx = 0$, with $s > 0$ sufficiently small.

Proof Since the outer inverse T_*^- is injective, the equation $Fx = 0$ is equivalent with $T_*^- Fx = 0$; and hence the statement follows from [19, Theorem 3]. $\square$

Regarding that here $\ker F'(x_*)$ has dimension $m - l = r_0$ we obtain $Fx = 0$, a local solution set of dimension r_0.

The situation in higher-index cases is more involved. If F is regular with tractability index $\mu \geq 2$ and $x_* \in \operatorname{dom} F$ is sufficiently smooth, then there are completely decoupling projector functions $Q_{*0}, \ldots Q_{*\mu-1}$ associated with the linearization $F'(x_*)$. The linear IVPs

$$F'(x_*)x = \underbrace{G_{*\mu} P_{*1} \cdots P_{*\mu-1} G_{*\mu}^{-1}}_{=:\mathcal{P}_*} y, \quad C_* x(t_a) = 0, \tag{4.35}$$

with $C_* \in \mathcal{L}(\mathbb{R}^m, \mathbb{R}^{m-l})$, $\ker C_* = \ker \Pi_{*\mu-1}(t_a)$, $l = \sum_{j=0}^{\mu-1}(m-r_j)$, are uniquely solvable for each arbitrary $y \in \mathcal{C}(\mathcal{I}, \mathbb{R}^m)$. The operator

$$T_*^- \in \mathcal{L}(\mathcal{C}(\mathcal{I}, \mathbb{R}^m), \mathcal{C}_D^1(\mathcal{I}, \mathbb{R}^m)),$$

$$T_*^- y := \text{ solution of the IVP (4.35)}, \quad y \in \mathcal{C}(\mathcal{I}, \mathbb{R}^m),$$

is actually a bounded outer inverse of $F'(x_*)$. It holds that

$$F'(x_*)T_*^- y = y, \quad T_*^- F'(x_*)T_*^- y = T_*^- y, \quad y \in \mathcal{C}(\mathcal{I}, \mathbb{R}^m).$$

Now, in the higher-index case, T_*^- is no longer injective, but

$$\ker T_*^- = \{y \in \mathcal{C}(\mathcal{I}, \mathbb{R}^m) : \mathcal{P}_* y = 0\}.$$

If $x_* \in \operatorname{dom} F$ lacks in smoothness, $x_* \notin C^m(\mathcal{I}, \mathbb{R}^m)$, then $F'(x_*)$ is at least approximately outer invertible (cf. Appendix 6.1.2) as the following lemma shows.

Lemma 4.9 *Let F be the differential-algebraic operator described in Sect. 4.1, let F be regular with tractability index $\mu \geq 2$, $x_* \in \operatorname{dom} F$. Then the derivative $F'(x_*)$ is approximately outer invertible with a constant bound function $\Gamma(\rho) = M$ and approximate outer inverses $T_\rho^{\simeq}$ such that (cf. (4.36))*

$$\ker T_\rho^{\simeq} = \ker \mathcal{P}_\rho.$$

Proof Because of the continuity of $F'(\cdot)$, to $\varepsilon > 0$ there exists a $\delta(\varepsilon) > 0$ such that

$$\|x - x_*\|_{C_D^1} \leq \delta(\varepsilon) \quad \Rightarrow \quad \|F'(x) - F'(x_*)\| \leq \varepsilon.$$

Let $\delta_0 > 0$ be sufficiently small. We consider an arbitrary element $x_\rho \in B(x_*, \delta_0) \cap C^m(\mathcal{I}, \mathbb{R}^m)$ and turn to the linearization $F'(x_\rho)$ which is fine. There are completely decoupling projector functions $Q_{\rho 0}, \ldots Q_{\rho \mu - 1}$ associated with the linearization $F'(x_\rho)$. The linear IVPs

$$F'(x_\rho)x = \underbrace{G_{\rho \mu} P_{\rho 1} \cdots P_{\rho \mu - 1} G_{\rho \mu}^{-1}}_{=:\mathcal{P}_\rho} y, \quad C_\rho x(t_a) = 0, \tag{4.36}$$

with $C_\rho \in \mathcal{L}(\mathbb{R}^m, \mathbb{R}^{m-l})$, $\ker C_\rho = \ker \Pi_{\rho \mu - 1}(t_a)$, $l = \sum_{j=0}^{\mu-1}(m - r_j)$, are uniquely solvable for each arbitrary $y \in \mathcal{C}(\mathcal{I}, \mathbb{R}^m)$. The operator

$$T_\rho^- \in \mathcal{L}(\mathcal{C}(\mathcal{I}, \mathbb{R}^m), \mathcal{C}_D^1(\mathcal{I}, \mathbb{R}^m)),$$

$$T_\rho^- y := \text{ solution of the IVP (4.36)}, \quad y \in \mathcal{C}(\mathcal{I}, \mathbb{R}^m),$$

is a bounded outer inverse of $F'(x_\rho)$. It holds that

$$F'(x_\rho)T_\rho^- y = \mathcal{P}_\rho y, \quad T_\rho^- F'(x_\rho)T_\rho^- y = T_\rho^- y, \quad y \in \mathcal{C}(\mathcal{I}, \mathbb{R}^m).$$

Owing to the construction of the matrix function sequences and the representation of the coefficients of the linearizations (cf. [55], also Appendix 6.1), there is a uniform upper bound $\|T_\rho^-\| \le M$ for all those reference functions x_ρ.

Next, for each $\rho \in (0, 1)$ we ensure $\delta(\rho/M) \le \delta_0$, and fix an element $x_\rho \in B(x_*, \delta(\rho/M)) \cap \mathcal{C}^m(\mathcal{I}, \mathbb{R}^m)$. We have then

$$\|(T_\rho^- F'(x_*)T_\rho^- - T_\rho^-)y\|_{\mathcal{C}_D^1} = \|(T_\rho^-(F'(x_*) - F'(x_\rho))T_\rho^- y\|_{\mathcal{C}_D^1} \le \rho \|T_\rho^- y\|.$$

This means that $T_\rho^{\simeq} := T_\rho^-$ represents the required approximate outer inverse of $F'(x_*)$, with the bound function $\Gamma(\rho) = M$. $\qquad \Box$

Theorem 4.10 *Let F be the differential-algebraic operator described in Sect. 4.1, let F be regular with tractability index $\mu \ge 2$, $x_* \in \mathrm{dom}\, F$, and $F(x_*) = 0$.*

Then, whenever $z_* \in \mathcal{C}_D^1(\mathcal{I}, \mathbb{R}^m))$ satisfies $\|z_*\|_{\mathcal{C}_D^1} = 1$ and $F'(x_*)z_* = 0$, there exists a solution $x(s) = x_* + sz_* + o(s)$ to the equation $T_\rho^- F x = 0$, with $s > 0$ sufficiently small, with an appropriate choice of $\rho = \rho(s) \downarrow 0$ as $s \downarrow 0$, which means

$$F(x(s)) \in \ker \mathcal{P}_{\rho(s)}, \quad s > 0 \quad \text{sufficiently small.}$$

If, moreover, x_* is sufficiently smooth so that the derivative $F'(x_*)$ is fine, then $x(s)$ solves the equation $T_*^- F x = 0$. It holds that

$$F(x(s)) \in \ker \mathcal{P}_*, \quad s > 0 \quad \text{sufficiently small.}$$

Proof The statement follows from [19, Theorem 3], cf. Theorem 6.8. $\qquad \Box$

Regarding that ker $F'(x_\rho)$ has dimension $m - l$ one obtains a solution set of dimension $m - l$ to equation $T_*^- Fx = 0$.

4.3.2   Well-Posed IVPs and BVPs with Regular Index-1 Operators

Any regular differential-algebraic operator F with tractability index $\mu \in \{0, 1\}$ is Fredholm, its Fréchet derivative $F'(x)$ is surjective, and ker $F'(x)$ has dimension $m - l = r_0 = \operatorname{rank} D$, thus the Fredholm index is $\operatorname{ind}_{fred} F'(x) = r_0$. As indicated by the Fredholm index, aiming at a well-posed equation $\mathcal{F}x = 0$, one has to complete the equation $Fx = 0$ by exactly r_0 appropriate boundary and initial conditions. If one adds more or fewer boundary conditions, the composed equation fails to be well-posed.

We concentrate on the index-1 case. Index-0 operators can be treated analogously, however, then we have $r_0 = m$, which makes this case less interesting and close to the well-known ODE theory.

The operator equation (see (4.11))

$$\mathcal{F}x = (Fx, b(x(t_a), x(t_e))) = (q, d) \tag{4.37}$$

describes the BVP

$$f((Dx)'(t), x(t), t) = q(t), \ t \in \mathcal{I} = [t_a, t_e], \quad b(x(t_a), x(t_e)) = d. \tag{4.38}$$

Let F be regular with tractability index 1, $x_* \in \operatorname{dom} F$. Owing to Theorem 4.2 and Corollary 4.3, if $\mathbf{e}_i \in \mathbb{R}^m$ denotes the ith unit vector, there exists exactly one $\xi_i \in \mathcal{C}_D^1(\mathcal{I}, \mathbb{R}^m)$ such that

$$F'(x_*)\xi_i = 0, \quad D(t_a)\xi_i(t_a) = D(t_a)\mathbf{e}_i, \quad i = 1, \dots, m.$$

Introduce the matrix-valued function

$$X_* := [\xi_1, \dots, \xi_m] \tag{4.39}$$

such that

$$A_*(DX_*)' + B_*X_* = 0, \quad D(t_a)(X_*(t_a) - I) = 0.$$

This yields the representation

$$\ker F'(x_*) = \{x = X_*c : c \in \mathbb{R}^m\}.$$

Notice that X_* is the so-called *maximal fundamental solution matrix normalized at* t_a, e.g.,[55]. The matrix function X_* has constant rank r_0, and it holds that ker $X_* = $ ker D.

The following theorem provides conditions for the map $\mathcal{F}$ to be a local diffeomorphism (cf. Theorem 6.7).

Theorem 4.11 *Let $\mathcal{F}$ be the composed operator (4.11) described in Sect. 4.1, with differential-algebraic operator F being regular with tractability index 1. Suppose further $x_* \in \mathrm{dom}\, F$, $\mathcal{F}(x_*) = 0$, and let the matrix*

$$S_* := b_a'(x_*(t_a), x_*(t_e))X_*(t_a) + b_e'(x_*(t_a), x_*(t_e))X_*(t_e) \in \mathcal{L}(\mathbb{R}^m, \mathbb{R}^{r_0})$$

satisfy the conditions

$$\mathrm{im}\, S_* = \mathbb{R}^{r_0}, \quad \ker S_* = \ker D(t_a). \tag{4.40}$$

Then, the equation $\mathcal{F}x = 0$ is well-posed around x_*; to each $(q, d) \in \mathcal{C}(\mathcal{I}, \mathbb{R}^m) \times \mathbb{R}^{r_0}$ being sufficiently small, the perturbed equation $\mathcal{F}x = (q, d)$ possesses exactly one solution $x(q, d)$ in the neighborhood of x_*, and the inequality

$$\|x(q, d) - x_*\|_{\mathcal{C}_D^1} \le \kappa(\|q\|_\infty + |d|)$$

is valid with a constant κ.

Proof We introduce the differentiable map $\mathcal{H}(x, q, d) := \mathcal{F}x - (q, d)$, $\mathcal{H}_x'(x, q, d) = \mathcal{F}'(x)$, $\mathcal{H}(x_*, 0, 0) = 0$. If $\mathcal{F}'(x_*)$ is a bijection, then the statement results from the standard implicit function theorem in Banach spaces. It remains to show the injectivity and surjectivity of the linearization $\mathcal{F}'(x_*)$.

If $\mathcal{F}'(x_*)z = 0$ then $F'(x_*)z = 0$, thus $z = X_*c$, with a certain $c \in \mathbb{R}^m$. On the other hand, then we have $b_a'(x_*(t_a), x_*(t_e))z(t_a) + b_e'(x_*(t_a), x_*(t_e))z(t_e) = 0$, therefore $S_*c = 0$. Condition (4.40) implies $c \in \ker D(t_a)$, and hence $z = 0$.

We already know that F is surjective. It remains to prove that the BVP

$$F'(x_*)z = 0, \quad b_a'(x_*(t_a), x_*(t_e))z(t_a) + b_e'(x_*(t_a), x_*(t_e))z(t_e) = d,$$

is solvable for each arbitrary $d \in \mathbb{R}^{r_0}$. The function $z_d := X_*S_*^+d$ satisfies the first equation and the boundary condition reads $b_a'(x_*(t_a), x_*(t_e))z_d(t_a) + b_e'(x_*(t_a), x_*(t_e))z_d(t_e) = S_*S_*^+ = d$. □

Often one deals with the simpler IVPs.

Corollary 4.12 *Let $\mathcal{F}$ be the composed operator (4.11) described in Sect. 4.1, with differential-algebraic operator F being regular with tractability index 1. Let the boundary condition simplify to the initial condition*

$$b(x(t_a), x(t_e)) := Cx(t_a) - h = 0, \tag{4.41}$$

whereby $C \in \mathcal{L}(\mathbb{R}^m, \mathbb{R}^{r_0})$, $h \in \mathbb{R}^{r_0}$.

Let $x_* \in$ dom F, $\mathcal{F}(x_*) = 0$, and

$$\ker C \cap \{w \in \mathbb{R}^m : B_*(t_a)w \in A_*(t_a)\} = \{0\}. \tag{4.42}$$

Then the IVP $\mathcal{F}x = 0$ is well-posed around x_*.

Proof $\Pi_{*can}(t_a) \in \mathcal{L}(\mathbb{R}^m)$ represents the projector given by

$$\operatorname{im} \Pi_{*can}(t_a) = \{w \in \mathbb{R}^m : B_*(t_a)w \in A_*(t_a)\},$$

$$\ker \Pi_{*can}(t_a) = \ker D(t_a).$$

$\Pi_{*can}(t_a)$ has rank r_0. Regarding that $X_*(t_a) = \Pi_{*can}(t_a)$ we arrive at $S_* = C\Pi_{*can}(t_a)$. It remains to check condition (4.40).

$S_*z = 0$ means $\Pi_{*can}(t_a)z \in \ker C$. Because of (4.42) it follows that $\Pi_{*can}(t_a)z = 0$, thus $D(t_a)z = D(t_a)\Pi_{*can}(t_a)z = 0$. Conversely, $z \in \ker D(t_a)$ implies $S_*z = C\Pi_{*can}(t_a)z = C\Pi_{*can}(t_a)D(t_a)^+D(t_a)z = 0$. This proves the relation $\ker S_* = \ker D(t_a)$. Finally, for reasons of dimensions, it holds that $\operatorname{im} S_* = \mathbb{R}^{r_0}$. $\square$

The decomposition

$$\{w \in \mathbb{R}^m : B_*(t_a)w \in A_*(t_a)\} \oplus \ker D(t_a) = \mathbb{R}^m, \tag{4.43}$$

which is associated with the projector $\Pi_{*can}(t_a)$, makes evident that, choosing the matrix C so that

$$\ker C = \ker D(t_a), \tag{4.44}$$

condition (4.42) is always valid. Otherwise, additional structural restrictions are required. The choice (4.44) is natural in the sense that it directly applies to the inherent dynamical part. However, in practice, it can be required to fix other components for different reasons.

Example 4.13 (Additional Structural Restriction) The differential-algebraic operator F associated with the semi-explicit system of $m_1 + m_2 = m$ equations

$$x_1'(t) - g_1(x_1(t), x_2(t), t) = 0,$$

$$g_2(x_1(t), x_2(t), t) = 0,$$

is regular with tractability index 1, if the partial Jacobian g_{2,x_2} is everywhere nonsingular. Then $r_0 = m_1$ results. Assume, for simplicity, $m_1 = m_2$ and set

$$C = \begin{bmatrix} 0 & I \end{bmatrix}.$$

Condition (4.42) means now that

$$\ker C \ \cap \ \{z \in \mathbb{R}^m : g_{*2,x_1}(t_a)z_1 + g_{*2,x_i}(t_a)z_2 = 0\}$$
$$= \{z \in \mathbb{R}^m : z_2 = 0, \ g_{*2,x_1}(t_a)z_1 = 0\} = \{0\}$$

is valid, with $g_{*2,x_i}(t_a) := g_{2,x_i}(x_{*1}(t_a), x_{*2}(t_a), t_a)$. Evidently, here condition (4.42) requires that also the matrix $g_{*2,x_1}(t_a)$ must be nonsingular. $\qquad\square$

4.3.3 Regular Index-2 Operators and Advanced Setting

Let the linear differential-algebraic operator $T \in \mathcal{L}_b(\mathcal{C}_D^1(\mathcal{I}, \mathbb{R}^m), \mathcal{C}(\mathcal{I}, \mathbb{R}^m))$,

$$Tx = A(Dx)' + Bx, \quad x \in \mathcal{C}_D^1(\mathcal{I}, \mathbb{R}^m), \tag{4.45}$$

be regular with tractability index 2. Then, by Theorem 4.2, the image $\operatorname{im} T$ is a nonclosed proper subset in $\mathcal{C}(\mathcal{I}, \mathbb{R}^m)$. We are interested in revealing the detailed structure of this subset, and, eventually, in modifying the image space to enforce surjectivity and the Fredholm property as in Sect. 2.4.1.

Choose $Q_0 = D^+D$, and let Q_1 denote the projector function onto $N_1 = \ker G_1$ along $S_1 = \{z \in \mathbb{R}^m : Bz \in \operatorname{im} G_1\}$. This provides a fine decoupling, and, in particular, $Q_1 = Q_1 G_2^{-1} B P_0$ (e.g.,[55, Section 2.4.3]). Compute for each arbitrary $x \in \mathcal{C}_D^1(\mathcal{I}, \mathbb{R}^m)$

$$D\Pi_0 Q_1 G_2^{-1} Tx = D\Pi_0 Q_1 G_2^{-1}\{G_1(D^-(Dx)' + Q_0 x) + BP_0 x\}$$
$$= D\Pi_0 Q_1 G_2^{-1} BP_0 x = D\Pi_0 Q_1 x = \underbrace{D\Pi_0 Q_1 D^- Dx}_{\in \mathcal{C}^1} \in \mathcal{C}^1(\mathcal{I}, \mathbb{R}^n),$$

such that $\operatorname{im} T \subseteq \mathcal{C}^{ind\,2}(\mathcal{I}, \mathbb{R}^m)$,

$$\mathcal{C}^{ind\,2}(\mathcal{I}, \mathbb{R}^m) := \{q \in \mathcal{C}(\mathcal{I}, \mathbb{R}^m) : D\Pi_0 Q_1 G_2^{-1} q \in \mathcal{C}^1(\mathcal{I}, \mathbb{R}^n)\}. \tag{4.46}$$

By applying the decoupling procedure to equation $Tx = q$ one shows solvability for each arbitrary $q \in \mathcal{C}^{ind\,2}(\mathcal{I}, \mathbb{R}^m)$, and hence

$$\operatorname{im} T = \mathcal{C}^{ind\,2}(\mathcal{I}, \mathbb{R}^m). \tag{4.47}$$

We equip the linear space $\mathcal{C}^{ind\,2}(\mathcal{I}, \mathbb{R}^m)$ with the norm

$$\|q\|_{ind\,2} := \|q\|_\infty + \|(D\Pi_0 Q_1 G_2^{-1} q)'\|_\infty, \quad x \in \mathcal{C}^{ind\,2}(\mathcal{I}, \mathbb{R}^m), \tag{4.48}$$

which yields a Banach space. We derive further that

$$\|Tx\|_{ind\,2} = \|Tx\|_\infty + \|(D\Pi_0 Q_1 G_2^{-1} Tx)'\|_\infty$$

$$\leq \kappa \|x\|_{\mathcal{C}_D^1} + \|(D\Pi_0 Q_1 D^- Dx)'\|_\infty \leq \kappa_{new} \|x\|_{\mathcal{C}_D^1}$$

for each arbitrary $x \in \mathcal{C}_D^1(\mathcal{I}, \mathbb{R}^m)$. This means that our differential-algebraic operator is bounded and surjective, thus normally solvable in the new setting $T \in \mathcal{L}(\mathcal{C}_D^1(\mathcal{I}, \mathbb{R}^m), \mathcal{C}^{ind\,2}(\mathcal{I}, \mathbb{R}^m))$. By Theorem 4.2, ker T has dimension $m - l = r_0 - (m - r_1)$ such that T is a Fredholm operator in the advanced setting, and $\mathrm{ind}_{fred}(T) = m - l$.

The composed operator $\mathcal{T} \in \mathcal{L}(\mathcal{C}_D^1(\mathcal{I}, \mathbb{R}^m), \mathcal{C}^{ind\,2}(\mathcal{I}, \mathbb{R}^m) \times \mathbb{R}^{m-l})$ associated to a respective IVP,

$$\mathcal{T}x = (Tx, Cx(t_a)), \quad x \in \mathcal{C}_D^1(\mathcal{I}, \mathbb{R}^m),$$

with $C \in \mathcal{L}(\mathbb{R}^m, \mathbb{R}^{m-l})$, ker $C = N_0(t_a) \oplus N_1(t_a)$, acts bijectively in Banach spaces. Therefore, the IVP $\mathcal{T}x = (q, d)$ is well-posed in the adapted setting. So far so good.

We turn to the nonlinear differential-algebraic operator F and ask whether we can modify Theorem 4.11 and Corollary 4.12 accordingly.

Suppose $x_* \in \mathrm{dom}\, F$, $Fx_* = 0$ and apply the matrix function sequence and the associated projector functions to the linearization $F(x_*)$. Below, the subscript $*$ indicates the resulting dependence of x_*.

We provide first an appropriate fundamental solution matrix $X_*(t)$. Choose a matrix $C_* \in \mathcal{L}(\mathbb{R}^m, \mathbb{R}^{m-l})$, $l = m - r_0 + m - r_1$, so that the condition ker $C_* = N_0(t_a) \oplus N_{*1}(t_a)$ is satisfied. Then there exists exactly one $\xi_i \in \mathcal{C}_D^1(\mathcal{I}, \mathbb{R}^m)$ such that

$$F'(x_*)\xi_i = 0, \quad C_*\xi_i(t_a) = C_*e_i, \quad i = 1, \ldots, m.$$

The matrix-valued function

$$X_* := [\xi_1, \ldots, \xi_m] \tag{4.49}$$

satisfies the IVP

$$A_*(DX_*)' + B_*X_* = 0, \quad C_*(X_*(t_a) - I) = 0,$$

and the representation

$$\ker F'(x_*) = \{x = X_*c : c \in \mathbb{R}^m\}.$$

Proposition 4.14 *Let $\mathcal{F}$ be the composed operator (4.11) described in Sect. 4.1, with differential-algebraic operator F being regular with tractability index 2, $l = m - r_0 + m - r_1$.*

Let $x_* \in \text{dom } F$, $\mathcal{F}(x_*) = 0$. Let there exist an open neighborhood $\mathcal{U}_{x_*}$ of x_* such that

$$Fx \in \text{im } F'(x_*), \quad x \in \mathcal{U}_{x_*}, \tag{4.50}$$

and let the matrix

$$S_* := b_a'(x_*(t_a), x_*(t_e))X_*(t_a) + b_e'(x_*(t_a), x_*(t_e))X_*(t_e) \in \mathcal{L}(\mathbb{R}^m, \mathbb{R}^{m-l})$$

satisfy the conditions

$$\text{im } S_* = \mathbb{R}^{m-l}, \quad \ker S_* = N_0(t_a) \oplus N_{*1}(t_a). \tag{4.51}$$

Then, the BVP $\mathcal{F}x = 0$ is well-posed around x_* in the advanced setting; to each $(q, d) \in \mathcal{C}_*^{ind\,2}(\mathcal{I}, \mathbb{R}^m) \times \mathbb{R}^{m-l}$ being sufficiently small, the perturbed equation $\mathcal{F}x = (q, d)$ possesses exactly one solution $x(q, d)$ in the neighborhood of x_*, and the inequality

$$\|x(q, d) - x_*\|_{\mathcal{C}_D^1} \leq \kappa(\|q\|_\infty + \|(D\Pi_0 Q_{*1}G_{*2}^{-1}q)'\|_\infty + |d|)$$

is valid with a constant κ.

Proof Regarding the advanced setting the statement is proved in the same way as Theorem 4.11. □

Corollary 4.15 *Let $\mathcal{F}$ be the composed operator (4.11) described in Proposition 4.14. Let the projector functions Q_{*0}, Q_{*1} provide a complete decoupling of the linearization $F'(x_*)$, $\Pi_{*can} := P_{*0}P_{*1}$. Let the boundary condition simplify to the initial condition*

$$b(x(t_a), x(t_e)) := Cx(t_a) - h = 0, \tag{4.52}$$

whereby $C \in \mathcal{L}(\mathbb{R}^m, \mathbb{R}^{m-l})$, $h \in \mathbb{R}^{m-l}$, and

$$\ker C \cap \text{im } \Pi_{*can}(t_a) = \{0\}. \tag{4.53}$$

Then the IVP $\mathcal{F}x = 0$ is well-posed around x_* in the advanced setting.

Proof $\Pi_{*can}(t_a)$ has rank $m - l$. Regarding that $X_*(t_a) = \Pi_{*can}(t_a)$ we arrive at $S_* = C\Pi_{*can}(t_a)$. It remains to check condition (4.51).

$S_* z = 0$ means $\Pi_{*can}(t_a)z \in \ker C$. Because of (4.49) it follows that $\Pi_{*can}(t_a)z = 0$, thus $z \in \ker \Pi_{*can}(t_a) = N_0(t_a) \oplus N_{*1}(t_a)$. Conversely, $z \in (N_0(t_a) \oplus N_{*1}(t_a))$ implies $S_* z = C\Pi_{*can}(t_a)z = 0$. This proves the relation $\ker S_* = N_0(t_a) \oplus N_{*1}(t_a)$. For reasons of dimensions, it holds that $\text{im } S_* = \mathbb{R}^{m-l}$. □

In the light of the fact that im $F'(\tilde{x}_*)$ may vary with the reference function $\tilde{x}_*$, see Example 4.7, Condition (4.50) limits the class of relevant DAEs. The following structural restriction ensures (4.50). Let x_* be the reference solution and let the relation

$$f(y,x,t) - f(0, P_0(t)x,t) \in \operatorname{im} G_{*1}(t), \quad y \in \mathbb{R}^n, x \in \mathcal{D}_F, t \in \mathcal{I}, \qquad (4.54)$$

be valid. Then it follows for $x \in \operatorname{dom} F$ that

$$D(t)\Pi_0(t)Q_{*1}(t)G_{*2}(t)^{-1} f((Dx)'(t), x(t), t)$$
$$= D(t)\Pi_0(t)Q_{*1}(t)G_{*2}(t)^{-1} f(0, P_0(t)x(t), t)), \quad t \in \mathcal{I}.$$

Assuming, additionally, that f has also a continuous partial derivative f_t, and regarding that $P_0 x = D^- Dx$ is continuously differentiable, we conclude

$$D\Pi_0 Q_{*1} G_{*2}^{-1} Fx = D\Pi_0 Q_{*1} G_{*2}^{-1} f(0, (P_0 x)(\cdot), \cdot) \in \mathcal{C}^1(\mathcal{I}, \mathbb{R}^n),$$
$$Fx \in \operatorname{im} F'(x_*).$$

Condition (4.54) does not apply to Example 4.7. Fortunately, it applies to the widely used Hessenberg size-2 systems as we demonstrate by the next example.

Example 4.16 (Hessenberg size-2 DAE) Given the special system of $m = m_1 + m_2$ equations

$$x_1'(t) + g_1(x_1(t), x_2(t), t) = 0,$$
$$g_1(x_1(t), t) = 0,$$

yielding

$$D = \begin{bmatrix} I & 0 \end{bmatrix} \in \mathcal{L}(\mathbb{R}^m, \mathbb{R}^{m_1}), \quad f(y,x,t) = \begin{bmatrix} I \\ 0 \end{bmatrix} y + \begin{bmatrix} g_1(x_1, x_2, t) \\ g_2(x_1, t) \end{bmatrix},$$

$$P_0 = \begin{bmatrix} I & 0 \\ 0 & 0 \end{bmatrix}, \quad G_1 = \begin{bmatrix} I & g_{1\,x_2} \\ 0 & 0 \end{bmatrix},$$

and hence

$$f(y,x,t) - f(0, P_0 x, t) = \begin{bmatrix} y + g_1(x_1, x_2, t) - g_1(x_1, 0, t) \\ 0 \end{bmatrix}.$$

im G_1 is independent of its arguments and condition (4.54) is fulfilled.   □

4.3.4 Newton–Kantorovich-Like Iterations

Suppose that the composed operator $\mathcal{F}$ associated with the BVP (4.12) is a local diffeomorphism at $x_* \in \text{dom}\,\mathcal{F}$ and $\mathcal{F}(x_*) = 0$, then the well-known Newton–Kantorovich iteration

$$x_{k+1} = x_k - \mathcal{F}'(x_k)^{-1}\mathcal{F}(x_k), \quad k \geq 0, \tag{4.55}$$

can be applied to approximate x_*. If the initial guess x_0 is sufficiently close to x_*, then these iterations are well-defined and x_k tends to x_*. Practically, one solves the linear equations

$$\mathcal{F}'(x_k)z = -\mathcal{F}(x_k), \quad k \geq 0, \tag{4.56}$$

and, having the solution z_{k+1} of the linear problem (4.56), one puts

$$x_{k+1} = x_k + z_{k+1}. \tag{4.57}$$

The linear problem (4.56) represents the linear BVP

$$f_y(\xi_k(t))(Dz)'(t) + f_x(\xi_k(t))z(t) = -f(\xi_k(t)), \quad t \in \mathcal{I},$$
$$b_a(x_k(t_a), x_k(t_e))z(t_a) + b_e(x_k(t_a), x_k(t_e))z(t_e) = -b(x_k(t_a), x_k(t_e)),$$

with $(\xi_k(t)) := ((Dx_k)'(t), x_k(t), t)$ and the first partial derivatives b_a, b_e of the function b with respect to its first and second arguments.

Mostly, a damping parameter is incorporated, and instead of (4.57) one applies

$$x_{k+1} = x_k + \alpha_{k+1}z_{k+1}, \quad \text{with} \quad \alpha_{k+1} \in (0, 1]. \tag{4.58}$$

Usually the damping parameter is chosen so that the residuum $\mathcal{F}(x_{k+1})$ becomes smaller in some sense, that is

$$\|\mathcal{F}(x_{k+1})\|_{res} < \|\mathcal{F}(x_k)\|_{res},$$

with a suitable measure of the residuum, for instance,

$$\|\mathcal{F}(x)\|_{res} := \|\mathcal{F}(x)\| = \|F(x)\|_\infty + |b(x(t_a), x(t_e))|$$
$$\text{and} \quad \|\mathcal{F}(x)\|_{res}^2 := \|F(x)\|_{L^2}^2 + |b(x(t_a), x(t_e))|^2.$$

Sufficient conditions for the composed operator $\mathcal{F}$ to be a local diffeomorphism are described in Sect. 4.3.2 for the index-1 case and in Sect. 4.3.3 for a class of index-2 problems.

Next we take a look at the differentiable functional

$$J(x) := \frac{1}{2}\|F(x)\|_{L^2}^2 + \frac{1}{2}|b(x(t_a), x(t_e))|^2, \quad x \in \text{dom}\,\mathcal{F}. \tag{4.59}$$

Of course, the problem to solve the equation $\mathcal{F}(x) = 0$ can be regarded as the problem to minimize this functional.

For $x \in \text{dom}\,\mathcal{F}$ and $z \in \mathcal{C}_D^1(\mathcal{I}, \mathbb{R}^m)$, the directional derivative reads

$$J'(x)z = (F'(x)z, F(x))_{L^2}$$
$$+ \langle b_a(x(t_a), x(t_e))z(t_a) + b_e(x(t_a), x(t_e))z(t_e), b(x(t_a), x(t_e)) \rangle.$$

If $x^0 \in \text{dom}\,\mathcal{F}$ is fixed, $\mathcal{F}(x^0) \neq 0$, and if there exists a solution z_N of the linear equation,

$$\mathcal{F}'(x^0)z = -\mathcal{F}(x^0), \quad k \geq 0, \tag{4.60}$$

then it results that

$$J'(x^0)z_N = -\|F(x^0)\|_{L^2}^2 - |b(x^0(t_a), x^0(t_e))|^2 < 0$$

thus $J(x^0 + \alpha z_N) < J(x^0)$ for all sufficiently small $\alpha > 0$. Therefore, the so-called Newton direction z_N serves as descent direction. Constructing a descent method by applying Newton directions is essentially the same as the damped Newton–Kantorovich iteration. This works supposing the conditions described in Sects. 4.3.2 and 4.3.3 are given, that is, for index-1 and a restricted class of index-2 problems.

For equations $\mathcal{F}(x) = 0$ involving higher index differential-algebraic operators F, there are two principal difficulties concerning Newton descent and Newton–Kantorovich iteration:

1. The linear equation (4.56) resp. (4.60) is essentially ill-posed and might not be solvable. Changing to LSS does not make great sense, since the linearizations $\mathcal{F}'(x)$ are not normally solvable.
2. For an essentially ill-posed problem a small residuum $\mathcal{F}(x_k)$ does not mean that x_k is close to a solution, see Example 2.3.

Among the methods for ill-posed problems one finds generalizations of Newton-like methods using outer inverses, see [64]. Instead of the unbounded inverse $\mathcal{F}(x_k)^{-1}$ in (4.55) one uses a bounded outer inverse. Theorem 4.5 provides such an outer inverse.

Let $\mathcal{A}(x) \in \mathcal{L}_b(X, Y)$ be an approximation of $\mathcal{F}'(x) \in \mathcal{L}_b(X, Y)$, with $X = \mathcal{C}_D^1(\mathcal{I}, \mathbb{R}^m)$, $Y = \mathcal{C}(\mathcal{I}, \mathbb{R}^m)$. Further, let $x_0 \in \text{dom}\,\mathcal{F}$ and $\Gamma \in \mathcal{L}_b(Y, X)$ be an outer inverse of $\mathcal{A}(x_0)$. Then, supposing that x_k is sufficiently close to x_0 (cf. Lemma 6.6), by

$$\mathcal{A}(x_k)^- := (I + \Gamma(\mathcal{A}(x_k) - \mathcal{A}(x_0))^{-1}\Gamma$$

we obtain a bounded outer inverse of $\mathcal{A}(x_k)$ such that

$$\ker \mathcal{A}(x_k)^- = \ker \Gamma, \quad \text{im} \, \mathcal{A}(x_k)^- = \text{im} \, \Gamma.$$

We refer to [64, Theorem 3.1] for conditions ensuring the sequence

$$x_{k+1} = x_k - \mathcal{A}(x_k)^- \mathcal{F}(x_k), \quad k \geq 0, \tag{4.61}$$

with initial guess x_0, to be well-defined and to converge to a solution of the equation

$$\Gamma \mathcal{F}(x) = 0. \tag{4.62}$$

Then the Eq. (4.62) possesses a unique solution in $\{x_0 + \text{im} \, \Gamma\} \cap B(x_0, \tau)$, τ sufficiently small.

4.4 Different Views on Constant-Rank Conditions

Regularity of differential-algebraic operators in Definitions 4.2 and 4.4 is supported by several constant-rank conditions. These definitions are compatible with the regularity notion for DAEs within the projector based analysis (e.g.,[55], also Appendix 6.3). In the DAE literature several different opinions concerning regularity of time-varying and nonlinear DAEs can be found, which all reflect and generalize regularity of matrix pencils, see [55] for a comprehensive discussion.

Much work concerning DAEs (e.g.,[8, 12]) is focused on problems

$$E(t)x'(t) + F(t)x(t) = q(t), \quad t \in \mathcal{I}, \tag{4.63}$$

being smoothly transformable into the so-called standard canonical form (SCF)

$$\begin{bmatrix} I & 0 \\ 0 & N(t) \end{bmatrix} x'(t) + \begin{bmatrix} W(t) & 0 \\ 0 & I \end{bmatrix} x(t) = g(t), \quad t \in \mathcal{I},$$

which generalizes the Weierstraß–Kronecker form of matrix pencils, (cf. (2.4)). The matrix function N is strictly lower or upper diagonal, but there is no restriction concerning the rank of $N(t)$. In the easier cases, if N is absent or vanishes identically, the matrix $E(t)$ has constant rank, and this is, in essence, in agreement with our notion of regular index-0 and index-1 problems. However, concerning nontrivial N, there is an ongoing controversy.

Supposing sufficiently smooth or even real analytic N and g, DAEs being transformable into SCF are solvable in $\mathcal{C}^1$, and the flow does not show critical behavior.

In contrast, the regularity concept within the framework of the tractability index consequently indicates all corresponding rank changes as critical points. This way it detects serious singularities of the flow such as bifurcations, which are ruled out by the SCF approach a priori, but also so-called *harmless critical points*, which do not affect the flow in smooth environments, see [23,55].

The next example explains the difficulty arising from harmless critical points in a rigorous functional-analytic setting.

Example 4.17 (DAE in SCF with Harmless Critical Point) The DAE

$$
\underbrace{\begin{bmatrix} 0 & 1 & \alpha & 0 \\ 0 & 0 & 0 & \beta \\ 0 & 0 & 0 & 1 \\ 0 & 0 & 0 & 0 \end{bmatrix}}_{=E} x'(t) + x(t) = q(t), \quad t \in \mathcal{I} := [-1,1],
$$

is already in SCF and quasi-regular in the sense of [55]. Let the coefficient functions $\alpha \in \mathcal{C}^1(\mathcal{I}, \mathbb{R})$ and $\beta \in \mathcal{C}(\mathcal{I}, \mathbb{R})$ be given so that

$$
t \in \mathcal{I}_- := [-1,0] \Rightarrow \alpha(t) + \beta(t) = 0, \quad t \in \mathcal{I}_+ := (0,1] \Rightarrow \alpha(t) + \beta(t) > 0.
$$

The DAE is associated with the differential-algebraic operator $\overset{\circ}{T} x = Ex' + x$, $x \in \mathcal{C}^1(\mathcal{I}, \mathbb{R}^4)$, and the closure of $\overset{\circ}{T} \in \mathcal{L}(\mathcal{C}(\mathcal{I}, \mathbb{R}^4), \mathcal{C}(\mathcal{I}, \mathbb{R}^4))$ is given as

$$
Tx = \underbrace{\begin{bmatrix} 1 & 0 \\ 0 & \beta \\ 0 & 1 \\ 0 & 0 \end{bmatrix}}_{=A} (\underbrace{\begin{bmatrix} 0 & 1 & \alpha & 0 \\ 0 & 0 & 0 & 1 \end{bmatrix}}_{=D} x)' + \underbrace{\begin{bmatrix} 1 & 0 & -\alpha' & 0 \\ 0 & 1 & 0 & 0 \\ 0 & 0 & 1 & 0 \\ 0 & 0 & 0 & 1 \end{bmatrix}}_{=B} x, \quad x \in \mathrm{dom}\, T = \mathcal{C}_D^1(\mathcal{I}, \mathbb{R}^4).
$$

Compute

$$
Q_0 = \begin{bmatrix} 1 & 0 & 0 & 0 \\ 0 & 0 & -\alpha & 0 \\ 0 & 0 & 1 & 0 \\ 0 & 0 & 0 & 0 \end{bmatrix}, \quad G_1 = \begin{bmatrix} 1 & 1 & \alpha - \alpha' & 0 \\ 0 & 0 & -\alpha & \beta \\ 0 & 0 & 1 & 1 \\ 0 & 0 & 0 & 0 \end{bmatrix}.
$$

We have here $r_0 = \mathrm{rank}\, G_0 = \mathrm{rank}\, E = 2$. The matrix $G_1(t)$ changes the rank at $t_c = 0$, we have $r_0 = 2$ on $\mathcal{I}_-$ and $r_0 = 3$ on $\mathcal{I}_+$. The operator T fails to be regular on $\mathcal{I}$ since t_c is a critical point, more precisely, a harmless critical point. On the other hand, considering the restrictions $T|_{\mathcal{I}_-}$ and $T|_{\mathcal{I}_+}$ corresponding to the restrictions of the functions x onto the subintervals $\mathcal{I}_-, \mathcal{I}_+$, we may check that both operators are regular, however with different characteristics. Namely, $T|_{\mathcal{I}_-}$ has tractability index $\mu = 2$ and characteristics $r_0 = 2, r_1 = 2, r_2 = 4$, whereas $T|_{\mathcal{I}_+}$ has tractability index $\mu = 3$ and characteristics $r_0 = 2, r_1 = 3, r_2 = 3, r_3 = 4$. We observe qualitatively different operator images:

$$\operatorname{im} T|_{\mathcal{I}_-} = \{q \in C(\mathcal{I}_-, \mathbb{R}^4) : q_2, q_4 \in C^1(\mathcal{I}_-, \mathbb{R})\},$$

$$\operatorname{im} T|_{\mathcal{I}_+} = \{q \in C(\mathcal{I}_+, \mathbb{R}^4) : q_2 - (\alpha + \beta)q_4', \ q_4 \in C^1(\mathcal{I}_+, \mathbb{R})\}.$$

In essence, on $\mathcal{I}_+$, one needs the additional derivative q_4''. This plays its role in rigorous input-output relations. It does not matter if one is only interested in the flow for smooth data. $\square$

In the framework of the projector based analysis (cf. [55]), the harmless critical points are compensated within so-called quasi-regular DAEs. However, neither for the concept of quasi-regular DAEs nor for the concepts associated with the SCF and derivative arrays functional-analytic interpretations seem to be available.

4.5   Notes and References

Remark 4.1 A DAE in *standard* form,

$$\mathfrak{f}(x'(t), x(t), t) = 0$$

can be brought to the form (4.1) by introducing the additional function $\chi = x'$ and regarding the extended system

$$x'(t) - \chi(t) = 0,$$

$$\mathfrak{f}(\chi(t), x(t), t) = 0.$$

However, we do not recommend this way, which would constrain us to C^1-solutions (cf. the discussion in Sect. 2.3). By this, for regular linear constant coefficient DAEs the index would be increased by 1. Fortunately, there are more appropriate reformulations for large classes of DAEs (e.g., [55]), and, what is more important, the most frequently applied classes of DAEs such as the semi-explicit ones are originally in the form (4.1).

Furthermore, as applied, e.g., in [32, 57, 58] as well as in Sect. 3.4, if $\ker \mathfrak{f}_{x^1}(x^1, x, t)$ is a C^1-subspace which is independent of x^1 and x, then there is a continuously differentiable projector valued function $D : \mathcal{I}_{\mathfrak{f}} \to \mathcal{L}(\mathbb{R}^m)$ such that

$$\mathfrak{f}(x^1, x, t) \equiv \mathfrak{f}(D(t)x^1, x, t), \quad \text{and}$$

$$\mathfrak{f}(x'(t), x(t), t) = \mathfrak{f}((Dx)'(t) - D'(t)x(t), x(t), t)$$

$$=: f((Dx)'(t), x(t), t), \quad \text{for } x \in C^1(\mathcal{I}, \mathbb{R}^m).$$

Moreover, DAEs

$$f((d(x(t),t))', x(t), t) = 0 \tag{4.64}$$

with properly involved nonlinear derivative term and border projector function $R \in \mathcal{C}^1(\mathcal{I}, \mathcal{L}(\mathbb{R}^n))$ (cf. [55,60]) can be treated as DAEs of the form (4.1) via the enlarged system

$$f((Ry)'(t), x(t), t) = 0,$$
$$y(t) - d(x(t), t) = 0.$$

This is of particular interest for quasi-linear DAEs

$$A(x(t), t)(d(x(t), t))' + b(x(t), t) = 0$$

which arise, e.g., in circuit simulation. We refer to [60] for detailed relations between (4.64) and its enlarged form.

Remark 4.2 Regularity of nonlinear differential-algebraic operators in the sense of Definition 4.4 is consistent with the definition of regular DAEs in [55] via Corollary 6.13, and hence, justified this way, see Appendix 6.3. In essence, F is a regular differential-algebraic operator if the set $\mathcal{D}_F \times \mathcal{I}$ totally belongs to a regularity region of the associated DAE.

Remark 4.3 It was pointed out, e.g., in [43,57] that operator equations representing higher-index DAEs are essentially ill-posed in their natural settings.

Lamour et al. [55, Section 3.9] addresses DAEs as operator equations and presents some parts of Sect. 4 concerning IVPs. It is pointed out that DAE solutions depend smoothly on a well-defined part of their initial value.

Let the differential-algebraic operator F be regular with index 1, $x_* \in \mathrm{dom}\, F$, $F(x_*) = 0, C \in \mathcal{L}(\mathbb{R}^m, \mathbb{R}^{r_0}), \mathrm{im}\, C = \mathbb{R}^{r_0}, \ker C = \ker D(t_a)$. Then the IVP

$$F(x) = 0, \quad Cx(t_a) = Cx_*(t_a) + z \tag{4.65}$$

is uniquely solvable for all sufficiently small $z \in \mathbb{R}^{r_0}$ (cf. Theorem 4.11). The solution is continuously differentiable with respect to z and the sensitivity matrix $X(t;z) := x'_z(t;z)$ satisfies the variational system

$$f_y(\eta(t;z))(DX)'(t;z) + f_x(\eta(t;z))X(t;z) = 0, \ t \in \mathcal{I}, \quad CX(t_a;z) = I_{r_0},$$

with $(\eta(t;z)) := ((Dx)'(t;z), x(t;z), t)$, that is,

$$F'(x(z))X(z) = 0, \quad CX(t_a;z) = I_{r_0}.$$

We emphasize that the initial condition, that is, the requirement for C, is in accordance with the Fredholm index of F. This statement would not longer hold for $C = I$.

For regular higher-index DAEs the situation is much more subtle. Since $F'(x_*)$ is no longer surjective and Fredholm, to apply the implicit function theorem in the advanced setting, we are obliged to assume the inclusion $F(x) \in \mathrm{im}\, F'(x_*)$ for all x in a neighborhood of the reference solution x_*. Furthermore, fewer initial conditions are allowed and the correct formulation of C depends on the reference solution.

For index-2 problems then the above statement concerning the IVP (4.65) and the variational system is valid with a full row-rank matrix C such that

$$\ker C = N_0(t_a) + N_{*1}(t_a).$$

The index $*$ indicates the dependence on the reference solution x_*. This is not too bad for $\mu = 2$, but worse for higher index, cf. Remark 4.5.

Remark 4.4 Analogous results concerning linear differential-algebraic operators as described in Sect. 4.2 can also be accomplished for different settings (cf. Remark 3.1), e.g.,

$$T \in \mathcal{L}_b(H_D^1(\mathcal{I}, \mathbb{R}^m), L^2(\mathcal{I}, \mathbb{R}^m)),$$

$$T \in \mathcal{L}_c(C(\mathcal{I}, \mathbb{R}^m), C(\mathcal{I}, \mathbb{R}^m)),$$

$$T \in \mathcal{L}_c(L^2(\mathcal{I}, \mathbb{R}^m), L^2(\mathcal{I}, \mathbb{R}^m)),$$

$$T \in \mathcal{L}_b(W_D^{1,\infty}(\mathcal{I}, \mathbb{R}^m), L^\infty(\mathcal{I}, \mathbb{R}^m)),$$

Hilbert spaces are favorable when looking for LSS, Moore–Penrose inverses, and pseudo-solutions. In any case, if T is regular with a higher index $\mu \geq 2$, then the equation $Tx = q$ is essentially ill-posed. In particular, then the Moore–Penrose inverse is unbounded, and a small residuum does not indicate a good approximation of a LSS, see Example 2.3.

Remark 4.5 The well-posed BVPs and IVPs in Sects. 4.3.2 and 4.3.3 are associated with composed maps $\mathcal{F}$ being local $\mathcal{C}^1$- diffeomorphisms at the solution x_*. Thereby, $Y = C(\mathcal{I}, \mathbb{R}^m) \times \mathbb{R}^{r_0}$ serves as image space in the index-1 case.

If the index is $\mu = 2$, we are already confronted with the more difficult advanced setting

$$Y_{*new} = C_*^{index\mu}(\mathcal{I}, \mathbb{R}^m) \times \mathbb{R}^{m-l}, \quad l = \sum_{i=0}^{\mu-1}(m - r_i).$$

Also in higher index cases $\mu > 2$ one can precisely describe the operator image $\mathrm{im}\, \mathcal{F}'(x_*)$ and introduce an appropriate norm $\| \cdot \|_{*index\mu}$ for obtaining the adapted Banach space Y_{*new} (cf. Sect. 2.4, also [55, Section 2.6.4]). Assuming again condition (4.50), that is,

$$Fx \in \text{im } F'(x_*), \quad x \in \mathcal{U}(x_*), \tag{4.66}$$

the composed operator $\mathcal{F}$ is a local $\mathcal{C}^1$- diffeomorphisms at the solution x_* and the equation $\mathcal{F}x = 0$ is well-posed in this adapted setting. This sounds fine, however, there are two principal concerns with this:

1. Though for $\mu = 2$ the adapted space Y_{*new} is transparent and can be seen as a reasonable compromise, in higher index cases the topology defined by the new norm $\| \cdot \|_{*index\mu}$ is far from meeting practical needs, since it is too strong as Examples 2.3 and 2.8 demonstrate.
2. Condition (4.66) is valid for quite a large class of index-2 DAEs including Hessenberg size-2 systems, which is very useful in optimal control. However, this condition is not given in many index-2 DAEs arising in circuit simulation, see [74] and Example 4.7. Moreover, for $\mu > 2$, no practical conditions ensuring (4.66) are in sight such as (4.54) ensuring (4.66) for $\mu = 2$.

Remark 4.6 Let the coefficients of the linear differential-algebraic operator $T \in \mathcal{L}(\mathcal{C}_D^1(\mathcal{I}, \mathbb{R}^m), C(\mathcal{I}, \mathbb{R}^m))$ be sufficiently smooth and T be regular with tractability index μ. Then the composed operator $\mathcal{T}$ associated with the IVP $Tx = q$, $Cx(t_a) = d$, with $\text{im } C = \mathbb{R}^{m-l}$, $\ker C = N_0(t_a) + \cdots + N_{\mu-1}(t_a)$, is injective and the inclusion $\mathcal{C}^{\mu-1}(\mathcal{I}, \mathbb{R}^m) \subset \text{im } \mathcal{T}$ is valid.

Set $Y_{new} := \text{im } \mathcal{T}$ and introduce on Y_{new} an appropriate norm $\| \cdot \|_{index\,\mu}$ to attain a Banach space (cf. [55, Section 3.9], also Sect. 2.4). Then the inequality

$$\|\mathcal{T}^{-1}(q, d)\|_\infty \leq \|\mathcal{T}^{-1}(q, d)\|_{\mathcal{C}_D^1} \leq \kappa(|d| + \|q\|_{index\,\mu}) \leq \tilde{\kappa}(|d| + \sum_{i=0}^{\mu-1} \|q^{(i)}\|_\infty),$$

is true for all $d \in \mathbb{R}^{m-l}$ and $q \in \mathcal{C}^{\mu-1}(\mathcal{I}, \mathbb{R}^m)$. This implies that a DAE associated with a regular differential-algebraic operator T has *perturbation index* μ (cf. [35]).

Let the nonlinear differential-algebraic operator F be regular with tractability index μ, $x_* \in \text{dom } F$, $Fx_* = 0$, and let the function f defining F as well as x_* be sufficiently smooth so that the linearization $F'(x_*)$ has sufficiently smooth coefficients for the inclusion $\mathcal{C}^{\mu-1}(\mathcal{I}, \mathbb{R}^m) \subset \text{im } F'(x_*)$ to hold. If, additionally, the condition (4.66) is satisfied, then the IVP $\mathcal{F}x = (q, Cx_*(t_a))$, with suitable matrix C, is uniquely solvable for all sufficiently small $q \in \mathcal{C}^{\mu-1}(\mathcal{I}, \mathbb{R}^m)$, and the inequality

$$\|x(q) - x_*\|_\infty \leq \|x(q) - x_*\|_{\mathcal{C}_D^1} \leq \kappa \|q\|_{*index\mu} \leq \tilde{\kappa} \sum_{i=0}^{\mu-1} \|q^{(i)}\|_\infty$$

follows, and hence, the DAE has perturbation index μ.

Since the unpleasant condition (4.66) is applied, this fails to confirm the conjecture in [55, p. 290] in its general form. It remains open whether this general conjecture can be verified.

Remark 4.7 Well-posed BVPs for regular index-1 DAEs and their discretizations are treated as operator equations, e.g., in [21, 22, 56–59].

The statements of Theorem 4.11 and Proposition 4.14 related to IVPs for quasi-linear index-1 and index-2 DAEs of the form (4.64) are proved in [60].

Remark 4.8 Least-squares collocation methods are known to belong to so-called regularizing algorithms for ill-posed problems (e.g. [34]). A first attempt to treat boundary value problems

$$E(t)x'(t) + F(t)x(t) = q(t), \; t \in [t_a, t_e], \quad C_a x(t_a) + C_e x(t_e) = d,$$

for higher-index time-varying linear DAEs by a least-squares collocation method is reported in [36]. Recent experiments lead us to expect some further progress on that score.

Let $\pi \subset [t_a, t_e]$ denote a finite set of points. A function $x_\pi : [t_a, t_e] \to \mathbb{R}^m$ is called a least-squares collocation solution of the BVP, if

$$E(t)x'_\pi(t) + F(t)x_\pi(t) = q(t), \; t \in \pi, \quad C_a x_\pi(t_a) + C_e x_\pi(t_e) = d, \qquad (4.67)$$

and x_π minimizes some scalar product norm among all solutions of (4.67). Reproducing kernels [2] serve as an essential tool for constructing and analyzing least-square collocation methods. The operator $\mathcal{T}$ associated with the BVP is given on the Sobolev space $H^k(\mathcal{I}, \mathbb{R}^m)$, $k \geq 2$. In order to reduce the computational expense, [36] concentrates on $k = 2$. $H^2(\mathcal{I}, \mathbb{R}^m)$ is endowed with the scalar product

$$(x, y)_{H^2} := \langle x(t_a), x(t_a) \rangle + \langle x(t_e), x(t_e) \rangle + (x'', y'')_{L^2}.$$

The so-called normal spline method in [31] repeats this approach for integro-differential equations,

$$E(t)x'(t) + F(t)x(t) - \int_{t_a}^{t_e} K(t, s)x(s)\mathrm{d}s = q(t), \; t \in \mathcal{I}, \; C_a x(t_a) + C_e x(t_e) = d.$$

As basic Sobolev space in [31] serves $H^k(\mathcal{I}, \mathbb{R}^m)$ equipped with the scalar product

$$(x, y)_{H^k} := \sum_{j=0}^{k-1} \langle x^{(j)}(t_a), x^{(j)}(t_a) \rangle + (x^{(k)}, y^{(k)})_{L^2}.$$

Remark 4.9 Convergence conditions for the Newton–Kantorovich iteration applied to well-posed BVPs for regular index-1 DAEs are derived in [59, 69]. Moreover, also for the class of regular index-2 DAEs described in Sect. 4.3.3, well-posedness is ensured by adapting the image-space and convergence conditions are provided. Practical computational experiments are reported in [69].

Further, [17, Section 2.2.10] is devoted to Newton–Kantorovich iterations via adapting the image spaces. However, it has been overlooked there that the images of the linearizations as well as the new advanced norms depend on the reference functions, see Remark 4.5.

Remark 4.10 The bounded outer inverse Γ in Theorem 4.5 was constructed first in [73, Chapter 4] for the index-3 case, aiming for a solvability result of the corresponding equation $\Gamma(\mathcal{F}x - (q, d)) = 0$ by applying [64, Theorem 3].

Concerning the computational treatment of DAEs by Newton-like iteration methods using outer inverses, as yet, there seems to be no practical experience and no efforts are reported.

Remark 4.11 By means of differentiating certain derivative-free equations of the DAE

$$A(Dx)' + Bx = q, \tag{4.68}$$

with m unknowns and $k \geq m$ equations, and adding the differentiated part to the given one, one can reduce the tractability index (see [55, Section 10.2] for the definition of the tractability index of nonregular DAEs). This allows to modify the essentially ill-posed problem $Tx = q$ to a prolongated system $\bar{T}x = \bar{q}$ having a normally solvable differential-algebraic operator $\bar{T}$ and dom $\bar{T}$ = dom T.

More precisely, if the DAE (4.68) has index $\mu \geq 2$ and G_μ has full column-rank, then the enlarged DAE

$$\begin{bmatrix} A \\ W_{\mu-1}BD^- \end{bmatrix}(Dx)' + \begin{bmatrix} B \\ (W_{\mu-1}BD^-)'D \ = \end{bmatrix}\begin{bmatrix} q \\ (W_{\mu-1}q)' \end{bmatrix}, \tag{4.69}$$

with the same m unknowns, but $k + m$ equations, has tractability index $\mu - 1$ owing to [55, Proposition 10.8].

In particular, starting from a regular DAE with tractability index $\mu \geq 2$, one can successively reduce the index and eventually arrives at an over-determined DAE with tractability index 1. The latter is associated with a normally solvable differential-algebraic operator. Possibly, this explains the advantage of overdetermined discretizations as used, e.g., in [28] for regular index-3 DAEs in Hessenberg form.

Remark 4.12 The present paper intends to provide a basic functional analysis for linear DAEs with continuous coefficients and nonlinear DAEs given by continuously differentiable data.

Appropriate modification of the linear theory for linear DAEs with integrable coefficients can be accomplished. A first approach can be found in [38], see also [55, Section 12.3]. Up to now, there is a lack of a comprehensive theory in this respect.

Furthermore, in [65–67, 81], quasi-linear DAEs

$$A(t)(Px)'(t) + B(t)x(t) + g(x(t), t) = 0, \quad \text{a.e. in } [t_a, t_e],$$

with a measurable and bounded function g, are treated as inclusions

$$A(t)(Px)'(t) + B(t)x(t) \in G(x(t), t), \quad \text{a.e. in } [t_a, t_e],$$

with Filippov functions G. Regularity with index 1 (transferability in [66, 67]) is adapted to the inclusion. Then, solvability in Filippov's sense is proved.

Linear DAEs with piecewise-smooth distributional coefficients are considered in [77] in order to manage DAEs whose coefficient matrices have jumps in view of system theoretical aspects. This approach is limited to linear DAE.

5   Regularization of Essentially Ill-Posed DAEs

Regularization in the context of ill-posed problems means the approximation by a certain family of well-posed problems. For instance, the well-known Tikhonov regularization utilizes the functional

$$J_\alpha(x) := \|Kx - q\|_Y^2 + \alpha \|x\|_X^2 \tag{5.1}$$

for the ill-posed equation $Kx = q$ stated in Hilbert spaces X and Y. The functional J_α is to be minimized for special sequences of the parameter $\alpha > 0$ (e.g., [24, 26, 78]).

The traditional nomenclature used in the theory of matrix pencils and subsequently in the DAE theory occupies the notion *regular* for special pencils and DAEs. However, those regular pencils and DAEs may induce ill-posed operator equations. This entails that also regular DAEs may need a further regularization in the scope of ill-posed problems. We describe in Sect. 5.2 the respective efforts.

Most common in DAE analysis and applications are index reduction procedures and transformations into special form so that the latter allows a safer numerical treatment. Actually, in this way, an ill-posed problem is modified into a well-posed one, too. This also justifies speaking of a *regularization*.

For instance, Remarks 3.9 and 4.11 describe prolongations of essentially ill-posed DAEs to normally solvable over-determined ones. Thereby, the eventually required derivatives are carried out analytically.

At most in the context of reduction procedures, one forms a derivative array system, also called a prolongated system, and then one looks for an embedded explicit ODE and derivative-free equations. This sounds simpler than it is! Of course, there are various ways to do this. In any case, again, all required derivatives are carried out analytically. We refer to [11, 17, 51, 70] for different overviews. As yet, to the author's knowledge, a systematical comparison of all these reduction procedures is missing. Below we describe the version of Chistyakov [13, 14] since it is basically a functional-analytic approach.

To the author's knowledge, in all these index reduction procedures the derivative array must be provided analytically. No errors are allowed in the prolongated system, since an appropriate sensitivity analysis is not available so far.

5.1   Chistyakov's Approach

An operator version of the reduction concept is developed in [13, 14, 17]. By means of prolongated systems, special left regularizers are arranged. Thereby several rank conditions play their role.

Without doubt, this is closely related to other reduction concepts aiming at regular (explicit or implicit) ODEs or index-1 DAEs such as those in [6, 51, 70]. However, the present treatise is not the right place to analyze the interrelations between the various reduction concepts which are developed to a large extent in parallel and without having notice of each other.

Given is the operator

$$T \in \mathcal{L}(X, Y), \quad Tx := Ex' + Fx, \quad x \in X, \tag{5.2}$$

with matrix coefficients $E(t)$, $F(t) \in \mathcal{L}(\mathbb{R}^m)$ sufficiently smooth on a neighborhood $\mathcal{I}_0$ of the interval $\mathcal{I} := [t_a, t_e]$. $E(t)$ is everywhere singular, which excludes regular ODEs. Denote

$$\rho := \max\{\text{rank } E(t) : t \in \mathcal{I}\} < m.$$

The role of the pre-image space X is assigned either to the function space $\mathcal{C}^\infty(\mathcal{I}, \mathbb{R}^m)$ or to $\mathcal{C}^1(\mathcal{I}, \mathbb{R}^m)$, whereas the role of the image space Y is assigned to $\mathcal{C}^\infty(\mathcal{I}, \mathbb{R}^m)$ and a factitious function space $\mathcal{C}^s_*(\mathcal{I}, \mathbb{R}^m)$, respectively. The latter one will be specified below, which requires full information about $T(\mathcal{C}^1(\mathcal{I}, \mathbb{R}^m))$. We refer to Sect. 2 for an interpretation of these settings.

It is investigated in [13, 14, 17] whether the nullspace of the operator T has finite dimension and whether T represents a Fredholm operator (called Noetherian there). The operator T is associated with the DAE

$$E(t)x'(t) + F(t)x(t) = q(t), \quad t \in \mathcal{I}, \tag{5.3}$$

as well as with the IVP

$$E(t)x'(t) + F(t)x(t) = q(t), \quad t \in \mathcal{I}, \quad x(t_a) = x_a \in \mathbb{R}^m. \tag{5.4}$$

We emphasize that $E(t)$ may change its rank here, cf. Sect. 4.4.

The DAE (5.3) as well as the coefficient pair $\{E, F\}$ is said to be SCF, if there exist a coordinate change $L \in C^1(\mathcal{I}, \mathcal{L}(\mathbb{R}^m))$ and a nonsingular scaling $K \in C(\mathcal{I}, \mathcal{L}(\mathbb{R}^m))$ converting the DAE coefficients to (e.g.,[6, Section 2.4.2])

$$LEK = \begin{bmatrix} I & \\ & N \end{bmatrix} \begin{matrix} \}m - l =: \delta \\ \}l \end{matrix} \quad , \quad LFK + LEK' = \begin{bmatrix} W & \\ & I \end{bmatrix} \begin{matrix} \}\delta \\ \}l \end{matrix} \quad .$$

Thereby, N is strictly upper (or lower) triangular. It may happen that $l = m, \delta = 0$, and then the so-called dynamic part is absent such that $LEK = N$, $LFK + LEK' = I$.

Evidently, if the coefficient pair $\{E, F\}$ is transformable into SCF, then the nullspace $\ker T$ has finite dimension, namely $\dim \ker T = \delta$. The opposite is true in a limited version ([17, p. 68]):

Proposition 5.1 *Suppose that* $E, F \in C^m(\mathcal{I}_0, \mathcal{L}(\mathbb{R}^m))$ *and* $\dim \ker T < \infty$.
Then there is a subinterval $[\bar{t}_a, \bar{t}_e] \subseteq \mathcal{I}$ *on which the DAE is transformable into SCF.*

The central ideas in [13, 14, 17] are *regularizers* and *solution representations of Cauchy type.* We quote the respective notions and results from [17] and illustrate it by an example.

Definition 5.1 The differential operator

$$Lz := \sum_{j=0}^{s} L_j z^{(j)}, \quad z \in C^s(\mathcal{I}, \mathbb{R}^m), \tag{5.5}$$

with coefficient functions $L_j \in C(\mathcal{I}, \mathcal{L}(\mathbb{R}^m))$, $j = 0, \ldots s$, is called a *left regularizer* (LR) of the operator T, if

$$(L \circ T)x = L(Tx) = x' + \tilde{F}x, \quad \text{for all } x \in C^{s+1}(\mathcal{I}, \mathbb{R}^m), \tag{5.6}$$

with a continuous matrix function $\tilde{F}$.

The minimal such index s is referred to a *left index* (in [13]: *unsolvability index*) of the operator T on the interval $\mathcal{I}$.

Example 5.2 (Left Regularizer) Consider the differential-algebraic operator $Tx := Ex' + Fx$ with constant coefficients

$$E = \begin{bmatrix} 0 & 1 & 0 \\ 0 & 0 & 0 \\ 0 & 0 & 1 \end{bmatrix}, \quad F = \begin{bmatrix} 1 & 0 & 0 \\ 0 & 1 & 0 \\ 0 & 1 & 0 \end{bmatrix}.$$

The matrix pair $\{E, F\}$ is regular with Kronecker index 2. At the beginning we show that there is no left regularizer with $s = 1$. On the contrary, if $Lz := L_1 z' + L_0 z$ is a left regularizer, then it holds that $L_1 E = 0$ and $L_0 E + L_1 F = I$, which implies $L_0 E^2 + L_1(FE - E) = E$. The latter relation reads in detail

$$L_0 \begin{bmatrix} 0 & 0 & 0 \\ 0 & 0 & 0 \\ 0 & 0 & 1 \end{bmatrix} + L_1 \begin{bmatrix} 0 & 0 & 0 \\ 0 & 0 & 0 \\ 0 & 0 & -1 \end{bmatrix} = \begin{bmatrix} 0 & 1 & 0 \\ 0 & 0 & 0 \\ 0 & 0 & 1 \end{bmatrix},$$

but this can never be valid.

Next we put $s = 2$ and $Lz := L_2 z'' + L_1 z' + L_0 z$. We have

$$L(Tx) = L_2 E x''' + (L_1 E + L_2 F)x'' + (L_0 E + L_1 F)x' + L_0 Fx.$$

For arbitrary $a, b, c \in \mathbb{R}$, the matrices

$$L_2 = \begin{bmatrix} 0 & -1 & 0 \\ 0 & 0 & 0 \\ 0 & 0 & 0 \end{bmatrix}, L_1 = \begin{bmatrix} 1 & 0 & 0 \\ 0 & c & 0 \\ 0 & 0 & 0 \end{bmatrix}, L_0 = \begin{bmatrix} 0 & a & 0 \\ 1 - c & b & 0 \\ 0 & 0 & 1 \end{bmatrix} \tag{5.7}$$

yield

$$L_2 E = 0, \ L_1 E + L_2 F = 0, \ L_0 E + L_1 F = I, \ L_0 F = \begin{bmatrix} 0 & a & 0 \\ 1 - c & b & 0 \\ 0 & 1 & 0 \end{bmatrix},$$

and hence, L is a left regularizer and the DAE has left index 2. More precisely, we are confronted with a family of left regularizers. Each of these left regularizers replaces the original DAE $Tx = q$ by the explicit ODE $(L \circ T)x = x' + L_0 Fx = Lq$; in more detail, the DAE

$$x_2' + x_1 = q_1$$
$$x_2 = q_2$$
$$x_3' + x_2 = q_3$$

is replaced by the explicit ODE

$$x_1' + a x_2 = a q_2 + q_1' - q_2''$$
$$x_2' + (1 - c)x_1 + b x_2 = (1 - c)q_1 + b q_2 + c q_2'$$
$$x_3' + x_2 = q_3.$$

The matrix $L_0 F$ has the eigenvalues 0 and $\frac{b}{2} \pm \sqrt{\frac{b^2}{4} + a(1 - c)}$. Therefore, for different values a, b, c, one obtains explicit ODEs which might feature quite different solution quality. This complicates the recognition of the original DAE solution in practice.   □

Definition 5.2 One says that the DAE (5.3) possesses a general solution of Cauchy type on $\mathcal{I}$, if the DAE (5.3) is solvable for each $q \in \mathcal{C}^{\rho+1}(\mathcal{I}, \mathbb{R}^m)$ and if there are a smooth matrix function X_δ with constant rank δ and a vector valued function φ such that any function x given by

$$x(t, c) = X_\delta(t)c + \varphi(t), \quad t \in \mathcal{I}, \ c \in \mathbb{R}^m, \tag{5.8}$$

represents a solution of the DAE, and, moreover, on any subinterval $[\bar{t}_a, \bar{t}_e] \subseteq \mathcal{I}$, there are no solutions other than restrictions of (5.8) onto $[\bar{t}_a, \bar{t}_e]$.

Proposition 5.3 *Let the operator T with coefficients $E, F \in \mathcal{C}^{2m}(\mathcal{I}_0, \mathcal{L}(\mathbb{R}^m))$ have a left regularizer with coefficients from $\mathcal{C}^{2s}(\mathcal{I}, \mathcal{L}(\mathbb{R}^m))$, whereby s is the left index of the operator T.*

Then, for $q \in \mathcal{C}^s(\mathcal{I}, \mathbb{R}^m)$, the equation $Tx = q$ has the general solution

$$x(t, c) = X_\delta(t)c + \int_{t_a}^t K(t, s)q(s)\mathrm{d}s + \sum_{j=0}^{s-1} C_j(t)q^j(t), \quad t \in \mathcal{I}, \ c \in \mathbb{R}^m, \tag{5.9}$$

which is a Cauchy type solution.

Moreover, supposing x_a is consistent, the IVP (5.4) is uniquely solvable, and the solution satisfies the inequalities

$$\|x\|_\infty \leq \kappa_1 \|x_a\| + \kappa_2 \|q\|_{\mathcal{C}^{s-1}},$$

$$\|x\|_{L^2} \leq \kappa_3 \|x_a\| + \kappa_4 \|q\|_{W_2^{s-1}}.$$

Theorem 5.4 *If the coefficients E, F of the operator T are real analytic, then the following statements are equivalent:*

1. *The operator T has on the interval $\mathcal{I}$ a left regularizer.*
2. *The DAE (5.3) has a general solution of Cauchy type on the interval $\mathcal{I}$.*
3. *The DAE (5.3) can be transformed on the interval $\mathcal{I}$ into canonical form by real analytic transforms L and K.*

By means of successive elimination and differentiation steps, left regularizers L can be stepwise constructed such that ([17, Section 2.1.5])

$$L = E_{[s]}^{-1}\Omega_s \cdots \Omega_1,$$

where $E_{[s]}$ is a nonsingular matrix function, s is the left index, and each factor Ω_i represents a special first order differential-algebraic operator.

Denote by $C_*^s(\mathcal{I}, \mathbb{R}^m)$ the completion of $C^s(\mathcal{I}, \mathbb{R}^m)$ with respect to the norm

$$\|q\|_* := \|q\|_\infty + \|\Omega_1 q\|_\infty + \cdots + \|\Omega_s \cdots \Omega_1 q\|_\infty, \quad q \in C^s(\mathcal{I}, \mathbb{R}^m),$$

We emphasize that the resulting Banach space strongly depends on the special problem, thus it is rather factitious, see Sects. 2.3 and 2.4.

Example 5.5 (Continuation of Example 5.2) Consider once more the matrix pair

$$E = \begin{bmatrix} 0 & 1 & 0 \\ 0 & 0 & 0 \\ 0 & 0 & 1 \end{bmatrix}, \; F = \begin{bmatrix} 1 & 0 & 0 \\ 0 & 1 & 0 \\ 0 & 1 & 0 \end{bmatrix}.$$

First we compute

$$K_1 = \begin{bmatrix} 1 & 0 & 0 \\ 0 & 0 & 1 \\ 0 & 1 & 0 \end{bmatrix}, \; K_1 E = \begin{bmatrix} 0 & 1 & 0 \\ 0 & 0 & 1 \\ 0 & 0 & 0 \end{bmatrix}, \quad \Omega_1 := \begin{bmatrix} 1 & 0 & 0 \\ 0 & 1 & 0 \\ 0 & 0 & \frac{d}{dt} \end{bmatrix} K_1,$$

$$\Omega_1 z = \begin{bmatrix} 0 & 0 & 0 \\ 0 & 0 & 0 \\ 0 & 1 & 0 \end{bmatrix} z' + \begin{bmatrix} 1 & 0 & 0 \\ 0 & 0 & 1 \\ 0 & 0 & 0 \end{bmatrix} z,$$

and further

$$\Omega_1(Tx) := \begin{bmatrix} 0 & 1 & 0 \\ 0 & 0 & 1 \\ 0 & 1 & 0 \end{bmatrix} x' + \begin{bmatrix} 1 & 0 & 0 \\ 0 & 1 & 0 \\ 0 & 0 & 0 \end{bmatrix} x =: E_{[1]} x' + F_{[1]} x.$$

The equation $\Omega_1(Tx) = \Omega_1 q$ reads in detail

$$x_2' + x_1 = q_1,$$
$$x_3' + x_2 = q_3,$$
$$x_2' = q_2',$$

which confirms a first index reduction; the pair $\{E_{[1]}, F_{[1]}\}$ is regular with Kronecker index 1. Next we perform

$$K_2 = \begin{bmatrix} 0 & 1 & 0 \\ 1 & 0 & -1 \\ 0 & 0 & 1 \end{bmatrix}, \; K_2 E_{[1]} = \begin{bmatrix} 0 & 0 & 1 \\ 0 & 0 & 0 \\ 0 & 1 & 0 \end{bmatrix}, \quad \Omega_2 := \begin{bmatrix} 1 & 0 & 0 \\ 0 & \frac{d}{dt} & 0 \\ 0 & 0 & 1 \end{bmatrix} K_2,$$

$$\Omega_2 z = \begin{bmatrix} 0 & 0 & 0 \\ 1 & 0 & -1 \\ 0 & 0 & 0 \end{bmatrix} z' + \begin{bmatrix} 0 & 1 & 0 \\ 0 & 0 & 0 \\ 0 & 0 & 1 \end{bmatrix} z,$$

$$\Omega_2 \Omega_1 z = \begin{bmatrix} 0 & 0 & 0 \\ 0 & 0 & -1 \\ 0 & 0 & 0 \end{bmatrix} z'' + \begin{bmatrix} 0 & 0 & 0 \\ 1 & 0 & 0 \\ 0 & 1 & 0 \end{bmatrix} z' + \begin{bmatrix} 0 & 0 & 1 \\ 0 & 0 & 0 \\ 0 & 0 & 0 \end{bmatrix} z,$$

and further

$$\Omega_2 \Omega_1 (Tx) := \begin{bmatrix} 0 & 0 & 1 \\ 1 & 0 & 0 \\ 0 & 1 & 0 \end{bmatrix} x' + \begin{bmatrix} 0 & 1 & 0 \\ 0 & 0 & 0 \\ 0 & 0 & 0 \end{bmatrix} x =: E_{[2]} x' + F_{[2]} x.$$

The equation $\Omega_2 \Omega_1 (Tx) = \Omega_2 \Omega_1 q$ reads in detail

$$x_3' + x_2 = q_3,$$
$$x_1' = q_1' - q_2'',$$
$$x_2' = q_2',$$

which confirms another index reduction. Obviously, $L := E_{[2]}^{-1} \Omega_2 \Omega_1$ is a left regularizer of the operator T. It coincides with the left regularizer given by (5.7) for $a = b = 0, c = 1$.

The norm

$$\|q\|_* := \|q\|_\infty + \|\Omega_1 q\|_\infty + \|\Omega_2 \Omega_1 q\|_\infty = \|q\|_\infty + \left\| \begin{bmatrix} q_1 \\ q_3 \\ q_2' \end{bmatrix} \right\|_\infty + \left\| \begin{bmatrix} q_3 \\ q_1' - q_2'' \\ q_2' \end{bmatrix} \right\|_\infty$$

defined on $\mathcal{C}^2(\mathcal{I}, \mathbb{R}^3)$ is equivalent to the norm

$$\|q\|_{**} := \|q\|_\infty + \|q_2'\|_\infty + \|q_1' - q_2''\|_\infty,$$

and the function space

$$\mathcal{C}_*^2(\mathcal{I}, \mathbb{R}^3) = \{q \in \mathcal{C}(\mathcal{I}, \mathbb{R}^3) : q_1 - q_2', q_2 \in \mathcal{C}^1(\mathcal{I}, \mathbb{R}^1)\} = T(\mathcal{C}^1(\mathcal{I}, \mathbb{R}^3))$$

results as completion of $\mathcal{C}^2(\mathcal{I}, \mathbb{R}^3)$ by this norm. □

If the operator T has a real-analytic coefficient pair $\{E, F\}$ then, for $X = \mathcal{C}^\infty(\mathcal{I}, \mathbb{R}^m)$, it holds that $\operatorname{im} T \subseteq \mathcal{C}^\infty(\mathcal{I}, \mathbb{R}^m)$. In contrast, if $X = \mathcal{C}^1(\mathcal{I}, \mathbb{R}^m)$ then $\operatorname{im} T \subseteq \mathcal{C}(\mathcal{I}, \mathbb{R}^m)$. Concerning both these settings, similar arguments apply as in the case of constant coefficients in Sect. 2.

Theorem 5.6 *Let the coefficients E, F of the operator T be real-analytic and let the associated operator $T \in \mathcal{L}(X, Y)$ have a left regularizer. Suppose either $X = Y = \mathcal{C}^\infty(\mathcal{I}, \mathbb{R}^m)$ or $X = \mathcal{C}^1(\mathcal{I}, \mathbb{R}^m)$, $Y = \mathcal{C}_*^s(\mathcal{I}, \mathbb{R}^m)$.*

1. *Then the operator T is Fredholm with $\alpha(T) = \delta$, $\beta(T) = 0$, $\text{ind}_{fred}(T) = \delta$.*
2. *The composed operator $\mathcal{T} \in \mathcal{L}(X, Y \times \mathbb{R}^m)$ associated with the IVP (5.4) is Fredholm with $\alpha(T) = 0$, $\beta(T) = m - \delta$, $\text{ind}_{fred}(T) = -(m - \delta)$.*

Once again, as already exposed in Sects. 2.2 and 2.4, in higher index cases, surjectivity is exacted by a special and also factitious setting.

As demonstrated in Example 5.5, the construction of left regularizers can be done stepwise by index reduction. One can also terminate the procedure earlier when either a regular index-1 DAE or a regular implicit ODE is achieved. This leads to the notion of *left semi-regularizers* ([17, p. 105]) which incorporates this idea.

Definition 5.3 The differential operator

$$Lz := \sum_{j=0}^{s} L_j z^{(j)}, \quad z \in \mathcal{C}^s(\mathcal{I}, \mathbb{R}^m), \tag{5.10}$$

with coefficient functions $L_j \in \mathcal{C}(\mathcal{I}, \mathcal{L}(\mathbb{R}^m))$, $j = 0, \dots s$, is called a *left semi-regularizer* (LSR) of the operator T, if

$$(L \circ T)x = \tilde{E}x' + \tilde{F}x, \quad \text{for all } x \in \mathcal{C}^{s+1}(\mathcal{I}, \mathbb{R}^m), \tag{5.11}$$

and $\tilde{E}, \tilde{F}$ are continuous matrix functions such that

$$\deg \det(\lambda \tilde{E}(t) + \tilde{F}(t)) = \text{rank} \, \tilde{E}(t) = \delta = \text{const}, \quad t \in \mathcal{I}. \tag{5.12}$$

The condition (5.12) is valid, exactly if the pair $\{\tilde{E}(t), \tilde{F}(t)\}$ is regular with Kronecker index 0 or 1, uniformly for all $t \in \mathcal{I}$. A DAE having coefficients which satisfy (5.12) is known to be a DAE with index 0 or 1. From this viewpoint the next assertion ([17, p. 105]) is self-explanatory.

Theorem 5.7 *If $E, F \in \mathcal{C}^{2s+3}(\mathcal{I}_0, \mathcal{L}(\mathbb{R}^m))$, then the operator T has a left regularizer if and only if it has a left semi-regularizer.*

5.2 Singular Perturbation and Tikhonov Regularization

Aiming for a regularization of regular higher-index DAEs, several parametrizations are investigated mainly in the early literature concerning DAEs, which was strongly affected by the singular perturbation theory.

The so-called *pencil regularization* [5, 7, 9, 18] which approximates the standard form DAE

$$Ex' + Fx = q \qquad (5.13)$$

by the regular implicit ODE

$$(E + \alpha F)x' + Fx = q,$$

with a small parameter $\alpha > 0$, is the earliest general such approach. Thereby, one has to assume regular local matrix pairs $\{E(t), F(t)\}$. This property is not given in general, but it is given, e.g., for DAEs in Hessenberg form and the class described in [54]. A review of convergence results is presented in [40]. Chistyakov [15] provides some additional discussion. Roughly speaking, if it works, then the pencil regularization leads to singular singularly perturbed equations.

Alternative approaches aiming at a regular index-1 DAE are studied, e.g., in [37, 39–44, 48]. For instance, the parametrizations

$$(E + \alpha FP)(Px)' + (F - EP')x = q \qquad (5.14)$$

and

$$(E + \alpha WFP)(Px)' + (F - EP')x = q, \qquad (5.15)$$

with P and W being projector functions along $\ker E$ and $\operatorname{im} E$, respectively, are used to approximate the equation

$$E(Px)' + (F - EP')x = q. \qquad (5.16)$$

The DAE (5.16) has a properly stated leading term; this version corresponds to the closure of the operator representing the DAE (5.13), cf. Sect. 2. The latter two parametrizations have appropriate physical interpretations for DAEs describing electrical networks [41, 48]. They lead to singularly perturbed index-1 DAEs, and they are less severe than the ODEs resulting from pencil regularization.

For special autonomous nonlinear DAEs

$$\mathfrak{f}(x', x) = 0, \qquad (5.17)$$

with constant $\ker \mathfrak{f}_{x'}$, such as Hessenberg form DAEs, in [40, 42] the parametrizations

$$\mathfrak{f}(x', x + \alpha x') = 0 \quad \text{and} \quad \mathfrak{f}(x', x + \alpha Px') = 0 \qquad (5.18)$$

are compared. Owing to the identity $\mathfrak{f}(x', x) \equiv \mathfrak{f}(Px', x))$, the second version is more favorable. Deep results concerning convergence as $\alpha \to 0$ and asymptotic expansions are obtained for regular index-2 and index-3 DAEs (5.13) and also for special classes of nonlinear DAEs in Hessenberg form, e.g., in [40–42].

In appropriate Hilbert space settings, these parametrizations yield regularizations in the sense of Tikhonov (see [26,76]). We quote a typical result for regular index-2 DAEs. Consider the operator $T \in \mathcal{L}(H_P^1(\mathcal{I}, \mathbb{R}^m), L^2(\mathcal{I}, \mathbb{R}^m))$ defined by

$$Tx := E(Px)' + (F - EP')x, \quad x \in H_P^1(\mathcal{I}, \mathbb{R}^m),$$

the composed operator $\mathcal{T} \in \mathcal{L}(H_P^1(\mathcal{I}, \mathbb{R}^m), L^2(\mathcal{I}, \mathbb{R}^m) \times \mathbb{R}^m)$,

$$\mathcal{T}x := (Tx, \, Cx(t_a)), \quad x \in H_P^1(\mathcal{I}, \mathbb{R}^m),$$

and further the operators

$$T_\alpha x := (E + \alpha FP)(Px)' + (F - EP')x, \quad x \in H_P^1(\mathcal{I}, \mathbb{R}^m),$$

$$\mathcal{T}_\alpha x := \begin{bmatrix} T_\alpha x \\ Cx(t_a) \\ (\Pi_0 Q_1)(t_a)x(t_a) \end{bmatrix}, \quad x \in H_P^1(\mathcal{I}, \mathbb{R}^m),$$

with $C \in \mathcal{L}(\mathbb{R}^m)$. We refer to Sect. 4 for the meaning of $\Pi_0 Q_1, \Pi_1$. Denote $M := \operatorname{im} C$ and $L := \operatorname{im} (\Pi_0 Q_1)(t_a)$.

Proposition 5.8 *Let T be regular with tractability index 2 and $\ker C = \ker \Pi_1(t_a)$. Then the following statements hold:*

1. *The operator $\mathcal{T}$ is injective.*
2. *The operator T_α is regular with tractability index 1 for each sufficiently small $\alpha > 0$.*
3. *The operator $\mathcal{T}_\alpha \in \mathcal{L}(H_P^1(\mathcal{I}, \mathbb{R}^m), L^2(\mathcal{I}, \mathbb{R}^m) \times M \times L)$ is a bijection for all sufficiently small $\alpha > 0$.*
4. *For $(q, d) \in \operatorname{im} \mathcal{T}$ and*

$$x_* := \mathcal{T}^{-1}(q, d) \quad x_\alpha := \mathcal{T}_\alpha^{-1}(q, d, (\Pi_1 Q_1 G_2^{-1} q)(t_a)),$$

   $\alpha \to 0$ implies $\|x_\alpha - x_*\|_{H_P^1} \to 0$ and $\|Px_\alpha - Px_*\|_{H^1} = O(\alpha)$.
5. *If $(q, d) \in L^2(\mathcal{I}, \mathbb{R}^m) \times M$, $q \notin \operatorname{im} T$, then*

$$\|x_\alpha\|_{H_P^1} \to \infty, \quad as \quad \alpha \to 0.$$

Under additional smoothness one obtains $\|x_\alpha - x_*\|_{H_P^1} = O(\alpha^{\frac{1}{2}})$ and even $\|x_\alpha - x_*\|_{H_P^1} = O(\alpha)$.

The convergence behavior is similar to Tikhonov regularization for integral equations of the first kind [34]. Motivated by this experience, in [25] nonlinear semi-explicit DAEs are treated by Tikhonov regularization in several settings.

In spite of all these contributions, as yet, there are no sufficiently matured procedures to solve BVPs and IVPs in higher-index DAEs relying on Tikhonov regularization.

6 Appendices

6.1 Functional-Analytic Basics and Notations

We are mainly interested in bounded operators acting in real Banach spaces and in closed operators acting in real Hilbert spaces. We refer to [20, 29, 45, 83, 84] for details regarding the material below.

6.1.1 Linear Operators in Normed Spaces

Let X and Y be normed linear spaces over $\mathbb{R}$. $\mathcal{L}(X, Y)$ denotes the set of all linear operators K mapping an individual *definition domain* dom $K \subseteq X$ into Y. We also shorten $\mathcal{L}(X) := \mathcal{L}(X, X)$. For each operator $K \in \mathcal{L}(X, Y)$ we introduce its *range* (also image) and *nullspace* (also kernel) as

$$\operatorname{im} K := \{y \in Y : \exists x \in X, \, y = Kx\}, \quad \ker K := \{x \in X : Kx = 0\}.$$

The operator K is said to be *densely defined* if dom K is dense in X and *densely solvable* if im K is dense in Y.

The *graph* of K is determined by

$$\operatorname{graph} K := \{(x, Kx) : x \in \operatorname{dom} K\} \subseteq X \times Y.$$

The sets $\mathcal{L}_b(X, Y)$ and $\mathcal{L}_c(X, Y)$ consist of all linear bounded and closed operators, respectively, such that the inclusions

$$\mathcal{L}_b(X, Y) \subseteq \mathcal{L}_c(X, Y) \subseteq \mathcal{L}(X, Y)$$

are given.

The operator $K \in \mathcal{L}(X, Y)$ is said to be *closed*, if for each sequence $x_n \to x_* \in X$, $x_n \in \operatorname{dom} K$ for all $n \in \mathbb{N}$, and $Kx_n \to y_* \in Y$ it results that $x_* \in \operatorname{dom} K$ and $Kx_* = y_*$.

The operator $K \in \mathcal{L}(X, Y)$ is closed, exactly if *graph K* is a closed subspace in $X \times Y$. The closed-graph theorem says that a closed linear operator K which maps all of a Banach space X (i.e., dom $K = X$) into a Banach space Y is bounded. In contrast, often one is confronted with the fact that dom K is merely a proper subset of X.

The operator $K \in \mathcal{L}(X, Y)$ is called *closable* if it admits a closed extension. The minimal closed extension is said to be the *closure* of K.

We equip the linear space dom $K =: X_K$ with the so-called *graph norm*

$$\|x\|_K := \|x\| + \|Kx\|, \quad \|x\| \leq \|x\|_K \text{ for all } x \in \text{dom } K.$$

Because of the evident inequality $\|Kx\| \leq \|x\|_K$ given for all $x \in X_K$, the operator K is bounded in this new setting (we keep the notation K), therefore

$$K \in \mathcal{L}_c(X, Y) \quad \text{implies} \quad K \in \mathcal{L}_b(X_K, Y).$$

If X and Y are Banach spaces then also X_K is a Banach space.

The operator $K \in \mathcal{L}_c(X, Y)$, where X and Y are normed spaces, is said to be a *Fredholm operator*, if it has a closed range im K and the dimensions dim ker $K =:$ $\alpha(K)$ and codim im $K =: \beta(K)$ are finite. Then, the difference

$$\alpha(K) - \beta(K) =: \text{ind}_{fred}(K)$$

is called the *Fredholm index*. The operator $K \in \mathcal{L}_c(X, Y)$ is said to be *semi-Fredholm*, if it has a closed range im K and either dim ker $K =: \alpha(K)$ or codim im $K =: \beta(K)$ is finite.

Often, Fredholm operators are ab initio supposed to act in Banach spaces and to be bounded, and then the closed-range property is not explicitly listed. Owing to Kato's theorem (e.g., [45, p. 310]), if im K possesses a finite codimension it is necessarily closed. Although similar arguments apply also to closed operators acting in Banach spaces, we impose an explicit listing of the closed-range property.

Sometimes one uses the name *Noetherian index* instead of Fredholm index, e.g., [13, 17].

A closed, densely defined operator K acting from the Banach space X into the Banach space Y is said to be *normally solvable* if im K is closed in Y (e.g., [47, p. 234]). Normally solvable operators comprise useful properties similar to Fredholm operators (cf. Theorem 6.2), and sometimes then the problem $Kx = q$ is said to be *well-posed in the sense of Fichera*. Well-posedness in Fichera's sense does not necessarily suppose either injectivity or surjectivity.

If K is a bounded bijection acting on Banach spaces, then K is a Fredholm operator with ind$_{fred}(K) = \alpha(K) = 0$ and the equation $Kx = q$ is well-posed in the sense of Hadamard.

We denote by $X^* := \mathcal{L}_b(X, \mathbb{R})$ the *dual* of the real normed space X. Equipped with its natural norm, X^* becomes a Banach space.

We set $\langle \eta, x \rangle := \eta(x)$ for every $x \in X$ and $\eta \in X^*$. The resulting bilinear form $\langle \cdot, \cdot \rangle$ indicates the duality pairing between X^* and X, also called *scalar product* between X^* and X.

For each densely defined operator $K \in \mathcal{L}(X, Y)$ there is the *dual operator* $K^* \in \mathcal{L}(Y^*, X^*)$ given by

$$\langle K^* g, x \rangle = \langle g, Kx \rangle \quad \text{for all} \quad x \in \text{dom } K, \ g \in \text{dom } K^*,$$

$$\text{dom } K^* = \{ g \in Y^* : \exists \eta_g \in X^* \text{ such that } \langle g, Kx \rangle = \langle \eta_g, x \rangle \ \forall x \in \text{dom } K \},$$

$$K^* g = \eta_g \quad \text{for } g \in \text{dom } K^*.$$

The dual operator of a bounded operator is also bounded and it holds that $\| K^* \| = \| K \|$.

We note that the dual operator K^* is also called the transposed and adjoint operator in the literature.

For each element x of the normed space X there exists an element z of the *bidual* space $X^{**} := (X^*)^*$ such that $\langle z, g \rangle = \langle g, x \rangle$ for all $g \in X^*$ and further $\| x \| = \| z \|$. This allows to assume the inclusion $X \subseteq X^{**}$. If $X = X^{**}$, then X is called a *reflexive* Banach space.

Proposition 6.1 *Let X and Y be Banach spaces, and $K \in \mathcal{L}(X, Y)$ be densely defined. Then the following hold:*

1. *The dual operator K^* is closed, $K^* \in \mathcal{L}_c(Y^*, X^*)$.*
2. *If X and Y are reflexive, then the dual operator K^* is likewise densely defined.*
3. *If X and Y are reflexive and the operator K is closable, then K^{**} represents the closure of K.*

Proofs can be found, e.g., in [47, Chapter III].

We quote the closed image theorem, e.g., [20, p. 348],[82, p. 205]:

Theorem 6.2 *Let X and Y be Banach spaces, and $K \in \mathcal{L}_c(X, Y)$ be densely defined. Then* im K *is closed in Y if and only if* im K^* *is closed in Y^*.*

In addition, under this hypothesis,

$$(\text{im } K)^\perp = \{ y \in Y : \langle y, g \rangle = 0 \ \forall g \in \ker K^* \} = \ker K^*,$$

$$\text{im } K^* = \{ \eta \in X^* : \langle x, \eta \rangle = 0 \ \forall x \in \ker K \} = (\ker K)^\perp,$$

$$\alpha(K) = \beta(K^*),$$

$$\alpha(K^*) = \beta(K).$$

A Hilbert space X is comfortable owing to the scalar product $(\cdot, \cdot)$ which is defined on $X \times X$. As before, the symbol $\langle \cdot, \cdot \rangle$ indicates the scalar product between X^* and X. If $\mathcal{J} \in \mathcal{L}_b(X, X^*)$ denotes the so-called duality map, then it holds that

$$\langle \mathcal{J}x, \xi \rangle = (x, \xi) \text{ for all } x, \xi \in X, \quad \text{and} \quad \| \mathcal{J}x \| = \| x \| \text{ for all } x \in X.$$

This feature allows to identify the Hilbert space and its dual, that is $X = X^*$. This implies reflexivity.

If X and Y are Hilbert spaces and the operator $K \in \mathcal{L}(X, Y)$ is densely defined, then K has the *adjoint operator* $K^* \in \mathcal{L}(Y, X)$ given by

$$(K^* y, x) = (y, Kx) \quad \text{for all} \quad x \in \operatorname{dom} K, \; y \in \operatorname{dom} K^*,$$

$$\operatorname{dom} K^* = \{y \in Y : \exists \eta_y \in X \text{ such that } (y, Kx) = (\eta_y, x) \; \forall x \in \operatorname{dom} K\},$$

$$K^* y = \eta_y \quad \text{for} \; y \in \operatorname{dom} K^*.$$

The adjoint operator is likewise densely defined, and it is closely related to the dual operator. If $\mathcal{J}_X$ and $\mathcal{J}_Y$ are the duality maps associated to X and Y, respectively, and if $K^* \in \mathcal{L}(Y^*, X^*)$ is the dual operator of the densely defined operator $K \in \mathcal{L}(X, Y)$, then the adjoint operator is given by $\mathcal{J}_X^{-1} K^* \mathcal{J}_Y$. We use the same symbol for the dual and the adjoint operators. No confusion will arise.

Moreover, for a linear densely defined operator K acting in Hilbert spaces X and Y, the *biadjoint* $K^{**} := (K^*)^* \in \mathcal{L}(X, Y)$ exists and is given by

$$(K^{**} x, y) = (x, K^* y) \quad \text{for all} \quad y \in \operatorname{dom} K^*, \; x \in \operatorname{dom} K^{**},$$

$$\operatorname{dom} K^{**} = \{x \in X : \exists \eta_x \in X \text{ such that } (x, K^* y) = (\eta_x, y) \; \forall y \in \operatorname{dom} K^*\},$$

$$K^* y = \eta_y \quad \text{for} \; y \in \operatorname{dom} K^*.$$

Owing to Proposition 6.1, the biadjoint K^{**} of a closable, densely defined operator $K \in \mathcal{L}(X, Y)$ acting in Hilbert spaces represents the closure of K.

6.1.2 Inner Inverses, Outer Inverses, Generalized Inverses, and Least-Squares Solutions

Here we adapt the terminology applied in [19,62,64] and collect a few of the results reported therein.

Let X and Y be Banach spaces, $K \in \mathcal{L}(X, Y)$, and $K^- \in \mathcal{L}(Y, X)$. If $KK^- K = K$ holds on $\operatorname{dom} K$, then the map K^- is said to be an *inner inverse* of K. If $K^- KK^- = K^-$ holds on $\operatorname{dom} K^-$, then the map K^- is said to be an *outer inverse* of K. If K^- is both an inner and an outer inverse of K, then it is called an *algebraic generalized inverse* of K.

If K^- is either an inner or an outer inverse of K, then both KK^- and $K^- K$ are linear idempotents, that is, *algebraic projectors*.

From the viewpoint of analysis, however, these algebraic constructions are of little use unless the resulting operators are continuous; hence our special interest is directed to results on bounded projectors and inverses.

The linear mapping K is called *approximately outer invertible*, if, for each $\rho \in (0, 1)$, there exists an operator $K_\rho^\simeq \in \mathcal{L}_b(Y, X)$ and a bound $\Gamma(\rho)$ such that

$$\|(K_\rho^\simeq K K_\rho^\simeq - K_\rho^\simeq)y\| \leq \rho \|K_\rho^\simeq y\| \quad \text{and} \quad \|K_\rho^\simeq y\| \leq \Gamma(\rho)\|y\|, \quad \text{for all} \ \ y \in Y.$$

Then each $K_\rho^\simeq$ is called an *approximate outer inverse of K, with bound function* $\Gamma(\rho)$.

In case of infinite-dimensional spaces, the symbols $\dotplus$ and $\oplus$ indicate algebraic direct sums and topological direct sums (with closed subspaces), respectively. In finite-dimensional spaces all subspaces are closed; then we only apply the symbol $\oplus$ for direct sums.

Topological direct sum decompositions of a Banach space are intimately connected with bounded projectors. The range of each bounded projector on a Banach space is a closed subspace. A closed subspace of a Banach space has a topological complement if and only if it is the range of some bounded projector.

A Hilbert space is more comfortable: any closed subspace M has a topological complement, and $M^\perp$ is one such complement; the orthogonal projector P onto a closed subspace is linear, idempotent and it holds that $P^* = P$ and $\|P\| = 1$.

For a densely defined operator (or even bounded operator) $K \in \mathcal{L}_c(X, Y)$ in Hilbert spaces, one can define the so-called *orthogonal generalized inverse*, denoted by $K^+ \in \mathcal{L}(Y, X)$, which is the operator version of the Moore–Penrose inverse given for matrices.

Let X and Y be Hilbert spaces, $K \in \mathcal{L}_c(X, Y)$, $\overline{\operatorname{dom} K} = Y$. Then $\ker K$ is closed. Let $\mathfrak{P} \in \mathcal{L}_b(X)$ and $\mathfrak{R} \in \mathcal{L}_b(Y)$ denote the orthoprojectors onto $(\ker K)^\perp$ and $\overline{\operatorname{im} K}$, respectively. It follows that $K = K\mathfrak{P} = \mathfrak{R}K$. One defines

$$K^+ y := (K|_{(\operatorname{dom} K) \cap (\ker K)^\perp})^{-1} \mathfrak{R}y \quad \text{for all} \ y \in \operatorname{im} K + (\operatorname{im} K)^\perp =: \operatorname{dom} K^+.$$

This definition implies the relations

$$KK^+ K = K \ \text{on dom} \ K,$$
$$K^+ KK^+ = K^+ \ \text{on dom} \ K^+,$$
$$KK^+ = \mathfrak{R}|_{\operatorname{dom} K+},$$
$$K^+ K = \mathfrak{P}|_{\operatorname{dom} K},$$

which show that K^+ is at the same time a particular algebraic generalized inverse.

The orthogonal generalized inverse K^+ of the densely defined closed operator K is also closed. KK^+ and $K^+ K$ are then densely defined and symmetric, but not necessarily closed. $K^+ K$ becomes closed exactly if K is bounded, and KK^+ is closed exactly if K^+ is bounded.

Furthermore, K^+ is bounded exactly if im K is closed in Y.

Let X be a vector space and Y be a Hilbert space, $K \in \mathcal{L}(X, Y)$, $y_* \in Y$. The element $x_* \in X$ is said to be a LSS of the equation $Kx = y_*$, if

$$\|Kx_* - y_*\| = \inf\{\|Kx - y_*\| : x \in \operatorname{dom} K\}.$$

An LSS is also named a *quasi solution*, e.g., [78].

Let $\mathfrak{R} \in \mathcal{L}_b(Y)$ again denote the orthoprojector onto $\overline{\operatorname{im} K}$. Then, the equation $Kx = y_*$ has a LSS exactly if $\mathfrak{R}y_* \in \operatorname{im} K$, i.e., $y_* \in \operatorname{im} K + (\operatorname{im} K)^\perp$. If X is normed, then one can look for a *LSS with minimal norm*. If X is also an Hilbert space, then the LSS with minimal norm is called a *pseudo solution*, e.g., [78].

There is a close relation between LSS and the orthogonal generalized inverse (cf. [62, Theorem 5.1]):

Theorem 6.3 *Let X and Y be Hilbert spaces, $K \in \mathcal{L}_c(X, Y)$ be densely defined, $y_* \in Y$. Then the following is true:*

1. *x_* is a LSS exactly if it satisfies the equation $K^*(Kx - y_*) = 0$.*
2. *For $y_* \in \operatorname{dom} K^+$, $K^+ y_*$ is the unique solution of minimal norm of the equation $Kx = \mathfrak{R}y_*$.*
3. *For $y_* \in \operatorname{dom} K^+$, $K^+ y_*$ is the unique LSS of minimal norm of the equation $Kx = y_*$.*
4. *For $y_* \in \operatorname{dom} K^+$, $K^+ y_*$ is the unique solution of minimal norm of the equation $K^*(Kx - y_*) = 0$.*
5. *If, additionally, K is bounded, then x_* is a LSS exactly if the normal equation $K^*Kx_* = K^*y_*$ is satisfied.*

We quote [62, Theorem 7.1] and parts of [62, Theorem 7.2]:

Theorem 6.4 *Let the operator $K \in \mathcal{L}(X, Y)$ acting in Banach spaces be bounded or closed with dense domain. Then, K has a bounded inner inverse exactly if ker K and im K are closed and have topological complements in X and Y respectively.*

Theorem 6.5 *For the operator $K \in \mathcal{L}_b(X, Y)$ acting in Hilbert spaces the following statements are equivalent:*

1. *K has a bounded inner inverse.*
2. *im K is closed.*
3. *K is normally solvable.*
4. *$\inf\{\|Kx - y\| : x \in X\}$ is attained for each $y \in Y$.*
5. *The orthogonal generalized (Moore–Penrose) inverse K^+ of K is bounded.*
6. *For some $\gamma > 0$, it holds that $\|Kx\| \geq \gamma \|x\|$ for all $x \in M$, where $X = \ker K \oplus M$.*

For the operator $K \in \mathcal{L}_b(X, Y)$ acting in Hilbert spaces and each arbitrary parameter value $\alpha > 0$, the operator $K^*K + \alpha I \in \mathcal{L}_b(X)$ is bijective and

$$\|(K^*K + \alpha I)^{-1}\| \le \frac{1}{\alpha}.$$

Then, the equation $K^*Kx + \alpha x = y$ is well-posed in Hadamard's sense, also if K fails to be injective and im K is nonclosed. The unique solution x_α of this equation is actually the only minimizer of Tikhonov's functional

$$J_\alpha(x) := \|Kx - y\|^2 + \alpha\|x\|^2, \quad x \in X.$$

This fact is utilized by the so-called *Tikhonov regularization* of ill-posed problems, see, e.g., [24, 76, 78].

Here we point out that the normal equation $K^*Kx = K^*y$ fails to be well-posed; it is again essentially ill-posed if im K is nonclosed.

Finally we quote the perturbation result [64, Lemma 2.2]:

Lemma 6.6 *Let X and Y be Banach spaces, $K \in \mathcal{L}_b(X, Y)$, and let $K^- \in \mathcal{L}_b(Y, X)$ be a bounded outer inverse of K. Let $B \in \mathcal{L}_b(X, Y)$ be such that*

$$\|K^-(B - K)\| < 1.$$

Then $B^- := (I + K^-(B-K))^{-1}K^-$ is a bounded outer inverse of B with $\ker B^- = \ker K^-$ and $\operatorname{im} B^- = \operatorname{im} K^-$. Moreover

$$\|B^- - K^-\| \le \frac{K^-(B - K)K^-}{1 - \|K^-(B - K)\|}.$$

6.1.3 Nonlinear Fréchet-Differentiable Operators

Let X and Y be Banach spaces, $U \subseteq X$ be an open subset. We consider the map $K : \operatorname{dom} K := U \to Y$. We say that K is *Fréchet-differentiable at $u \in U$*, if there exists an operator $A \in \mathcal{L}_b(X, Y)$ such that, if we set

$$R(h) = K(u + h) - K(u) - A(h),$$

there results $R(h) = o(\|h\|)$, that is $\frac{\|R(h)\|}{\|h\|} \to 0$ as $h \to 0$. Such an A will be called Fréchet-differential of K at u and denoted by $A = dK(u)$.

Let K be differentiable at all $u \in U$. Then the map $K' : U \to \mathcal{L}_b(X, Y)$, $K'(u) := dK(u)$, $u \in U$, is called the *Fréchet-derivative* of K. $K'(u)$ is also called the *linearization* of K at u.

If the derivative K' is continuous as a map from U to $\mathcal{L}_b(X, Y)$, then we will say that K is a C^1- *operator*, see e.g., [1, Section 1.1].

Let X and Y be Banach spaces. The $\mathcal{C}^1$-operator K acting between X and Y, with definition domain dom K open in X, is called a *Fredholm operator* exactly if for each $x \in$ dom K the linearization $K'(x) \in \mathcal{L}_b(X, Y)$ is a Fredholm operator, see e.g., [84, p. 317].

If dom $K = X$ then $\text{ind}_{fred}(K'(x))$ is independent of x, and then one sets

$$\text{ind}_{fred}(K) := \text{ind}_{fred}(K'(x)),$$

see [84, Section 5.15].

We quote the local inverse mapping theorem from [84, p. 259].

Theorem 6.7 *Let X, Y be Banach spaces, $r \in \mathbb{N}$, and $\mathcal{U}(x_0) \subseteq X$ be an open neighborhood of $x_0 \in X$. Let $K : \mathcal{U}(x_0) \subseteq X \to Y$ be a $\mathcal{C}^r$ map. Then K is a local $\mathcal{C}^r$-diffeomorphism at x_0 if and only if $K'(x_0) : X \to Y$ is bijective.*

Often, classical *well-posedness in Hadamard's sense* of the nonlinear operator equation $Kx = p$, with K acting in Banach spaces X, Y, is unrealistic. Instead of this global requirement, one applies its local version which actually means that K is a local $\mathcal{C}^1$- diffeomorphism at the wanted solution:

Let X, Y be Banach spaces. The equation $Kx = 0$, with a Fréchet-differentiable operator $K : \text{dom } K \subseteq X \to Y$, is said to be *well-posed locally around* $x_* \in$ dom K, with $Kx_* = 0$, if the Fréchet derivative $K'(x_*) \in \mathcal{L}(X, Y)$ is a homeomorphism. Then, owing to the classical implicit function theorem in Banach spaces (e.g., [1]), for each sufficiently small $q \in Y$, in the neighborhood of x_* exists exactly one solution $x(q)$ to the equation $Kx = q$, and the inequality

$$\|x(q) - x_*\| \leq \kappa \|q\|$$

is given with a constant κ.

Regular higher-index differential-algebraic operators (in their natural Banach spaces) are Fréchet-differentiable, however, im $K'(x_*)$ is a nonclosed subset in Y, such that K fails to be Fredholm and, to make matters worse, $K'(x)$ is no longer normally solvable. One can treat equation $Kx = 0$ as an ill-posed problem applying the respective methods (e.g., [19, 24, 34, 62–64, 76, 78]). The following implicit function theorem is a consequence of [19, Theorem 3].

Theorem 6.8 *Let X and Y be Banach spaces, Let the function $K : X \to Y$ be strongly Fréchet-differentiable at $x_* \in X$, with $K(x_*) = 0$. Let the Fréchet derivative $K'(x_*) \in \mathcal{L}_b(X, Y)$ be approximately outer invertible, with approximate outer inverses $K_\rho^\simeq$ and bound function $\Gamma(\rho) = h_0 \rho^{-\gamma}$, where $\gamma < 1$. Then, whenever $z_* \in X$ satisfies $K'(x_*)z_* = 0$, and $\|z_*\| = 1$, there exists a solution $x = x_* + tz_* + o(t)$ to $K(x) \in \ker K_\rho^\simeq$.*

6.2 Matrix Functions, Varying Subspaces, and Special Function Spaces

We identify matrices $M \in \mathbb{R}^{n \times m}$ and their associated operators $M \in \mathcal{L}(\mathbb{R}^m, R^n)$.

By $\mathcal{C}(\Omega, \mathbb{R}^s)$ and $\mathcal{C}^k(\Omega, \mathbb{R}^s)$ we denote the linear spaces of continuous and k-times continuously differentiable functions defined on $\Omega \subseteq \mathbb{R}^n$ with values in $\mathbb{R}^s$, $k \in \mathbb{N}$ and $k = \infty$.

We apply the usual notations $L^2(\mathcal{I}, \mathbb{R}^n)$ and $H^1(\mathcal{I}, \mathbb{R}^n)$, with a compact interval $\mathcal{I}$, for the respective Lebesgue and Sobolev spaces.

By $|\cdot|$ we denote absolute values as well as norms of vectors and matrices. In contrast, $\|\cdot\|$ is a function or operator norm. At episodes when several norms are to be distinguished, we use specific indices.

Let $s, n \geq 1$, $k \geq 0$ be integers, $\Omega \subseteq \mathbb{R}^s$ be an connected set, and let $S(z) \subseteq \mathbb{R}^n$ be a subspace for all $z \in \Omega$. We say that $S(\cdot)$ is a $\mathcal{C}^k$-subspace in R^n, if there exists a projector-valued function $P \in \mathcal{C}^k(\Omega, L(\mathbb{R}^s, \mathbb{R}^s))$ such that im $P(z) = S(z)$ for all $z \in \Omega$. Equivalently, $S(\cdot)$ is a $\mathcal{C}^k$-subspace, if the orthoprojector function onto $S(\cdot)$ belongs to the class $\mathcal{C}^k$.

A $\mathcal{C}^k$-subspace has constant dimension on Ω. It has further local $\mathcal{C}^k$-bases. For $s = 1$, there is even a global $\mathcal{C}^k$-basis, see [55, Appendix A].

Lemma 6.9 *Given are two matrix functions $M \in \mathcal{C}^1(\mathcal{I}, \mathcal{L}(\mathbb{R}^m, \mathbb{R}^s))$ and $\tilde{M} \in \mathcal{C}^1(\mathcal{I}, \mathcal{L}(\mathbb{R}^m, \mathbb{R}^{\tilde{s}}))$, both with constant rank on the interval $\mathcal{I}$. Additionally, let the condition*

$$\ker \tilde{M}(t) = \ker M(t), \quad \text{for all} \quad t \in \mathcal{I}.$$

be satisfied. Then the following becomes valid:

1. $\ker \tilde{M}(\cdot) = \ker M(\cdot)$ is a $\mathcal{C}^1$-subspace varying in $\mathbb{R}^m$.
2. The function spaces

$$\mathcal{C}_M^1(\mathcal{I}, \mathbb{R}^m) := \{x \in \mathcal{C}(\mathcal{I}, \mathbb{R}^m) : Mx \in \mathcal{C}^1(\mathcal{I}, \mathbb{R}^s)\},$$

$$H_M^1(\mathcal{I}, \mathbb{R}^m) := \{x \in L^2(\mathcal{I}, \mathbb{R}^m) : Mx \in H^1(\mathcal{I}, \mathbb{R}^s)\},$$

coincide with

$$\mathcal{C}_{\tilde{M}}^1(\mathcal{I}, \mathbb{R}^m) := \{x \in \mathcal{C}(\mathcal{I}, \mathbb{R}^m) : \tilde{M}x \in \mathcal{C}^1(\mathcal{I}, \mathbb{R}^{\tilde{s}})\},$$

$$H_{\tilde{M}}^1(\mathcal{I}, \mathbb{R}^m) := \{x \in L^2(\mathcal{I}, \mathbb{R}^m) : \tilde{M}x \in H^1(\mathcal{I}, \mathbb{R}^{\tilde{s}})\},$$

respectively.
3. Let $\mathcal{I}$ be compact. Equipped with the norm

$$\|x\|_{\mathcal{C}_M^1} := \|x\|_\infty + \|(Mx)'\|_\infty, \quad x \in \mathcal{C}_M^1(\mathcal{I}, \mathbb{R}^m),$$

the function space $\mathcal{C}_M^1(\mathcal{I}, \mathbb{R}^m)$ becomes a Banach space. The norms $\|\cdot\|_{\mathcal{C}_M^1}$ and $\|\cdot\|_{\mathcal{C}_{\tilde{M}}^1}$ are equivalent.

4. Let $\mathcal{I}$ be compact. Equipped with the scalar product

$$(x, \xi)_{H_M^1} := (x, \xi)_{L^2} + ((Mx)', (M\xi)')_{L^2}, \quad x \in H_M^1(\mathcal{I}, \mathbb{R}^m), \qquad (6.1)$$

the function space $H_M^1(\mathcal{I}, \mathbb{R}^m)$ becomes a Hilbert space. The associated norms $\|\cdot\|_{H_M^1}$ and $\|\cdot\|_{H_{\tilde{M}}^1}$ are equivalent.

Proof Statement (1) is a consequence of [55, Lemma A.15].

2. Owing to the constant-rank properties the Moore–Penrose inverses M^+ and $\tilde{M}^+$ are also continuously differentiable. It holds that $\tilde{M}^+ \tilde{M} = M^+ M$.

   For each $x \in \mathcal{C}_M^1(\mathcal{I}, \mathbb{R}^m)$, we have $x \in \mathcal{C}(\mathcal{I}, \mathbb{R}^m)$ and $\tilde{M}x = \tilde{M}\tilde{M}^+ \tilde{M}x = \tilde{M}M^+ Mx \in \mathcal{C}^1(\mathcal{I}, \mathbb{R}^{\tilde{s}})$, such that $x \in \mathcal{C}_{\tilde{M}}^1(\mathcal{I}, \mathbb{R}^m)$, and hence $\mathcal{C}_M^1(\mathcal{I}, \mathbb{R}^m) = \mathcal{C}_{\tilde{M}}^1(\mathcal{I}, \mathbb{R}^m)$.

   Similarly, for each $x \in H_M^1(\mathcal{I}, \mathbb{R}^m)$, we have $x \in L^2(\mathcal{I}, \mathbb{R}^m)$ and $\tilde{M}x = \tilde{M}\tilde{M}^+ \tilde{M}x = \tilde{M}M^+ Mx \in H^1(\mathcal{I}, \mathbb{R}^{\tilde{s}})$, such that $x \in H_{\tilde{M}}^1(\mathcal{I}, \mathbb{R}^m)$, and hence $H_M^1(\mathcal{I}, \mathbb{R}^m) = H_{\tilde{M}}^1(\mathcal{I}, \mathbb{R}^m)$.

3. The given expression defines a norm in fact. The norm equivalence results from the inequality

$$\|x\|_{\mathcal{C}_M^1} := \|x\|_\infty + \|(Mx)'\|_\infty = \|x\|_\infty + \|(M\tilde{M}^+ \tilde{M}x)'\|_\infty$$

$$= \|x\|_\infty + \|(M\tilde{M}^+)'\tilde{M}x + M\tilde{M}^+(\tilde{M}x)'\|_\infty$$

$$\leq \kappa(\|x\|_\infty + \|(\tilde{M}x)'\|_\infty) = \kappa\|x\|_{\mathcal{C}_{\tilde{M}}^1}.$$

For proving the completeness, we consider the sequence $\{x_l\}$ being a Cauchy sequence in $\mathcal{C}_M^1$. Then, $\{x_l\}$ and $\{Mx_l\}$ are Cauchy sequences in $\mathcal{C}$, therefore there is a $x_* \in \mathcal{C}$ such that $\|x_l - x_*\|_\infty \to 0$ and $\|Mx_l - Mx_*\|_\infty \to 0$. Furthermore, $\{Mx_l\}$ is a Cauchy sequence in $\mathcal{C}^1$, and there is a $y_* \in \mathcal{C}^1$ so that $\|Mx_l - y_*\|_{\mathcal{C}^1} \to 0$. This implies $Mx_* = y_* \in \mathcal{C}^1$, thus $x_* \in \mathcal{C}_M^1$ as well as $\|x_l - x_*\|_{\mathcal{C}_M^1} \to 0$.

4. Formula (6.1) defines a scalar product and induces a norm on H_M^1. Further we can proceed analogously to (3). $\qquad\square$

Lemma 6.10 *Let $V \in \mathcal{C}(\mathcal{I}, \mathcal{L}(\mathbb{R}^m))$ be a projector-valued function on the compact interval $\mathcal{I}$.*

1. *Then, for $X = \mathcal{C}(\mathcal{I}, \mathbb{R}^m)$ and $X = L^2(\mathcal{I}, \mathbb{R}^m)$, the relation*

$$(\mathcal{V}x)(t) := V(t)x(t), \quad t \in \mathcal{I}, \quad resp.\ a.e.\ in\ \mathcal{I}, \qquad (6.2)$$

defines a projector $\mathcal{V} \in \mathcal{L}_b(X)$ such that the topological direct sum decomposition

$$\ker \mathcal{V} \oplus \operatorname{im} \mathcal{V} = X \tag{6.3}$$

is valid.

2. *If the additional matrix-valued function $D \in \mathcal{C}^1(\mathcal{I}, \mathcal{L}(\mathbb{R}^m, \mathbb{R}^n))$ is given, and*

$$DV(I - D^+ D) = 0, \quad DVD^+ \in \mathcal{C}^1(\mathcal{I}, \mathcal{L}(\mathbb{R}^n)), \tag{6.4}$$

then statement (1) is also valid for $X = \mathcal{C}_D^1(\mathcal{I}, \mathbb{R}^m)$ and $X = H_D^1(\mathcal{I}, \mathbb{R}^m)$.

Proof 1. $\mathcal{V}$ is idempotent since $V(t)^2 = V(t)$ for all t. Since $V(t)$ and $I - V(t)$ are uniformly bounded on $\mathcal{I}$, the projectors $\mathcal{V}, I - \mathcal{V} \in \mathcal{L}(X)$ are bounded. Then, their nullspaces are closed, hence (6.3) is valid.

2. For $x \in \mathcal{C}_D^1$ it holds that $D\mathcal{V}x = DVD^+Dx \in \mathcal{C}^1$ and $(D\mathcal{V}x)' = (DVD^+)'Dx + DVD^+(Dx)'$. We have $\mathcal{V}x \in \mathcal{C}_D^1$ in fact, and $\|\mathcal{V}x\|_{\mathcal{C}_D^1} \leq \gamma \|x\|_{\mathcal{C}_D^1}$.

As bounded operators, $\mathcal{V}, I - \mathcal{V}$ have closed nullspaces. Analogous arguments apply for $X = H_D^1(\mathcal{I}, \mathbb{R}^m)$.

□

We emphasize that $\operatorname{im} \mathcal{V}$ and $\ker \mathcal{V}$ are infinite-dimensional subspaces of the function space X, though $\operatorname{im} V(t), \ker V(t) \subseteq \mathbb{R}^m$ are finite-dimensional.

Often, if no confusion looms, we use the letter V instead of $\mathcal{V}$.

6.3 Basics Concerning Regular DAEs

We collect basic facts on the DAE

$$f((Dx)'(t), x(t), t) = 0, \tag{6.5}$$

which exhibits the involved derivative by means of an extra matrix-valued function D. The function $f : \mathbb{R}^n \times \mathcal{D}_f \times \mathcal{I}_f \longrightarrow \mathbb{R}^m$, $\mathcal{D}_f \times \mathcal{I}_f \subseteq \mathbb{R}^m \times \mathbb{R}$ open is continuous and has continuous partial derivatives f_y and f_x with respect to the first two variables $y \in \mathbb{R}^n$, $x \in \mathcal{D}_f$. The partial Jacobian $f_y(y, x, t)$ is everywhere singular. The matrix function $D : \mathcal{I}_f \to \mathcal{L}(\mathbb{R}^m, \mathbb{R}^n)$ is continuously differentiable and $D(t)$ has constant rank r on the given interval $\mathcal{I}_f$. Then, $\operatorname{im} D$ is a $\mathcal{C}^1$-subspace in $\mathbb{R}^n$. We refer to [55] for proofs, motivation, and more details.

6.3.1 Regular DAEs, Regularity Regions

Definition 6.1 The DAE (6.5) has a *properly involved derivative*, also called *properly stated leading term*, if ker f_y is another C^1-subspace varying in $\mathbb{R}^n$, and the transversality condition

$$\ker f_y(y, x, t) \oplus \operatorname{im} D(t) = \mathbb{R}^n, \quad (y, x, t) \in \mathbb{R}^n \times D_f \times \mathcal{I}_f, \tag{6.6}$$

is valid.

Below, we always assume the DAE (6.5) to have a properly stated leading term. To simplify matters we further assume the nullspace ker $f_y(y, x, t)$ to be independent of y. Then, the transversality condition (6.6) pointwise induces a projector matrix $R(x, t) \in \mathcal{L}(\mathbb{R}^n)$, the so-called *border projector*, such that

$$\operatorname{im} R(x, t) = \operatorname{im} D(t), \quad \ker R(x, t) = \ker f_y(y, x, t), \quad (y, x, t) \in \mathbb{R}^n \times \mathcal{D}_f \times \mathcal{I}_f. \tag{6.7}$$

Since both subspaces im D and ker f_y are C^1-subspaces, the border projector function $R : \mathcal{D}_f \times \mathcal{I}_f \to \mathcal{L}(\mathbb{R}^n)$ is continuously differentiable, see [55, Lemma A.20].

Note that, if the subspace ker $f_y(y, x, t)$ actually depends on y. then we can modify the DAE by letting $\tilde{f}(y, x, t) := f(D(t)D(t)^+ y, x, t)$ such that ker $\tilde{f}_y(y, x, t) = (\operatorname{im} D(t))^\perp$ solely depends on t.

Next we depict the notion of regularity regions of a DAE (6.5). For these aims we introduce *admissible matrix function sequences* and associated projector functions (cf. [55]). Denote

$$A(x^1, x, t) := f_y(D(t)x^1 + D'(t)x, x, t) \in \mathcal{L}(\mathbb{R}^n, \mathbb{R}^m),$$

$$B(x^1, x, t) := f_x(D(t)x^1 + D'(t)x, x, t) \in \mathcal{L}(\mathbb{R}^m),$$

$$G_0(x^1, x, t) := A(x^1, x, t)D(t) \in \mathcal{L}(\mathbb{R}^m),$$

$$B_0(x^1, x, t) := B(x^1, x, t) \in \mathcal{L}(\mathbb{R}^m) \quad \text{for } x^1 \in \mathbb{R}^m, x \in \mathcal{D}_f, t \in \mathcal{I}_f.$$

The transversality condition (6.6) implies ker $G_0(x^1, x, t) = \ker D(t)$. We introduce projector-valued functions $Q_0, P_0, \Pi_0 \in C(\mathcal{I}_f, \mathcal{L}(\mathbb{R}^m))$ such that for all $t \in \mathcal{I}_f$

$$\operatorname{im} Q_0(t) = N_0(t) := \ker D(t), \quad \Pi_0(t) := P_0(t) := I - Q_0(t). \tag{6.8}$$

Since D has constant rank, the orthoprojector function onto N_0 is as smooth as D. Therefore, as Q_0 we can choose the orthoprojector function onto N_0 which is even continuously differentiable. Next we determine the generalized inverse $D(x,t)^-$ of $D(t)$ pointwise for all arguments by

$$D(x,t)^- D(t) D(x,t)^- = D(x,t)^-,$$
$$D(t) D(x,t)^- D(t) = D(t),$$
$$D(x,t)^- D(t) = P_0(t),$$
$$D(t) D(x,t)^- = R(x,t).$$

The resulting function D^- is continuous, if P_0 is continuously differentiable then so is D^-.

Definition 6.2 Let the DAE (6.5) have a properly involved derivative, the set $\mathcal{G} \subseteq \mathcal{D}_f \times \mathcal{I}_f$ be open connected.

For the given level $\kappa \in \mathbb{N}$, we call the sequence $G_0, \ldots, G_\kappa$ an *admissible matrix function sequence* associated with the DAE (6.5) on the set $\mathcal{G}$, if it is built pointwise for all $(x,t) \in \mathcal{G}$ and all arising $x^j \in \mathbb{R}^m$ by the rule:
set $G_0 := AD$, $B_0 := B$, $N_0 := \ker G_0$,
for $i \geq 1$:

$$G_i := G_{i-1} + B_{i-1} Q_{i-1}, \tag{6.9}$$

$$N_i := \ker G_i, \quad \widehat{N}_i := (N_0 + \cdots + N_{i-1}) \cap N_i,$$

find a complement X_i such that $N_0 + \cdots + N_{i-1} = \widehat{N}_i \oplus X_i$,

choose a projector Q_i such that $\operatorname{im} Q_i = N_i$ and $X_i \subseteq \ker Q_i$,

set $P_i := I - Q_i$, $\Pi_i := \Pi_{i-1} P_i$,

$$B_i := B_{i-1} P_{i-1} - G_i D^- (D \Pi_i D^-)' D \Pi_{i-1}, \tag{6.10}$$

and, additionally,

(a) the matrix function G_i has constant rank r_i on $\mathbb{R}^{mi} \times \mathcal{G}$, $i = 0, \ldots, \kappa$,
(b) the intersection $\widehat{N}_i$ has constant dimension $u_i := \dim \widehat{N}_i$ there,
(c) the product function Π_i is continuous and $D \Pi_i D^-$ is continuously differentiable on $\mathbb{R}^{mi} \times \mathcal{G}$, $i = 0, \ldots, \kappa$.

The projector functions $Q_0, \ldots, Q_\kappa$ linked with an admissible matrix function sequence are said to be *admissible* themselves.

An admissible matrix function sequence $G_0, \ldots, G_\kappa$ is said to be *regular admissible*, if

$$\widehat{N}_i = \{0\} \quad \text{for all} \quad i = 1, \ldots, \kappa.$$

Then, also the projector functions $Q_0, \ldots, Q_\kappa$ are called *regular admissible*.

The numbers $r_0 = \text{rank}\,G_0, \ldots, r_\kappa = \text{rank}\,G_\kappa$ and $u_1, \ldots, u_\kappa$ are called *characteristic values* of the DAE on $\mathcal{G}$.

To shorten the wording we often speak simply of *admissible projector functions*, having in mind the admissible matrix function sequence built with these admissible projector functions. Admissible projector functions are always cross-linked with their matrix function sequence. Changing a projector function yields a new matrix function sequence.

We refer to [55] for many useful properties of the admissible matrix function sequences. A special instance of a sequence is given in Example 4.1. It always holds that

$$r_0 \leq \cdots \leq r_{\kappa-1} \leq r_\kappa.$$

The notion of *characteristic values* makes sense, since these values are independent of the special choice of admissible projector functions and invariant under regular transformations.

In case of a linear constant coefficient DAE, the construct simplifies to a sequence of matrices. In particular, the second term in the definition of B_i disappears. It has long been known [33] that a pair $\{E, F\}$ of $m \times m$ matrices E, F is regular with Kronecker index μ exactly if an admissible sequence of matrices starting with $G_0 = AD = E$, $B_0 := F$ yields

$$r_0 \leq \cdots \leq r_{\mu-1} < r_\mu = m. \tag{6.11}$$

Thereby, neither the factorization nor the special choice of admissible projectors matter. The characteristic values describe the structure of the Weierstraß–Kronecker form (2.4): we have $l = \sum_{j=0}^{\mu-1}(m - r_j)$ and the nilpotent part N contains altogether $s = m - r_0$ Jordan blocks, among them $r_i - r_{i-1}$ Jordan blocks of order i, $i = 1, \ldots, \mu$, see [55, Corollary 1.32].

For linear DAEs with time-varying coefficients, the term $(\cdot)'$ in (6.10) means the derivative in time, and all matrix functions are functions in time. In general, the term $(\cdot)'$ in (6.10) stands for the total derivative in jet variables and then the matrix function G_i depends on the basic variables $(x, t) \in \mathcal{G}$ and, additionally, on the jet variables $x^1, \ldots, x^{i+1} \in \mathbb{R}^m$. Owing to the total derivative $(D\Pi_i D^-)'$ the new variable $x^{i+2} \in \mathbb{R}^m$ comes in at this level, see [55, Section 3.2].

Owing to the constant-rank conditions, the terms $D\Pi_i D^-$ are basically continuous. It may happen, for making these terms continuously differentiable, that the data function f must satisfy additional smoothness requirements. A precise description of those smoothness is much too involved and an overall sufficient condition, say $f \in C^m$, is much too superficial. To indicate that there might be additional smoothness demands we restrict ourselves to the wording f *is sufficiently smooth*.

The next definition ties regularity up to the inequalities (6.11) and so generalizes regularity of matrix pencils for time-varying linear DAEs as well as for nonlinear DAEs. We emphasize that regularity is supported by several constant-rank conditions.

Definition 6.3 Let the DAE (6.5) have a properly involved derivative. Let $\mathcal{G} \subseteq \mathcal{D}_f \times \mathcal{I}_f$ be an open, connected subset. The DAE (6.5) is said to be

1. *regular on $\mathcal{G}$ with tractability index 0*, if $r_0 = m$;
2. *regular on $\mathcal{G}$ with tractability index μ*, if an admissible matrix function sequence exists such that (6.11) is valid on $\mathcal{G}$;
3. *regular on $\mathcal{G}$*, if it is, on $\mathcal{G}$, regular with any index (i.e., case (1) or (2) applies).

The open connected subset $\mathcal{G}$ is called a *regularity region* or *regularity domain*.

A point $(\bar{x}, \bar{t}) \in \mathcal{D}_f \times \mathcal{I}_f$ is a *regular point*, if there is a regularity region $\mathcal{G} \ni (\bar{x}, \bar{t})$.

If $\mathcal{D} \subseteq \mathcal{D}_f$ is an open subset and $\mathcal{I} \subseteq \mathcal{I}_f$ is a compact subinterval, then the DAE (6.5) is said to be regular on $\mathcal{D} \times \mathcal{I}$, if there is a regularity region $\mathcal{G}$ such that $\mathcal{D} \times \mathcal{I} \subset \mathcal{G}$.

Example 6.11 (Regularity Regions) We write the DAE

$$x_1'(t) + x_1(t) = 0,$$
$$x_2(t)x_2'(t) - x_3(t) = 0,$$
$$x_1(t)^2 + x_2(t)^2 - 1 - \gamma(t) = 0,$$

in the form (6.5), with $n = 2$, $m = k = 3$,

$$f(y, x, t) = \begin{bmatrix} y_1 + x_1 \\ x_2 y_2 - x_3 \\ x_1^2 + x_2^2 - \gamma(t) - 1 \end{bmatrix}, \quad f_y(y, x, t) = \begin{bmatrix} 1 & 0 \\ 0 & x_2 \\ 0 & 0 \end{bmatrix},$$

$$D(t) = \begin{bmatrix} 1 & 0 & 0 \\ 0 & 1 & 0 \end{bmatrix},$$

for $y \in \mathbb{R}^2$, $x \in \mathcal{D}_f = \mathbb{R}^3$, $t \in \mathcal{I}_f = \mathbb{R}$.

The derivative is properly involved on the open subsets $\mathbb{R}^2 \times \mathcal{G}_+$ and $\mathbb{R}^2 \times \mathcal{G}_-$, $\mathcal{G}_+ := \{x \in \mathbb{R}^3 : x_2 > 0\} \times \mathcal{I}_f$, $\mathcal{G}_- := \{x \in \mathbb{R}^3 : x_2 < 0\} \times \mathcal{I}_f$. We have there

$$G_0 = AD = \begin{bmatrix} 1 & 0 & 0 \\ 0 & x_2 & 0 \\ 0 & 0 & 0 \end{bmatrix}, \quad B_0 = \begin{bmatrix} 1 & 0 & 0 \\ 0 & x_2^1 & -1 \\ 2x_1 & 2x_2 & 0. \end{bmatrix}.$$

Letting

$$Q_0 = \begin{bmatrix} 0 & 0 & 0 \\ 0 & 0 & 0 \\ 0 & 0 & 1 \end{bmatrix}, \quad \text{yields} \quad G_1 = \begin{bmatrix} 1 & 0 & 0 \\ 0 & 2x_2 & -1 \\ 0 & 0 & 0 \end{bmatrix}.$$

G_1 is singular but has constant rank. Since $N_0 \cap N_1 = \{0\}$ we find a projector function Q_1 such that $N_0 \subseteq \ker Q_1$. We choose

$$Q_1 = \begin{bmatrix} 0 & 0 & 0 \\ 0 & 1 & 0 \\ 0 & \frac{1}{x_2} & 0 \end{bmatrix}, P_1 = \begin{bmatrix} 1 & 0 & 0 \\ 0 & 0 & 0 \\ 0 & -\frac{1}{x_2} & 1 \end{bmatrix}, \Pi_1 = \begin{bmatrix} 1 & 0 & 0 \\ 0 & 0 & 0 \\ 0 & 0 & 0 \end{bmatrix}, D\Pi_1 D^- = \begin{bmatrix} 1 & 0 \\ 0 & 0 \end{bmatrix},$$

and obtain $B_1 = B_0 P_0 Q_1$, and then

$$G_2 = \begin{bmatrix} 1 & 0 & 0 \\ 0 & 2x_2 + x_2^1 & -1 \\ 0 & 2x_2 & 0 \end{bmatrix}.$$

The matrix $G_2 = G_2(x^1, x, t)$ is nonsingular for all arguments (x^1, x, t) with $x_2 \neq 0$. The admissible matrix function sequence terminates at this level. The open connected subsets $\mathcal{G}_+$ and $\mathcal{G}_-$ are regularity regions, here both with characteristics $r_0 = 2, r_1 = 2, r_2 = 3$, and tractability index $\mu = 2$.   □

For regular DAEs, all intersections $\widehat{N}_i$ are trivial ones, thus $u_i = 0$, $i \geq 1$. Namely, because of the inclusions

$$\widehat{N}_i \subseteq N_i \cap N_{i+1} \subseteq N_{i+1} \cap N_{i+2} \subseteq \cdots \subseteq N_{\mu-1} \cap N_\mu,$$

for reaching a nonsingular G_μ, which means $N_\mu = \{0\}$, it is necessary to have $\widehat{N}_i = \{0\}$, $i \geq 1$. This is a useful condition for checking regularity in practice.

Observe that each open connected subset of a regularity region is again a regularity region. A regularity region consist of regular points having uniform characteristics. The union of regularity regions is, if it is connected, a regularity region, too. Further, the nonempty intersection of two regularity regions is also a regularity region. Only regularity regions with uniform characteristics may yield nonempty intersections. *Maximal regularity regions* are then bordered by so-called critical points. Solutions may cross the borders of maximal regularity regions and undergo there bifurcations et cetera, see examples in [53, 55, 61]. Much further research is needed to elucidate these phenomena.

6.3.2   The Structure of Linear DAEs

The general DAE (6.5) captures linear DAEs

$$A(t)(Dx)'(t) + B(t)x(t) - q(t) = 0 \tag{6.12}$$

as $f(y, x, t) := A(t)y + B(t)x - q(t)$, $t \in \mathcal{I}_f$. Now, admissible matrix function sequences depend only on time t; and hence, we speak of *regularity intervals* instead of regions. A regularity interval is open by definition. We say that the linear DAE with properly leading term is *regular on the compact interval* $[t_a, t_e]$, if there is an accommodating regularity interval, or equivalently, if all points of $[t_a, t_e]$ are regular.

If the linear DAE is regular on the interval $\mathcal{I}$, then it is also regular on each subinterval of $\mathcal{I}$ with the same characteristics. This sounds as a triviality; however, there is a continuing profound debate about some related questions, cf. Sect. 4.4.

If the linear DAE (6.12) is regular on the interval $\mathcal{I}$, then (see [55, Section 2.4]) it can be decoupled by admissible projector functions into an IERODE

$$u' - (D\Pi_{\mu-1}D^-)'u + D\Pi_{\mu-1}G_\mu^{-1}B_\mu D^- u = D\Pi_{\mu-1}G_\mu^{-1}q \tag{6.13}$$

and a triangular subsystem of several equations including differentiations

$$\begin{bmatrix} 0 & \mathcal{N}_{01} & \cdots & \mathcal{N}_{0,\mu-1} \\ & 0 & \ddots & \vdots \\ & & \ddots & \mathcal{N}_{\mu-2,\mu-1} \\ & & & 0 \end{bmatrix} \begin{bmatrix} 0 \\ (Dv_1)' \\ \vdots \\ (Dv_{\mu-1})' \end{bmatrix}$$

$$+ \begin{bmatrix} I & \mathcal{M}_{01} & \cdots & \mathcal{M}_{0,\mu-1} \\ & I & \ddots & \vdots \\ & & \ddots & \mathcal{M}_{\mu-2,\mu-1} \\ & & & I \end{bmatrix} \begin{bmatrix} v_0 \\ v_1 \\ \vdots \\ v_{\mu-1} \end{bmatrix} + \begin{bmatrix} \mathcal{H}_0 \\ \mathcal{H}_1 \\ \vdots \\ \mathcal{H}_{\mu-1} \end{bmatrix} D^- u = \begin{bmatrix} \mathcal{L}_0 \\ \mathcal{L}_1 \\ \vdots \\ \mathcal{L}_{\mu-1} \end{bmatrix} q.$$

The subspace im $D\Pi_{\mu-1}$ is an invariant subspace for the IERODE (6.13).

This structural decoupling is associated with the decomposition

$$x = D^- u + v_0 + v_1 + \cdots + v_{\mu-1}.$$

The coefficients are continuous and explicitly given in terms of an admissible matrix function sequence as

$$\mathcal{N}_{01} := -Q_0 Q_1 D^-$$
$$\mathcal{N}_{0j} := -Q_0 P_1 \cdots P_{j-1} Q_j D^-, \qquad\qquad j = 2, \ldots, \mu - 1,$$

$$\mathcal{N}_{i,i+1} := -\Pi_{i-1}Q_iQ_{i+1}D^-,$$

$$\mathcal{N}_{ij} := -\Pi_{i-1}Q_iP_{i+1}\cdots P_{j-1}Q_jD^-, \qquad j = i+2,\ldots,\mu-1,$$
$$i = 1,\ldots,\mu-2,$$

$$\mathcal{M}_{0j} := Q_0P_1\cdots P_{\mu-1}\mathcal{M}_jD\Pi_{j-1}Q_j, \qquad j = 1,\ldots,\mu-1,$$

$$\mathcal{M}_{ij} := \Pi_{i-1}Q_iP_{i+1}\cdots P_{\mu-1}\mathcal{M}_jD\Pi_{j-1}Q_j, \quad j = i+1,\ldots,\mu-1,$$
$$i = 1,\ldots,\mu-2,$$

$$\mathcal{L}_0 := Q_0P_1\cdots P_{\mu-1}G_\mu^{-1},$$

$$\mathcal{L}_i := \Pi_{i-1}Q_iP_{i+1}\cdots P_{\mu-1}G_\mu^{-1}, \qquad i = 1,\ldots,\mu-2,$$

$$\mathcal{L}_{\mu-1} := \Pi_{\mu-2}Q_{\mu-1}G_\mu^{-1},$$

$$\mathcal{H}_0 := Q_0P_1\cdots P_{\mu-1}\mathcal{K}\Pi_{\mu-1},$$

$$\mathcal{H}_i := \Pi_{i-1}Q_iP_{i+1}\cdots P_{\mu-1}\mathcal{K}\Pi_{\mu-1}, \qquad i = 1,\ldots,\mu-2,$$

$$\mathcal{H}_{\mu-1} := \Pi_{\mu-2}Q_{\mu-1}\mathcal{K}\Pi_{\mu-1},$$

with

$$\mathcal{K} := (I - \Pi_{\mu-1})G_\mu^{-1}B_{\mu-1}\Pi_{\mu-1} + \sum_{l=1}^{\mu-1}(I - \Pi_{l-1})(P_l - Q_l)(D\Pi_l D^-)'D\Pi_{\mu-1},$$

$$\mathcal{M}_j := \sum_{k=0}^{j-1}(I - \Pi_k)\{P_k D^-(D\Pi_k D^-)' - Q_{k+1}D^-(D\Pi_{k+1}D^-)'\}D\Pi_{j-1}Q_lD^-,$$

$$l = 1,\ldots,\mu-1.$$

The IERODE is always uncoupled from the second subsystem, but the latter is tied to the IERODE (6.13) if among the coefficients $\mathcal{H}_0,\ldots,\mathcal{H}_{\mu-1}$ there is at least one which does not vanish. One speaks about a *fine decoupling*, if $\mathcal{H}_1 = \cdots = \mathcal{H}_{\mu-1} = 0$, and about a *complete decoupling*, if $\mathcal{H}_0 = 0$, additionally. A complete decoupling is given exactly if the coefficient $\mathcal{K}$ vanishes identically.

If the DAE (6.12) is regular and the original data are sufficiently smooth, then the DAE (6.12) is called *fine*. For fine DAEs, fine and complete decouplings always exist and can be constructed, see [55, Section 2.4.3]. Example 4.1 shows an instance of completely decoupling projectors.

It is noteworthy that, if $Q_0,\ldots,Q_{\mu-1}$ generate a complete decoupling for a constant coefficient DAE $Ex'(t) + Fx(t) = 0$, then $\Pi_{\mu-1}$ is the spectral projector of the matrix pencil $\{E, F\}$. This way, the projector function $\Pi_{\mu-1}$ associated with a complete decoupling of a fine time-varying DAE represents the generalization of the spectral projector.

6.3.3 Linearizations

Given is a reference function $x_* \in \mathcal{C}_D^1(\mathcal{I}_*, \mathbb{R}^m)$ on an individual interval $\mathcal{I}_* \subseteq \mathcal{I}_f$, whose values belong to $\mathcal{D}_f$. For each such reference function we may consider the linearization of the (6.5) along x_*, that is, the linearized DAE

$$A_*(t)(Dx)'(t) + B_*(t)x(t) = q(t), \quad t \in \mathcal{I}_*, \tag{6.14}$$

with coefficients

$$A_*(t) := f_y((Dx_*)'(t), x_*(t), t), \quad B_*(t) := f_x((Dx_*)'(t), x_*(t), t), \quad t \in \mathcal{I}_*.$$

The linear DAE (6.14) inherits from the nonlinear DAE (6.5) the properly stated leading term.

We denote by $\mathcal{C}_{ref}^m(\mathcal{G})$ the set of all $\mathcal{C}^m$ functions x_*, defined on individual intervals $\mathcal{I}_{x_*}$, and with graph in $\mathcal{G}$, that is, $(x_*(t), t) \in \mathcal{G}$ for $t \in \mathcal{I}_{x_*}$. Clearly, then we have also $x_* \in \mathcal{C}_D^1(\mathcal{I}_{x_*}, \mathbb{R}^m)$. By the smoothness of the reference functions x_* and the function f we ensure that also the coefficients A_* and B_* are sufficiently smooth for regularity.

Next we adapt the necessary and sufficient regularity condition from [55, Theorem 3.33] to our somewhat simpler situation.

Theorem 6.12 *Let the DAE (6.5) have a properly involved derivative and let f be sufficiently smooth. Let $\mathcal{G} \subseteq \mathcal{D}_f \times \mathcal{I}_f$ be an open connected set. Then the following statements are valid:*

1. *The DAE (6.5) is regular on $\mathcal{G}$ if the linearized DAE (6.14) along each arbitrary reference function $x_* \in \mathcal{C}_{ref}^m(\mathcal{G})$ is regular, and vice versa.*
2. *If the DAE (6.5) is regular on $\mathcal{G}$ with tractability index μ and characteristic values $r_0 \leq \cdots \leq r_{\mu-1} < r_\mu = m$, then all linearized DAEs (6.14) along reference functions $x_* \in \mathcal{C}_{ref}^m(\mathcal{G})$ are regular with uniform index μ and characteristics $r_0 \leq \cdots \leq r_{\mu-1} < r_\mu = m$.*
3. *If all linearized DAEs (6.14) along reference functions $x_* \in \mathcal{C}_{ref}^m(\mathcal{G})$ are regular, then they have uniform index and characteristics, and the nonlinear DAE (6.5) is also regular on $\mathcal{G}$, with the same index and characteristics.*

Corollary 6.13 *Let the DAE (6.5) have a properly involved derivative and let f be sufficiently smooth. Let $\mathcal{D} \subseteq \mathcal{D}_f$ be an open connected set and $\mathcal{I} \subset \mathcal{I}_f$ be a compact interval. Then the following statements are valid:*

1. *The DAE (6.5) is regular on $\mathcal{D} \times \mathcal{I}$ if the linearized DAE (6.14) along each arbitrary reference function $x_* \in \mathcal{C}^m(\mathcal{I}, \mathbb{R}^m)$ with values in $\mathcal{D}$ is regular, and vice versa.*
2. *If the DAE (6.5) is regular on $\mathcal{D} \times \mathcal{I}$ with tractability index μ and characteristic values $r_0 \leq \cdots \leq r_{\mu-1} < r_\mu = m$, then all linearized DAEs (6.14) along reference functions $x_* \in \mathcal{C}^m(\mathcal{I}, \mathbb{R}^m)$ with values in $\mathcal{D}$ are regular with uniform index μ and characteristics $r_0 \leq \cdots \leq r_{\mu-1} < r_\mu = m$.*

3. *If all linearized DAEs (6.14) along reference functions $x_* \in C^m(\mathcal{I}, \mathbb{R}^m)$ with values in $\mathcal{D}$ are regular, then they have uniform index and characteristics, and the nonlinear DAE (6.5) is also regular on $\mathcal{D} \times \mathcal{I}$, with the same index and characteristics.*

Proof Statement (1) is a consequence of Statements (2) and (3).

Statement (2) follows from the construction of the admissible matrix function sequences. Namely, for each $x_* \in C^m(\mathcal{I}, \mathbb{R}^m)$, with values in $\mathcal{D}$, we have

$$G_0(x'_*(t), x_*(t), t) =: G_{*0}(t),$$

$$B_{i-1}(x_*^{(i+1)}(t), \cdots, x'_*(t), x_*(t), t) =: B_{*i-1}(t),$$

$$G_i(x_*^{(i+1)}(t), \cdots, x'_*(t), x_*(t), t) =: G_{*i}(t), \quad t \in \mathcal{I}, \quad i = 1, \ldots, \mu,$$

which represents an admissible matrix function sequence for the linearized along x_* DAE.

Statement (3) is proved along the lines of [55, Theorem 3.33] by means of so-called widely orthogonal projector functions. The proof given in [55] also works if one supposes solely compact individual intervals $\mathcal{I}_{x_*}$.

By Lemma 6.14 below, each reference function given on an individual compact interval can be extended to belong to $x_* \in C^m(\mathcal{I}, \mathbb{R}^m)$, with values in $\mathcal{D}$. $\square$

Lemma 6.14 *Let $\mathcal{D} \subseteq \mathbb{R}^m$ be an open set and $\mathcal{I} \subset \mathbb{R}$ be a compact interval. Let $\mathcal{I}_* \subset \mathcal{I}$ be a compact subinterval and $s \in \mathbb{N}$.*

Then, for each function $x_* \in C^s(\mathcal{I}_*, \mathbb{R}^m)$, with values in $\mathcal{D}$, there is an extension $f \in C^s(\mathcal{I}, \mathbb{R}^m)$, with values in $\mathcal{D}$.

Proof It suffices to verify the statement for the case $\mathcal{I} = [t_a, t_e]$, $\mathcal{I}_* = [t_a, t_0]$, $t_0 < t_e$. We put

$$f(t) := x_*(t), \quad t \in [t_a, t_0],$$

$$f(t) := x_0 + e^{-\alpha(t-t_0)} p(t), \quad p(t) := (t - t_0) p_1 + \cdots + \frac{1}{s!}(t - t_0)^s p_s, \quad t \in (t_0, t_e].$$

Letting $x_0 := x_*(t_0)$ we have a continuous function f. We derive for $t > t_0$ and $j = 1, \ldots, s$:

$$f^{(j)}(t) = e^{-\alpha(t-t_0)} p^{(j)}(t) - \sum_{i=1}^{j-1} \alpha^{j-i} \binom{j}{i} f^{(i)}(t) - \alpha^j \binom{j}{0} (f(t) - x_0).$$

Fixing successively the coefficients p_j, for $j = 1, \ldots, s$, by

$$p_j = x_*^{(j)}(t_0) + \sum_{i=1}^{j-1} \alpha^{j-i} \binom{j}{i} x_*^{(i)}(t_0),$$

we ensure that $f \in C^s(\mathcal{I}, \mathbb{R}^m)$. It remains to show that the values of f remain in $\mathcal{D}$. We compute for $t > t_0$ and positive, sufficiently large α:

$$|f(t) - x_0| = |e^{-\alpha(t-t_0)} p(t)|$$

$$= |(t - t_0)e^{-\alpha(t-t_0)} p_1 + \cdots + \frac{1}{s!}(t - t_0)^s e^{-\alpha(t-t_0)} p_s|$$

$$= \frac{1}{\alpha} |\alpha(t - t_0)e^{-\alpha(t-t_0)} p_1 + \cdots + \frac{1}{s!}\alpha^s (t - t_0)^s e^{-\alpha(t-t_0)} \frac{1}{\alpha^{s-1}} p_s| \leq \frac{C}{\alpha}.$$

The last inequality holds true, since the expressions $\alpha^j (t-t_0)^j e^{-\alpha(t-t_0)}$ and $|\frac{1}{\alpha^{j-1}} p_j|$ are bounded by a constant C.

Together with $x_0 \in \mathcal{D}$ there is a ball $B(x_0, \varepsilon) \subset \mathcal{D}$. Choosing a sufficiently large α we arrive at $\frac{C}{\alpha} < \varepsilon$ and we are done.   □

6.4   List of Symbols and Abbreviations

$\mathcal{L}(X, Y)$	Set of linear operators from X to Y
$\mathcal{L}(X)$	$= \mathcal{L}(X, X)$
$\mathcal{L}(\mathbb{R}^m, \mathbb{R}^n)$	is identified with $\mathbb{R}^{n \times m}$
$\mathcal{L}_b(X, Y)$	Set of bounded linear operators from X to Y
$\mathcal{L}_c(X, Y)$	Set of closed linear operators from X to Y
X^*	Dual space
K^*	Dual and adjoint operator
K^-	Outer, inner, and generalized inverses
K^+	Orthogonal generalized (Moore–Penrose) inverse
dom K	Definition domain of the map K
ker K	Nullspace (kernel) of the operator K
im K	Image (range) of the operator K
ind $\{E, F\}$	Kronecker index of the matrix pair $\{E, F\}$
ind $_{fred}(K)$	Fredholm index of the operator K
$\langle \cdot, \cdot \rangle$	Scalar product in $\mathbb{R}^m$, dual pairing
$(\cdot, \cdot)$	Scalar product in function spaces
$|\cdot|$	Vector and matrix norms
$\|\cdot\|$	Norms on function spaces, operator norms
$\dot{+}$	Algebraic direct sum
$\oplus$	Topological direct sum
DAE	Differential-algebraic equation
ODE	Ordinary differential equation
IVP	Initial value problem
BVP	Boundary value problem
LSS	Least squares solution

References

1. Ambrosetti, A., Prodi, G.: A Primer in Nonlinear Analysis, Cambridge studies in advanced mathematics, vol. 34. Cambridge University Press, Cambridge (1995)
2. Aronszajn, N.: Theory of reproducing kernels. Trans. Am. Math. Soc. **68**, 337–404 (1950)
3. Baumanns, S.: Coupled Electromagnetic Field/Circuit Simulation. Ph.D. thesis, Universität zu Köln, Juni 2012. Logos, Berlin (2012)
4. Biegler, L., Campbell, S.L., Mehrmann, V.: Control and optimization with differential-algebraic constraints. SIAM, Philadelphia (2011)
5. Boyarintsev, Y.E.: Regular and Singular Systems of Linear Ordinary Differential Equations. Nauka (Sibirskoe otdelenie), Novosibirsk (1980, in Russian)
6. Brenan, K.E., Campbell, S.L., Petzold, L.R.: The Numerical Solution of Initial Value Problems in Ordinary Differential-Algebraic Equations. North Holland, New York (1989)
7. Campbell, S.L.: Singular Systems of Differential Equations II. Research Notes in Mathematics. Pitman, Marshfield (1982)
8. Campbell, S.L.: One canonical form for higher index linear time varying singular systems. Circuits Syst. Signal Process. **2**, 311–326 (1983)
9. Campbell, S.L.: Regularization of linear time varying singular systems. Automatica **20**, 365–370 (1984)
10. Campbell, S.L., Gear, C.W.: The index of general nonlinear daes. Numer. Math. **72**, 173–196 (1995)
11. Campbell, S.L., Kunkel, P., Mehrmann, V.: Regularization of linear and nonlinear descriptor systems. In: Biegler, L.T., Campbell, S.L., Mehrmann, V. (eds.) Control and Optimization with Differential-Algebraic Constraints, Advances in Design and Control, pp. 17–36. SIAM, New York (2012)
12. Chistyakov, V.F.: Vyrozhdennye sistemy obyknovennykh differentsial'nykh uravnenij, chapter 3: K metodam resheniya singul'yarnykh linejnykh sistem obyknovennykh differentsial'nykh uravnenij, pp. 37–65. Nauka, Novosibirsk (1982, in Russian, edited by Yu. E. Boyarintsev)
13. Chistyakov, V.F.: On Noetherian index of differential/algebraic systems. Sib. Math. J. **34**(3), 583–592 (1993)
14. Chistyakov, V.F.: Algebro-Differential'nye Operatory s Konechnomernym Yadrom. Nauka, Novosibirsk (1996, in Russian)
15. Chistyakov, V.F.: Regularization of differential-algebraic equations. Comput. Math. Math. Phys. **51**(12), 2052–2064 (2011). Original Russian text published in Zhurnal Vycislitel'noi Matematiki i Matematicheskoi Fiziki, vol. 51, pp. 2181–2193 (2011)
16. Chistyakov, V.F., Chistyakova, E.V.: Application of the least squares method to solving linear differential-algebraic equations. Numer. Anal. Appl. **6**(1), 77–90 (2013). Original Russian text published in Sibirskii Zhurnal Vycislitel'noi Matematiki, vol. 16, pp. 81–95 (2013)
17. Chistyakov, V.F., Shcheglova, A.A.: Izbrannye glavy teorii algebro-differential'nykh sistem. Nauka, Novosibirsk (2003, in Russian)
18. Cobb, D.: On the solutions of linear differential equations with singular coefficients. J. Differ. Equ. **46**, 310–323 (1982)
19. Craven, B.D., Nashed, M.Z.: Generalized implicit function theorems when the derivative has no bounded inverse. Nonlinear Anal. Theory Methods Appl. **6**(4), 375–387 (1982)
20. Dautray, R., Lions, J.-L.: Functional and Variational Methods. Mathematical Analysis and Numerical Methods for Science and Technology, vol. 2. Springer, Berlin/Heidelberg (1988)
21. Degenhardt, A.: A collocation method for boundary value problems of transferable differential-algebraic equations. Preprint (Neue Folge) 182, Humboldt-Universität zu Berlin, Sektion Mathematik (1988)
22. Degenhardt, A.: Collocation for transferable differential-algebraic equations. In: Griepentrog, E., Hanke, M., März, R. (eds.) Berlin Seminar on Differential-Algebraic Equations, vol. 92-1, pp. 83–104 (1992)

23. Dokchan, R.: Numerical Intergration of Differential-Algebraic Equations with Harmless Critical Points. Ph.D. thesis, Humboldt-University of Berlin (2011)
24. Engl, H.W., Groetsch, C.W. (eds.): Inverse and Ill-Posed Problems, Notes and Reports in Mathematics in Sciences and Engineering, vol. 4. Academic, Boston/Orlando/London (1987)
25. Engl, H.W., Hanke, M., Neubauer, A.: Tikhonov regularization of nonlinear differential-algebraic equations. In: Sabatier, P.C. (ed.) Inverse Methods in Action, pp. 92–105. Springer, Berlin/Heidelberg (1990)
26. Engl, H.W., Hanke, M., Neubauer, A.: Regularization of Inverse Problems. Mathematics and its Application. Kluwer Academic, Dordrecht (2000)
27. Favini, A., Yagi, A.: Degenerate Differential Equations in Banach Spaces. Pure and Applied Mathematics. Marcel Dekker, New York (1999)
28. Führer, C., Leimkuhler, B.J.: Numerical solution of differential-algebraic equations for constrained mechanical motion. Numer. Math. **59**, 55–69 (1991)
29. Gajewski, H., Gröger, K., Zacharias, K.: Nichtlineare Operatorgleichungen und Operatordifferentialgleichungen. Akademie, Berlin (1974)
30. Gantmacher, F.R.: Matrizenrechnung I+II. VEB Deutcher Verlag der Wissenchaften, Berlin (1970)
31. Gorbunov, V.K., Petrischev, V.V., Sviridov, V.Y.: Development of normal spline method for linear integro-differential equations. In: Sloot, P.M.A., Abramson, D., Bogdanov, A., Dongarra, J.J., Zomaya, A., Gorbachev, Y. (eds.) International Conference Computational Science—ICCS 2003, Lecture Notes in Computer Science, vol. 2658, pp. 492–499. Springer, Berlin/Heidelberg (2003)
32. Griepentrog, E., März, R.: Differential-Algebraic Equations and Their Numerical Treatment. Teubner-Texte zur Mathematik No. 88. BSB B.G. Teubner Verlagsgesellschaft, Leipzig (1986)
33. Griepentrog, E., März, R.: Basic properties of some differential-algebraic equations. Zeitschrift für Analysis und ihre Anwendungen **8**(1), 25–40 (1989)
34. Groetsch, C.W.: The theory of Tikhonov regularization for Fredholm equations of the first kind. Pitman, London (1984)
35. Hairer, E., Lubich, Ch., Roche, M.: The Numerical Solution of Differential-Algebraic Equations by Runge–Kutta Methods. Lecture Notes in Mathematics, vol. 1409. Springer, Heidelberg (1989)
36. Hanke, M.: On a least-squares collocation method for linear differential-algebraic equations. Numer. Math. **54**, 79–90 (1988)
37. Hanke, M.: Beiträge zur Regularisierung von Randwertaufgaben für Algebro-Differentialgleichungen mit höherem Index. Dissertation(B), Habilitation, Humboldt-Universität zu Berlin, Institut für Mathematik (1989)
38. Hanke, M.: Linear differential-algebraic equations in spaces of integrable functions. J. Differ. Equ. **79**, 14–30 (1989)
39. Hanke, M.: On the regularization of index 2 differential-algebraic equations. J. Math. Anal. Appl. **151**, 236–253 (1990)
40. Hanke, M.: Regularization methods for higher index differential-algebraic equations. In: Griepentrog, E., Hanke, M., März, R. (eds.) Berlin Seminar on Differential-Algebraic Equations, vol. 92-1, pp. 105–141 (1992)
41. Hanke, M.: Asymptotic expansions for regularization methods of linear fully implicit differential-algebraic equations. Zeitschrift für Analysis und ihre Anwendungen **13**, 513–535 (1994)
42. Hanke, M.: Regularization of differential-algebraic equations revisited. Math. Nachr. **174**, 159–183 (1995)
43. Hanke, M., März, R., Neubauer, A.: On the regularization of linear differential-algebraic equations. In: Engl, H.W., Groetsch, C.W. (eds.) Inverse and Ill-posed Problems, Notes and Reports in Mathematics in Science and Engineering, vol. 4, pp. 523–540. Academic, Orlando (1987)
44. Hanke, M., März, R., Neubauer, A.: On the regularization of a certain class of nontransferable differential-algebraic equations. J. Differ. Equ. **73**(1), 119–132 (1988)

45. Heuser, H.: Funktionalanalysis. Mathematische Leitfäden. B.G.Teubner Stuttgart (1992)
46. Ilchmann, A., Reis, T. (eds.): Surveys in Differential-Algebraic Equations I. Differential-Algebraic Equations Forum. Springer, Heidelberg/New York/Dordrecht/London (2013)
47. Kato, T.: Perturbation theory for linear operators. Springer, Berlin/Heidelberg/New York (1995). Reprint of the 1980 edition
48. Knorrenschild, M.: Regularisierung von differentiell-algebraischen Systemen—theoretische und numerische Aspekte. Ph.D. thesis, Rheinisch-Westfälische Technische Hochschule (1988)
49. Kronecker, L.: Gesammelte Werke, volume III, chapter Reduktion der Scharen bilinearer Formen, pp. 141–155. Akad. d. Wiss. Berlin (1890)
50. Kunkel, P., Mehrmann, V.: Generalized inverses of differential-algebraic operators. SIAM J. Matrix Anal. Appl. **17**, 426–442 (1996)
51. Kunkel, P., Mehrmann, V.: Differential-Algebraic Equations: Analysis and Numerical Solution. EMS Publishing House, Zürich (2006)
52. Kurina, G.A.: Singular perturbations of control problems with equation of state not solved for the derivative (a survey). J. Comput. Syst. Sci. Int. **31**(6), 17–45 (1993)
53. Lamour, R., März, R.: Detecting structures in differential algebraic equations: computational aspects. J. Comput. Appl. Math **236**(16), 4055–4066 (2012). Special Issue: 40 years of Numeric Mathematics
54. Lamour, R., März, R.: Differential-algebraic equations with regular local matrix pencils. Vestnik YuUrGU. Seriya Matematicheskoe modelirovanie i programmipovanie **6**(4), 39–47 (2013)
55. Lamour, R., März, R., Tischendorf, C.: Differential-Algebraic Equations: A Projector Based Analysis. Differential-Algebraic Equations Forum. Springer, Berlin/Heidelberg/New York/-Dordrecht/London (2013) (Series Editors: A. Ilchman, T. Reis)
56. März, R.: On difference and shooting methods for boundary value problems in differential-algebraic equations. ZAMM **64**(11), 463–473 (1984)
57. März, R.: On correctness and numerical treatment of boundary value problems in DAEs. Zhurnal Vychisl. Matem. i Matem. Fiziki **26**(1), 50–64 (1986)
58. März, R.: Numerical methods for differential-algebraic equations. Acta Numer. **1**, 141–198 (1992)
59. März, R.: On linear differential-algebraic equations and linearizations. Appl. Numer. Math. **18**, 267–292 (1995)
60. März, R.: Nonlinear differential-algebraic equations with properly formulated leading term. Technical Report 2001–3, Humboldt-Universität zu Berlin, Institut für Mathematik (2001)
61. März, R.: Notes on linearization of daes and on optimization with differential-algebraic constraints. In: Biegler, L.T., Campbell, S.L., Mehrmann, V. (eds.) Control and Optimization with Differential-Algebraic Constraints, Advances in Design and Control, pp. 37–58. SIAM, New York (2012)
62. Nashed, M.Z.: Inner, outer, and generalized inverses in banach and hilbert spaces. Numer. Funct. Anal. Optimiz. **9**, 261–325 (1987)
63. Nashed, M.Z.: A new approach to classification and regularization of ill-posed operator equations. In: Engl, H.W., Groetsch, C.W. (eds.) Inverse and Ill-posed Problems, Notes and Reports in Mathematics in Science and Engineering, vol. 4, pp. 53–75. Academic, New York (1987)
64. Nashed, M.Z., Chen, X.: Convergence of Newton-like methods for singular operator equations using outer inverses. Numer. Math. **66**, 235–257 (1993)
65. Niepage, D.: On the existence and approximation and approximate solution of discontinuous differential-algebraic systems. In: Griepentrog, E., Hanke, M., März, R. (eds.) Berlin Seminar on Differential-Algebraic Equations, vol. 92-1, pp. 179–194 (1992)
66. Niepage, H.-D.: A convergence and existence result for differential-algebraic inclusions. Numer. Funct. Anal. Optimiz. **9**, 1221–1250 (1987–1988)
67. Niepage, H.-D.: On the numerical solution of differential-algebraic equations with discontinuities. In: Strehmel, K. (ed.) Numerical Treatment of Differential Equations. Teubner, Leipzig (1990)

68. Nittka, R., Sauter, M.: Sobolev gradients for differential algebraic equations. Electron. J. Differ. Equ. **2008**(42), 1–31 (2008)
69. Petry, T.: Realisierung des Newton-Kantorovich-Verfahrens für nichtlineare Algebro-Differentialgleichungen mittels Abramov-Transfer. Ph.D. thesis, Humboldt-Universität zu Berlin, Juni 1998. Logos, Berlin (1998)
70. Rabier, P.J., Rheinboldt, W.C.: Theoretical and numerical analysis of differential-algebraic equations. In: Ciarlet, P.G. et al. (eds.) Handbook of Numerical Analysis, Techniques of Scientific Computing (Part 4), vol. VIII, pp. 183–540. North Holland/Elsevier, Amsterdam (2002)
71. Reis, T.: Systems Theoretic Aspects of PDAEs and Applications to Electrical Circuits. Ph.D. thesis, Technische Universiät Kaiserslautern (2006)
72. Riaza, R.: Differential-Algebraic Systems. Analytical Aspects and Circuit Applications. World Scientific, River Edge (2008)
73. Schumilina, I.: Charakterisierung der Algebro-Differentialgleichungen mit Traktabilitätsindex 3. Ph.D. thesis, Humboldt-Universität zu Berlin (2004)
74. Schwarz, D.E.: Consistent initialization for index-2 differential algebraic equations and its application to circuit simulation. Ph.D. thesis, Mathemematisch-Naturwissenschaftliche Fakultät II, Humboldt-Universität zu Berlin, Juli (2000)
75. Seufer, I.: Generalized inverses of differential-algebraic equations and their discretization. Ph.D. thesis, Technische Universität Berlin (2006)
76. Tikhonov, A.N., Arsenin, V.Y.: Methods for the solution of ill-posed problems. Nauka, Moskva (1974, in Russian.)
77. Trenn, S.: Distributional Differential Algebraic Equations. Ph.D. thesis, TU Ilmenau (2009)
78. Vajnikko, G.M., Veretennikov, A.Y.: Iteratsionnye protsedury v nekorrektnykh zadachakh. Nauka, Moskva (1986, in Russian)
79. Voigtmann, S.: General Linear Methods for Integrated Circiut Design. Ph.D. thesis, Humboldt-Universität zu Berlin, Juni 2006. Logos, Berlin (2006)
80. Weierstraß, K.: Gesammelte Werke, volume II, chapter Zur Theorie der bilinearen und quadratischen Formen, pp. 19–44. Akad. d. Wiss. Berlin (1868)
81. Wendt, W.: On a differential-algebraic inclusion model for LRS-networks. In: Griepentrog, E., Hanke, M., März, R. (eds.) Berlin Seminar on Differential-Algebraic Equations, vol. 92-1, pp. 195–218 (1992)
82. Yosida, K.: Functional Analysis, 6th edn. Springer, New York (1980)
83. Zeidler, E.: Applied Functional Analysis. Applications to Mathematical Physics. Applied Mathematical Sciences, vol. 108. Springer, New York (1995)
84. Zeidler, E.: Applied Functional Analysis. Main Principles and Their Applications. Applied Mathematical Sciences, vol. 109. Springer, New York (1995)
85. Zhuk, S.M.: Closedness and normal solvability of an operator generated by a degenerate linear differential equation with variable coefficients. Nonlinear Oscillations **10**(4), 469–486 (2007)

Algebraic Theory of Linear Systems: A Survey

Werner M. Seiler and Eva Zerz

Abstract An introduction into the algebraic theory of several types of linear systems is given. In particular, linear ordinary and partial differential and difference equations are covered. Special emphasis is given to the formulation of formally well-posed initial value problem for treating solvability questions for general, i.e. also under- and over-determined, systems. A general framework for analysing abstract linear systems with algebraic and homological methods is outlined. The presentation uses throughout Gröbner bases and thus immediately leads to algorithms.

Keywords Algebraic methods • Autonomy and controllability • Behavioural approach • Gröbner bases • Initial value problems • Linear systems • Over- and under-determined systems • Symbolic computation

Mathematics Subject Classification (2010): 13N10, 13P10, 13P25, 34A09, 34A12, 35G40, 35N10, 68W30, 93B25, 93B40, 93C05, 93C15, 93C20, 93C55

1 Introduction

We survey the algebraic theory of linear differential algebraic equations and their discrete counterparts. Our focus is on the use of methods from symbolic computation (in particular, the theory of Gröbner bases is briefly reviewed in an Appendix) for studying structural properties of such systems, e.g. autonomy and controllability, which are important concepts in systems and control theory. Moreover, the formulation of a well-posed initial value problem is a fundamental issue with differential algebraic equations, as it leads to existence and uniqueness

W.M. Seiler (✉)
Institut für Mathematik, Universität Kassel, 34109 Kassel, Germany
e-mail: seiler@mathematik.uni-kassel.de

E. Zerz
Lehrstuhl D für Mathematik, RWTH Aachen, 52062 Aachen, Germany
e-mail: eva.zerz@math.rwth-aachen.de

© Springer International Publishing Switzerland 2015
A. Ilchmann, T. Reis (eds.), *Surveys in Differential-Algebraic Equations II*,
Differential-Algebraic Equations Forum, DOI 10.1007/978-3-319-11050-9_5

theorems, and Gröbner bases provide a unified approach to tackle this question for both ordinary and partial differential equations, and also for difference equations.

Here are the key ideas of the algebraic approach: Given a linear differential or difference equation, we first identify a ring $\mathscr{D}$ of operators and a set $\mathscr{F}$ of functions where the solutions are sought. For instance, the equation $\ddot{f} + f = 0$ can be modelled by setting $\mathscr{D} = \mathbb{R}[\partial]$ (the ring of polynomials in the indeterminate ∂ with real coefficients) and $\mathscr{F} = \mathscr{C}^{\infty}(\mathbb{R}, \mathbb{R})$. The operator $d = \partial^2 + 1 \in \mathscr{D}$ acts on the function $f \in \mathscr{F}$ and the given equation takes the form $df = 0$. Partial differential equations with constant coefficients can be described by $\mathscr{D} = \mathbb{R}[\partial_1, \ldots, \partial_r]$ and, say, $\mathscr{F} = \mathscr{C}^{\infty}(\mathbb{R}^r, \mathbb{R})$. Similarly, difference equations such as the Fibonacci equation $f(t + 2) = f(t + 1) + f(t)$ can be put into this framework by setting $\mathscr{D} = \mathbb{R}[\sigma]$, where σ denotes the shift operator defined by $(\sigma f)(t) = f(t + 1)$. Then $d = \sigma^2 - \sigma - 1 \in \mathscr{D}$ acts on $f \in \mathscr{F} = \mathbb{R}^{\mathbb{N}_0}$, which is the set of all functions from $\mathbb{N}_0$ to $\mathbb{R}$. Again, this can easily be extended to partial difference equations by admitting several shift operators. The situation becomes more complicated when variable coefficients are involved, because then the coefficients do not necessarily commute with the operators ∂_i or σ_i, respectively. However, this setting can still be modelled by using appropriate noncommutative operator rings $\mathscr{D}$ such as the Weyl algebra (the ring of linear differential operators with polynomial coefficients). The function set $\mathscr{F}$ is supposed to have the structure of a left $\mathscr{D}$-module. This means that we may apply the operators ∂_i or σ_i arbitrarily often, and that for $f \in \mathscr{F}$, any df belongs again to $\mathscr{F}$, where $d \in \mathscr{D}$. Thus, the set of smooth functions or the set of distributions are the prototypes of such function sets in the continuous setting.

Having identified a suitable pair $(\mathscr{D}, \mathscr{F})$, one may just as well treat nonscalar equations $Df = 0$ with a matrix $D \in \mathscr{D}^{g \times q}$ and a vector $f \in \mathscr{F}^q$, where $(Df)_i = \sum_{j=1}^q D_{ij} f_j$ as usual. The set $S = \{f \in \mathscr{F}^q \mid Df = 0\}$ is the *solution set*, in $\mathscr{F}^q$, of the linear system of equations $Df = 0$. Associated with the system is the *row module* $N = \mathscr{D}^{1 \times g} D$ consisting of all $\mathscr{D}$-linear combinations of the rows of the matrix D and the *system module* $M = \mathscr{D}^{1 \times q}/N$, the corresponding factor module. Any $f \in \mathscr{F}^q$ gives rise to a $\mathscr{D}$-linear map $\phi(f) : \mathscr{D}^{1 \times q} \to \mathscr{F}$ which maps $d \in \mathscr{D}^{1 \times q}$ to $df = \sum_{j=1}^q d_j f_j \in \mathscr{F}$. Now if $f \in S$ is an arbitrary solution, then any vector $d \in N$ in the row module belongs to the kernel of $\phi(f)$. Thus $\phi(f)$ induces a well-defined $\mathscr{D}$-linear map $\psi(f) : M \to \mathscr{F}$ on the system module. An important observation by Malgrange [32] says that there is a bijection (actually even an isomorphism of Abelian groups with respect to addition) between the solution set S and the set of all $\mathscr{D}$-linear maps from the system module M to $\mathscr{F}$, that is,

$$S = \{f \in \mathscr{F}^q \mid Df = 0\} \cong \mathrm{Hom}_{\mathscr{D}}(M, \mathscr{F}), \quad f \mapsto \psi(f).$$

One of the nice features of this correspondence is the fact that it separates the information contained in the system S into a purely algebraic object (the system module M, which depends only on the chosen operator ring and the matrix representing the system) and an analytic object (the function set $\mathscr{F}$). Thus the study of the system module is important for all possible choices of $\mathscr{F}$. This makes it possible to consider S for successively larger function sets (smooth functions, distributions, hyperfunctions, etc.) as proposed by the "algebraic analysis" school of Sato, Kashiwara et al. (see, e.g., [21, 22, 34] and references therein).

This contribution is structured as follows. In the next three sections three particularly important classes of linear systems are studied separately: ordinary differential equations, difference equations and partial differential equations. The main emphasis here lies on an existence and uniqueness theory via the construction of formally well-posed initial value problems. Section 5 shows how the concept of an index of a differential algebraic equation can be recovered in the algebraic theory. Then Sect. 6 provides a general algebraic framework for studying abstract linear systems in a unified manner, using a common language for all the classes of linear systems considered in this paper. Here the main emphasis lies on systems theoretic aspects such as autonomy and controllability. The algebraic characterization of these properties is used throughout the paper, thus sparing the necessity of individual proofs for each system class. Finally, an appendix briefly recapitulates Gröbner bases as the main algorithmic tool for algebraic systems.

2 Linear Ordinary Differential Equations

Consider a linear ordinary differential equation

$$c_n(t)\frac{d^n f}{dt^n}(t) + \ldots + c_1(t)\frac{df}{dt}(t) + c_0(t)f(t) = 0. \tag{2.1}$$

The time-varying coefficients c_i are supposed to be real-meromorphic functions. Thus, the differential equation is defined on $\mathbb{R} \setminus \mathbb{E}_c$, where $\mathbb{E}_c$ is a discrete set (the collection of all poles of the functions c_i), and it is reasonable to assume that any solution f is smooth on the complement of some discrete set $\mathbb{E}_f \subset \mathbb{R}$.

For the algebraic approach, it is essential to interpret the left-hand side of (2.1) as the result of applying a differential operator to f. For this, let $\mathbb{k}$ denote the field of meromorphic functions over the reals. Let $\mathscr{D}$ denote the ring of linear ordinary differential operators with coefficients in $\mathbb{k}$, that is, $\mathscr{D}$ is a polynomial ring over $\mathbb{k}$ in the formal indeterminate ∂ which represents the derivative operator, i.e. $\partial \in \mathscr{D}$ acts on f via $\partial f = \frac{df}{dt}$. Then (2.1) takes the form $df = 0$, where

$$d = c_n \star \partial^n + \ldots + c_1 \star \partial + c_0 \in \mathscr{D}. \tag{2.2}$$

Due to the Leibniz rule $\frac{d}{dt}(cf) = c\frac{df}{dt} + \frac{dc}{dt}f$, the multiplication $\star$ in $\mathscr{D}$ satisfies

$$\partial \star c = c \star \partial + \frac{dc}{dt} \quad \text{for all } c \in \mathbb{k}.$$

For simplicity, we will write $\partial c = c\partial + \dot{c}$ below. Thus $\mathscr{D}$ is a noncommutative polynomial ring in which the indeterminate ∂ commutes with a coefficient c

according to this rule. In the language of Ore algebras (see Appendix), we have
$\mathscr{D} = \Bbbk[\partial; \mathrm{id}, \frac{d}{dt}]$.

The algebraic properties of $\mathscr{D}$ can be summarized as follows: The ring $\mathscr{D}$ is a left
and right Euclidean domain, that is, the product of two nonzero elements is nonzero,
and we have a left and right division with remainder. The Euclidean function is given
by the degree which is defined as usual, that is, $\deg(d) = \max\{i \mid c_i \neq 0\}$ for $d \neq 0$
as in (2.2). Thus $\mathscr{D}$ is also a left and right principal ideal domain, and it possesses a
skew field $\mathscr{K}$ of fractions $k = ed^{-1}$ with $e, d \in \mathscr{D}$ and $d \neq 0$. Therefore, the *rank*
of a matrix with entries in $\mathscr{D}$ can be defined as usual (i.e. as the dimension of its row
or column space over $\mathscr{K}$ [26]). Moreover, $\mathscr{D}$ is also a simple ring, that is, it has only
trivial two-sided ideals. These properties of $\mathscr{D}$ (see [7, 13]) imply that every matrix
$E \in \mathscr{D}^{g \times q}$ can be transformed into its Jacobson form [17] by elementary row and
column operations. Thus there exist invertible $\mathscr{D}$-matrices U and V such that

$$UEV = \begin{bmatrix} D & 0 \\ 0 & 0 \end{bmatrix} \tag{2.3}$$

with $D = \mathrm{diag}(1, \ldots, 1, d) \in \mathscr{D}^{r \times r}$ for some $0 \neq d \in \mathscr{D}$, where r is the
rank of E (over $\mathscr{K}$). The matrix on the right-hand side of (2.3) is called the
Jacobson normal form of E. (For its computation, see e.g. [30].) The existence
of this noncommutative analogue of the Smith form makes the algebraic theory of
linear ordinary differential equations over $\Bbbk$ very similar to the constant coefficient
case (for this, one mainly uses the fact that D is a diagonal matrix and not its
special form given above). The main analytical difference is that over $\Bbbk$, one has
to work locally due to the presence of singularities of the coefficients and solutions.
Therefore, let $\mathscr{F}$ denote the set of functions that are smooth up to a discrete set of
exception points. Then $\mathscr{F}$ is a left $\mathscr{D}$-module. In [57], it was shown that $\mathscr{F}$ is even an
injective cogenerator (see Sect. 6.1 for the definition). Thus the algebraic framework
outlined in Sect. 6 can be applied to systems of linear ordinary differential equations
which take the form $Ef = 0$, where E is a $\mathscr{D}$-matrix and f is a column vector with
entries in $\mathscr{F}$.

Let $S = \{f \in \mathscr{F}^q \mid Ef = 0\}$ for some $E \in \mathscr{D}^{p \times q}$. Due to the Jacobson
form, one may assume without loss of generality that E has full row rank. Two
representation matrices E_i of S, both with full row rank, differ only by a unimodular
left factor, that is, $E_2 = UE_1$ for some invertible matrix U [57]. According to
Theorem 6.1, the system S is autonomous (i.e. it has no free variables) if and
only if any representation matrix E has full column rank. Combining this with the
observation from above, we obtain that an autonomous system always possesses a
square representation matrix with full rank. Given an arbitrary representation matrix
E with full row rank, we can select a square submatrix P of E of full rank. Up to
a permutation of the columns of E, we have $E = [-Q, P]$. Partitioning the system
variables accordingly, the system law $Ef = 0$ reads $Py = Qu$, where u is a vector of
free variables, that is,

$$\forall u \in \mathscr{F}^{q-p} \exists y \in \mathscr{F}^p : Py = Qu \tag{2.4}$$

and it is maximal in the sense that $Py = 0$ defines an autonomous system. The number $m := q - p$, where $p = \text{rank}(E)$, is sometimes called the *input dimension* of S. With this terminology, a system is autonomous if and only if its input dimension is zero. To see that (2.4) holds, note that the homomorphism (of left $\mathcal{D}$-modules) $\mathcal{D}^{1 \times p} \to \mathcal{D}^{1 \times p}$, $x \mapsto xP$ is injective, because P has full row rank. Since $\mathcal{F}$ is an injective $\mathcal{D}$-module, the induced homomorphism (of Abelian groups w.r.t. addition) $\mathcal{F}^p \to \mathcal{F}^p$, $y \mapsto Py$ is surjective. This implies (2.4).

Such a representation $Py = Qu$ is called an *input-output decomposition* of S (with input u and output y). Note that it is not unique since it depends on the choice of the p linearly independent columns of E that form the matrix P. Once a specific decomposition $E = [-Q, P]$ is chosen, the input u is a free variable according to (2.4). For a fixed input u, any two outputs $y, \tilde{y}$ belonging to u must satisfy $P(y - \tilde{y}) = 0$. Since $Py = 0$ is an autonomous system, none of the components of y is free, and each component y_i of y can be made unique by an appropriate choice of initial conditions.

Consider an autonomous system, that is, $Py = 0$, where $P \in \mathcal{D}^{p \times p}$ has full rank. Our goal is to formulate a well-posed initial value problem. For this, one computes a minimal Gröbner basis of the row module of P with respect to an ascending POT term order (see Example 7.1), for instance

$$\partial^n \mathbf{e}_i \prec_{\text{POT}} \partial^m \mathbf{e}_j \quad \Leftrightarrow \quad i < j \text{ or } (i = j \text{ and } n < m), \tag{2.5}$$

where $\mathbf{e}_i$ denotes the ith standard basis row. According to Definition 7.6, a Gröbner basis G is called *minimal* if for all $g \neq h \in G$, $\text{lt}(g)$ does not divide $\text{lt}(h)$. This means that none of the elements of a minimal Gröbner basis is superfluous, that is, we have $\langle \text{lt}(G \setminus \{g\}) \rangle \subsetneq \langle \text{lt}(G) \rangle$ for all $g \in G$. In the next paragraph, we show that—possibly after re-ordering the generators—the result of this Gröbner basis computation is a lower triangular matrix $P' = UP$ with nonzero diagonal entries, where U is invertible.

To see this, let G be a minimal Gröbner basis of the row module of P. The minimality of G implies that there exist no $g \neq h \in G$ with $\text{lt}(g) = \partial^n \mathbf{e}_i$ and $\text{lt}(h) = \partial^m \mathbf{e}_i$. So for every $1 \leq i \leq p$ there is at most one $g_i \in G$ with $\text{lt}(g_i) = \partial^{n_i} \mathbf{e}_i$ for some n_i. By the choice of the POT order (2.5), the last $p - i$ components of g_i must be zero. On the other hand, since $\mathcal{D}^{1 \times p} \to \mathcal{D}^{1 \times p}$, $x \mapsto xP$ is injective, the row module of P is isomorphic to the free $\mathcal{D}$-module $\mathcal{D}^{1 \times p}$ and hence, it cannot be generated by less than p elements. Thus $G = \{g_1, \ldots, g_p\}$ and the matrix $P' \in \mathcal{D}^{p \times p}$ that has g_i as its ith row is lower triangular with nonzero diagonal entries.

The fact that P and P' have the same row module implies that $P' = UP$ with an invertible matrix U. Clearly, $Py = 0$ holds if and only if $P'y = 0$. Let $\rho_i := \deg(P'_{ii})$ for all $1 \leq i \leq p$. Then there exists an exception set $\mathbb{E}$ such that for all open intervals $I \subset \mathbb{R} \setminus \mathbb{E}$ and all $t_0 \in I$, the differential equation $Py = 0$ together with the initial data $y_i^{(j_i)}(t_0)$ for $1 \leq i \leq p$ and $0 \leq j_i < \rho_i$ determines $y|_I$ uniquely. This also shows that the set of solutions to $Py = 0$ on such an interval is a finite-dimensional real vector space (of dimension $\rho = \sum_{i=1}^p \rho_i$). The number

ρ is also equal to the degree of the polynomial d that appears in the Jacobson form $D = \mathrm{diag}(1, \ldots, 1, d)$ of P.

Example 2.1 Consider [24, Example 3.1]

$$\begin{bmatrix} -t & t^2 \\ -1 & t \end{bmatrix} \dot{f} = \begin{bmatrix} -1 & 0 \\ 0 & -1 \end{bmatrix} f.$$

Writing the system in the form $Ef = 0$, the operator matrix E acting on $f = [f_1, f_2]^T$ is given by

$$E = \begin{bmatrix} -t\partial + 1 & t^2\partial \\ -\partial & t\partial + 1 \end{bmatrix} \in \mathscr{D}^{2 \times 2}.$$

We compute a Gröbner basis of the row module of E with respect to the term order in (2.5) and obtain $E' = [1, -t]$. Indeed, E and E' have the same row module, as

$$E' = [1, -t]E \quad \text{and} \quad E = \begin{bmatrix} -t\partial + 1 \\ -\partial \end{bmatrix} E'.$$

The system S given by $Ef = 0$ is not autonomous, since $\mathrm{rank}(E) = \mathrm{rank}(E') = 1$, and thus, E does not have full column rank. In fact, the connection between the matrices E and E' shows that

$$S = \{ f \in \mathscr{F}^2 \mid Ef = 0 \} = \{ [tf_2, f_2]^T \mid f_2 \in \mathscr{F} \},$$

that is, S has an image representation and is therefore controllable. One may interpret f_2 as the system's input and f_1 as its output (or conversely). In this example, the output is uniquely determined by the input. The Jacobson form of E is $\mathrm{diag}(1, 0)$.

Example 2.2 Consider [24, Example 3.2]

$$\begin{bmatrix} 0 & 0 \\ 1 & -t \end{bmatrix} \begin{bmatrix} \dot{f}_3 \\ \dot{f}_4 \end{bmatrix} = \begin{bmatrix} -1 & t \\ 0 & 0 \end{bmatrix} \begin{bmatrix} f_3 \\ f_4 \end{bmatrix} + \begin{bmatrix} f_1 \\ f_2 \end{bmatrix}.$$

Writing the system in the form $Ef = 0$, where $f = [f_1, \ldots, f_4]^T$, one gets

$$E = \begin{bmatrix} -1 & 0 & 1 & -t \\ 0 & -1 & \partial & -t\partial \end{bmatrix} \in \mathscr{D}^{2 \times 4}.$$

Proceeding as above, we obtain

$$U = \begin{bmatrix} -t\partial + 1 & t \\ -\partial & 1 \end{bmatrix} \quad \text{and} \quad E' = UE = \begin{bmatrix} t\partial - 1 & -t & 1 & 0 \\ \partial & -1 & 0 & 1 \end{bmatrix}.$$

We may choose f_1 and f_2 as inputs, and then the outputs

$$f_3 = -t\dot{f}_1 + f_1 + tf_2 \quad \text{and} \quad f_4 = -\dot{f}_1 + f_2$$

are uniquely determined according to $E'f = 0$. Note that $E = [-I, E_1]$ with an invertible matrix E_1, where $U = E_1^{-1}$, and $E' = UE = [-E_1^{-1}, I]$. Thus

$$S = \{f \in \mathscr{F}^4 \mid Ef = 0\} = \{\begin{bmatrix} E_1 \\ I \end{bmatrix} u \mid u \in \mathscr{F}^2\} = \{\begin{bmatrix} I \\ E_1^{-1} \end{bmatrix} v \mid v \in \mathscr{F}^2\}.$$

Again, S has an image representation and is therefore controllable. The Jacobson form of E is $[I, 0] \in \mathscr{D}^{2 \times 4}$.

Example 2.3 Consider the system

$$\dot{f}_1 + tf_2 = 0, \quad \dot{f}_2 + tf_1 = 0.$$

Writing the system in the form $Ef = 0$, the operator matrix E acting on $f = [f_1, f_2]^T$ is given by

$$E = \begin{bmatrix} \partial & t \\ t & \partial \end{bmatrix} \in \mathscr{D}^{2 \times 2}.$$

Proceeding as above, we obtain

$$U = \begin{bmatrix} \partial - \frac{1}{t} & -t \\ \frac{1}{t} & 0 \end{bmatrix} \quad \text{and} \quad E' = UE = \begin{bmatrix} \partial^2 - \frac{1}{t}\partial - t^2 & 0 \\ \frac{1}{t}\partial & 1 \end{bmatrix}.$$

Clearly, the system S given by $Ef = 0$ is autonomous and has vector space dimension 2. The Jacobson form of E is $\mathrm{diag}(1, \partial^2 - \frac{1}{t}\partial - t^2)$.

3 Linear Difference Equations

Consider a linear ordinary difference equation

$$c_n f(t+n) + \ldots + c_1 f(t+1) + c_0 f(t) = 0 \quad \text{for all } t \in \mathbb{N}_0.$$

The coefficients c_i are supposed to be elements of a commutative quasi-Frobenius ring R (see e.g. [25]). This means that (i) R is Noetherian, and (ii) $\mathrm{Hom}_R(\cdot, R)$ is an exact functor. For instance, all fields are quasi-Frobenius rings, but also the residue class rings $\mathbb{Z}/k\mathbb{Z}$ for an integer $k \geq 2$. The ring of operators is $\mathscr{D} = R[\sigma]$, a univariate polynomial ring with coefficients in R. The action of $\sigma \in \mathscr{D}$ on a sequence $f : \mathbb{N}_0 \to R$ is given by the left shift $(\sigma f)(t) = f(t + 1)$. We set $\mathscr{F} = R^{\mathbb{N}_0}$, which is the set of all functions from $\mathbb{N}_0$ to R. Then $\mathscr{F}$ is a left $\mathscr{D}$-module and an injective cogenerator [31, 36, 58].

Example 3.1 Consider the Fibonacci equation

$$f(t + 2) = f(t + 1) + f(t)$$

over R, which can be written as $df = 0$ with $d = \sigma^2 - \sigma - 1 \in R[\sigma]$. Over the real numbers, one of its solutions is the famous Fibonacci sequence $0, 1, 1, 2, 3, 5, 8, \ldots$ Over a finite ring however, its solutions are periodic functions, because there exists a positive integer p such that $\sigma^2 - \sigma - 1$ divides $\sigma^p - 1$. (This is due to the fact that the element $\bar{\sigma} \in S := R[\sigma]/\langle \sigma^2 - \sigma - 1 \rangle$ belongs to the group of units of the finite ring S and hence, it has finite order.) Thus any solution to $df = 0$ must satisfy $(\sigma^p - 1)f = 0$, that is, $f(t + p) = f(t)$ for all t.

The discrete setting can easily be generalized to the multivariate situation. A linear partial difference equation takes the form

$$\sum_{\nu \in \mathbb{N}_0^r} c_\nu f(t + \nu) = 0 \quad \text{for all } t \in \mathbb{N}_0^r,$$

where only finitely many of the coefficients $c_\nu \in R$ are nonzero. The relevant operator ring is $\mathscr{D} = R[\sigma_1, \ldots, \sigma_r]$, where σ_i acts on $f : \mathbb{N}_0^r \to R$ via $(\sigma_i f)(t) = f(t + \mathbf{e}_i)$. We also use the multi-index notation $(\sigma^\nu f)(t) = f(t + \nu)$ for $t, \nu \in \mathbb{N}_0^r$. Let $\mathscr{F}$ denote the set of all functions from $\mathbb{N}_0^r$ to R. Then $\mathscr{F}$ is again a left $\mathscr{D}$-module and an injective cogenerator [31, 36, 58]. Finally, let $E \in \mathscr{D}^{g \times q}$ be given and consider $S = \{f \in \mathscr{F}^q \mid Ef = 0\}$. The system S is autonomous (i.e., it has no free variable) if and only if there exists a $\mathscr{D}$-matrix X such that $XE = \text{diag}(d_1, \ldots, d_q)$ with $0 \neq d_i \in \mathscr{D}$ for all i. In general, the input number (or "input dimension") of S is defined as the maximal m for which there exists a permutation matrix Π such that $S \to \mathscr{F}^m$, $f \mapsto [I_m, 0]\Pi f$ is surjective. Partitioning $\Pi f =: \begin{bmatrix} u \\ y \end{bmatrix}$ accordingly, this means that for all "inputs" $u \in \mathscr{F}^m$, there exists an "output" $y \in \mathscr{F}^{q-m}$ such that $f \in S$. Via the injective cogenerator property, the input number m is also the largest number such that there exists a permutation matrix Π with $\mathscr{D}^{1 \times g} E \Pi \cap (\mathscr{D}^{1 \times m} \times \{0\}) = \{0\}$. For simplicity, suppose that the columns of E have already been permuted such that $\mathscr{D}^{1 \times g} E \cap (\mathscr{D}^{1 \times m} \times \{0\}) = \{0\}$, where m is the input number of S. Let $E = [-Q, P]$ be partitioned accordingly such that the system law reads $Py = Qu$ with input u and output y. Moreover, the system given by $Py = 0$ is autonomous (otherwise, we'd have a contradiction to the maximality of m). By construction, we have $\ker(\cdot P) \subseteq \ker(\cdot Q)$ and this guarantees that $P\mathscr{F}^{q-m} \supseteq Q\mathscr{F}^m$. If R is a domain, then $\mathscr{D}$ has a quotient field $\mathscr{K}$, and we may simply set $m := q - \text{rank}(E)$ and choose P as a submatrix of E

whose columns are a basis of the column space $E\mathcal{K}^q$ of E. Then $Q = PH$ holds for some $\mathcal{K}$-matrix H, which clearly implies $\ker(\cdot P) \subseteq \ker(\cdot Q)$.

Example 3.2 Let $R = \mathbb{Z}/8\mathbb{Z}$ and consider the system given by

$$4f_1(t) = 2f_2(t + 1)$$

for all $t \in \mathbb{N}_0$. Then $E = [-4, 2\sigma]$. Since $\ker(\cdot 2\sigma) = \operatorname{im}(\cdot 4) \subseteq \ker(\cdot 4)$, we may choose $Q = 4$ and $P = 2\sigma$, that is, we may interpret $u := f_1$ as an input, and $y := f_2$ as an output. For any choice of u, the left hand side of the system law is in $\{0, 4\}$, and hence, the equation is always solvable for $y(t + 1)$. The autonomous system $2y(t + 1) = 0$ consists of all sequences with $y(t) \in \{0, 4\}$ for all $t \geq 1$ (with $y(0)$ being arbitrary). Conversely, f_2 is not a free variable, since the system law implies (by multiplying both sides by 2) that $4f_2(t + 1) = 0$ for all $t \in \mathbb{N}_0$, and hence $f_2(t) \in \{0, 2, 4, 6\}$ for all $t \geq 1$.

For the rest of this section, let $R = \Bbbk$ be a field. Let $Py = 0$ define an autonomous system, that is, let P have full column rank q. Then none of the components of y is free, and it can be made unique by an appropriate choice of initial conditions. The theory of Gröbner bases can be used to set up a well-posed initial value problem. For this, one computes a Gröbner basis G of the row module $N = \mathcal{D}^{1 \times g} P$ of P. If $d \in \mathcal{D}^{1 \times q}$ has leading term $\sigma^\mu e_i$, then we write $\operatorname{lead}(d) = (\mu, i) \in \mathbb{N}_0^r \times \{1, \ldots, q\}$. For a set $D \subseteq \mathcal{D}^{1 \times q}$ with $D \setminus \{0\} \neq \emptyset$, we put $\operatorname{lead}(D) := \{\operatorname{lead}(d) \mid 0 \neq d \in D\}$. With this notation, the fact that G is a Gröbner basis of N reads

$$\operatorname{lead}(N) = \operatorname{lead}(G) + (\mathbb{N}_0^r \times \{0\}).$$

Define

$$\Gamma := \mathbb{N}_0^r \times \{1, \ldots, q\} \setminus \operatorname{lead}(N).$$

Then the initial value problem $Py = 0$, $y|_\Gamma = z$ (that is, $y_i(\mu) = z(\mu, i)$ for all (μ, i) in Γ) has a unique solution $y \in \mathcal{F}^q$ for every choice of the initial data $z \in \Bbbk^\Gamma$ [36]. The solution y can be computed recursively, proceeding in the specific order on $\mathbb{N}_0^r \times \{1, \ldots, q\}$ that was used to compute the Gröbner basis. (Clearly, ordering this set is equivalent to ordering $\{\sigma^\mu e_i \mid \mu \in \mathbb{N}_0^r, 1 \leq i \leq q\}$.) For $(\mu, i) \in \operatorname{lead}(N)$, there exists a $g \in G$ such that $(\mu, i) = \operatorname{lead}(g) + (\nu, 0) = \operatorname{lead}(\sigma^\nu g)$, and thus, the value $y_i(\mu)$ can be computed from values $y_j(\kappa)$ with $(\kappa, j) < (\mu, i)$, that is, from values that are already known. Uniqueness follows by induction, since at each (μ, i), we may assume that all consequences d of the system law (that is, all equations $dy = 0$, where $d \in N$) with $\operatorname{lead}(d) < (\mu, i)$ are satisfied by the values that are already known. On the other hand, the values of y on Γ are unconstrained by the system law. More formally, the unique solvability of the initial value problem can be shown as follows: There is a $\Bbbk$-vector space isomorphism between S and $\operatorname{Hom}_\Bbbk(\mathcal{D}^{1 \times q}/N, \Bbbk)$. Clearly, a linear map on a vector space is uniquely determined by choosing the image of a basis. However, the set $\{[\sigma^\mu \mathbf{e}_i] \mid (\mu, i) \in \Gamma\}$ is indeed a $\Bbbk$-basis of $\mathcal{D}^{1 \times q}/N$. This shows that each element of S is uniquely determined by fixing its values on Γ. Note that the set Γ is infinite, in general. We remark that Γ is finite if and only if the system module is finite-dimensional as a $\Bbbk$-vector space [54].

Example 3.3 Let $\Bbbk = \mathbb{R}$ and consider the autonomous system given by

$$y(t_1 + 2, t_2) + y(t_1, t_2 + 2) = 0$$
$$y(t_1 + 3, t_2) + y(t_1, t_2 + 3) = 0$$

for all $t = (t_1, t_2) \in \mathbb{N}_0^2$. A Gröbner basis of $N = \langle \sigma_1^2 + \sigma_2^2, \sigma_1^3 + \sigma_2^3 \rangle$ with respect to the lexicographic order with $\sigma_1 > \sigma_2$ is given by $G = \{\sigma_1^2 + \sigma_2^2, \sigma_1 \sigma_2^2 - \sigma_2^3, \sigma_2^4\}$. Therefore, each solution $y : \mathbb{N}_0^2 \to \mathbb{R}$ is uniquely determined by its values on the complement of $\mathrm{lead}(N) = \{(2, 0), (1, 2), (0, 4)\} + \mathbb{N}_0^2$, that is, on the set $\Gamma = \{(0, 0), (0, 1), (0, 2), (0, 3), (1, 0), (1, 1)\}$. Thus, the solution set is a real vector space of dimension $|\Gamma| = 6$.

4   Linear Partial Differential Equations

The theory of partial differential equations shows some notable differences compared to ordinary differential equations. For general systems (i.e. including under- or over-determined ones), it is much harder to prove the existence of at least formal solutions, as now integrability conditions may be of higher order than the original system.[1] The reason is a simple observation. In systems of ordinary differential equations, only one mechanism for the generation of integrability conditions exist: (potentially after some algebraic manipulations) the system contains equations of different order and differentiation of the lower-order ones may lead to new equations. In the case of partial differential equations, such differences in the order of the individual equations are less common in practice. Here the dominant mechanism for the generation of integrability conditions are (generalized) cross-derivatives and these lead generally to equations of higher order. A comprehensive discussion of general systems of partial differential equations and the central notions of involution and formal integrability can be found in the monograph [52]. Within this article, we will consider exclusively linear systems where again standard Gröbner basis techniques can be applied. A somewhat more sophisticated approach using the formal theory of differential equations and involutive

---

[1]For arbitrary systems, not even an a priori bound on the maximal order of an integrability condition is known. In the case of linear equations, algebraic complexity theory provides a double exponential bound which is, however, completely useless for computations, as for most systems appearing in applications it grossly overestimates the actual order.

bases is contained in [16]; it provides intrinsic results independent of the used coordinates.

Example 4.1 We demonstrate the appearance of integrability conditions with a system of two second-order equations for one unknown function u of three independent variable (x, y, z) due to Janet [18, Example 47]:

$$\begin{bmatrix} \partial_z^2 + y\partial_x^2 \\ \partial_y^2 \end{bmatrix} f = 0. \tag{4.1}$$

It hides two integrability conditions of order 3 and 4, respectively, namely

$$\partial_x^2 \partial_y f = 0, \qquad \partial_x^4 f = 0. \tag{4.2}$$

It is important to note that they do not represent additionally imposed equations but that *any* solution of (4.1) will automatically satisfy (4.2).

A systematic derivation of these conditions is easily possible with Gröbner bases. As we have only one unknown function in the system, the row module becomes here the row ideal $I \lhd \mathscr{D} = \mathbb{R}(x, y, z)[\partial_x, \partial_y, \partial_z]$ in the ring of linear differential operators with rational coefficients generated by the operators $D_1 = \partial_z^2 + y\partial_x^2$ and $D_2 = \partial_y^2$. We use the reverse lexicographic order with $\partial_z \succ \partial_y \succ \partial_x$ and follow the Buchberger Algorithm 6 for the construction of a Gröbner basis. The S-polynomial (see (7.5)) of the two given operators is $S(D_1, D_2) = \partial_y^2 \cdot D_1 - \partial_z^2 \cdot D_2 = y\partial_x^2\partial_y^2 + 2\partial_x^2\partial_y$. Reduction modulo D_2 eliminates the first term so that $D_3 = \partial_x^2\partial_y$ and we have found the first integrability condition. The second one arises similarly from the S-polynomial $S(D_1, D_3) = \partial_x^2\partial_y \cdot D_1 - \partial_z^2 \cdot D_3$ leading to the operator $D_4 = \partial_x^4$ after reduction, whereas the S-polynomial $S(D_2, D_3)$ immediately reduces to zero. Since also all S-polynomials $S(D_i, D_4)$ reduce to zero, the set $\{D_1, D_2, D_3, D_4\}$ represents a Gröbner basis of the ideal I and there are no further hidden integrability conditions.

There are two reasons why one should make the integrability conditions (4.2) explicit. First of all, only after the construction of all hidden conditions, we can be sure that the system (4.1) indeed possesses solutions; in general, a condition like $1 = 0$ can be hidden which shows that the system is inconsistent.[2] Secondly, the knowledge of these conditions often significantly simplifies the integration of the system and provides information about the size of the solution space. As we will show later in this section, in our specific example one can immediately recognize from the combined system (4.1, 4.2) that the solution space of our problem is finite-dimensional (more precisely, 12-dimensional)—something rather unusual for partial differential equations. In fact, once (4.2) is taken into account, one easily

---

[2]Here we are actually dealing with the special case of a homogeneous linear system where consistency simply follows from the fact that $u = 0$ is a solution.

determines the general solution of the system, a polynomial with 12 arbitrary parameters:

$$
\begin{aligned}
f(x, y, z) = &-a_1 xyz^3 + a_1 x^3 z - 3a_2 xyz^2 - \frac{1}{3}a_3 yz^3 + a_2 x^3 + \\
&a_3 x^2 z - a_4 yz^2 + a_5 xyz + a_4 x^2 + a_6 xy + a_7 yz + \\
&a_8 xz + a_9 x + a_{10} y + a_{11} z + a_{12}.
\end{aligned}
\tag{4.3}
$$

An important notion for the formulation of existence and uniqueness results for differential equations is *well-posedness*. According to an informal definition due to Hadamard, an initial or boundary value problem is well-posed, if (i) a solution exists for arbitrary initial or boundary data, (ii) this solution is unique, and (iii) it depends continuously on the data. For a mathematically rigorous definition, one would firstly have to specify function spaces for the data and the solution and secondly define topologies on these spaces in order to clarify what continuous dependency should mean. In particular this second point is highly non-trivial and application dependent. For this reason, we will use in this article a simplified version which completely ignores (iii) and works with formal power series.

We will consider here exclusively initial value problems, however in a more general sense as usual. For notational simplicity, we assume in the sequel that there is only one unknown function f. Since we work with formal power series solutions, we further assume that some expansion point $\hat{\mathbf{x}} = (\hat{x}_1, \ldots, \hat{x}_n)$ has been chosen. For any subset $X' \subseteq X$ of variables, we introduce the $(n - |X'|)$-dimensional coordinate space $H_{X'} = \{x_i = \hat{x}_i \mid x_i \in X'\}$ through $\hat{\mathbf{x}}$. An initial condition then prescribes some derivative $f_\mu = \partial^{|\mu|} f / \partial x^\mu$ for a multi index $\mu \in \mathbb{N}_0^n$ of the unknown function f restricted to such a coordinate space $H_{X'}$. If several conditions of this kind are imposed with coordinate spaces $H_1, H_2, \ldots$, then one obtains an initial value problem in a strict sense only, if (after a suitable renumbering) the coordinate spaces form a chain $H_1 \subseteq H_2 \subseteq \cdots$. However, in general a formally well-posed initial value problem in this strict sense will exist only after a linear change of coordinates. Therefore, we will not require such a chain condition.

Definition 4.1 An initial value problem for a differential equation is *formally well-posed*, if it possesses a unique formal power series solution for arbitrary formal power series as initial data.

In principle, we approach the algorithmic construction of formally well-posed initial value problems for a given linear partial differential operator L in the same manner as described in the last section for linear difference operators: we compute a Gröbner basis of the row module N—which here is actually an ideal $I \trianglelefteq \mathscr{D}$ because of our assumption that there is only one unknown function—and then use the complement of the leading module. However, in order to be able to translate this complement into initial conditions, we need a complementary decomposition of it (see Definition 7.5) which can be constructed with the help of Algorithm 5 in the Appendix. In fact, it was precisely this problem of constructing formally well-posed problems which lead to the probably first appearance of such decompositions in the works of Riquier [45] and Janet [18].

A leading "term" can now be identified with a derivative f_μ. All derivatives in lt I are traditionally called *principal derivatives*, all remaining ones *parametric derivatives*. Assume that we want to compute a formal power series solution with expansion point $\hat{\mathbf{x}}$. Making the usual ansatz

$$f(\mathbf{x}) = \sum_{\mu \in \mathbb{N}_0^n} \frac{a_\mu}{\mu!}(\mathbf{x} - \hat{\mathbf{x}})^\mu, \tag{4.4}$$

we may identify by Taylor's theorem the coefficient a_μ with the value of the derivative f_μ at $\hat{\mathbf{x}}$.

Let the leading term of the scalar equation $Df = 0$ be f_μ. Because of the monotonicity of term orders, the leading term of the differentiated equation $\partial_i Df = 0$ is $f_{\mu+1_i} = \partial_i f_\mu$. Hence by differentiating the equations in our system sufficiently often, we can generate for any principal derivative f_μ a differential consequence of our system with f_μ as leading term. Prescribing initial data such that unique values are provided for all parametric derivatives (and no values for any principal derivative), we obtain a formally well-posed initial value problem, as its unique power series solution is obtained by determining each principal derivative via the equation that has it as leading term. Such initial data can be systematically constructed via complementary decompositions of the leading ideal.

A complementary decomposition of the monomial ideal lt I is defined by pairs (f_ν, X_ν) where f_ν is a parametric derivative and X_ν the associated set of multiplicative variables. Differentiating f_ν arbitrarily often with respect to variables contained in X_ν, we always obtain again a parametric derivative. Denoting by $\overline{X}_\nu = X \setminus X_\nu$ the set of all *non*-multiplicative variables of f_ν, we associate with the pair (f_ν, X_ν) the initial condition that f_ν restricted to the coordinate space $H_\nu = H_{\overline{X}_\nu}$ is some prescribed function $\rho_\nu(X_\nu)$. In this way, each complementary decomposition induces an initial value problem.[3]

Theorem 4.2 ([52, Theorem 9.3.5]) *For any complementary decomposition of* lt I, *the above constructed initial value problem for the linear differential operator* D *is formally well-posed.*

If both the coefficients of D and the prescribed initial data ρ_ν are analytic functions and a degree compatible term order has been used, then *Riquier's Theorem* [37,45] even guarantees the convergence of the unique formal power series solution to the thus constructed initial value problem. Under these assumptions, we therefore obtain an existence and uniqueness theorem in the analytic category.

---

[3]An initial value problem in the strict sense is obtained, if one starts with a complementary Rees decomposition (see Definition 7.5).

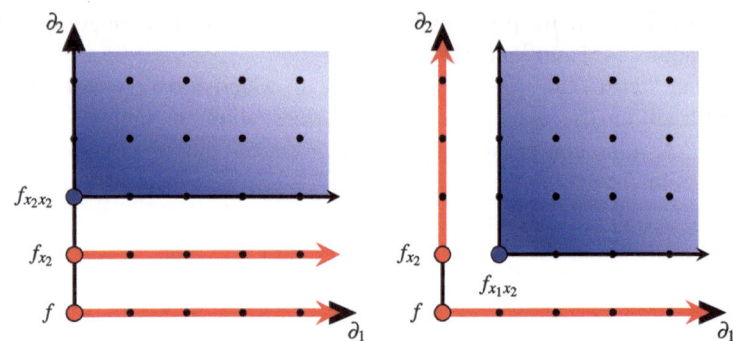

Fig. 1 Complementary decompositions for the wave equation

Example 4.3 Let us begin with a scalar first-order equation like the simple advection equation $(\partial_2 - \partial_1)f = 0$. If we assume that for the chosen term order f_{x_2} is the leading derivative, then it is easy to see that a complementary decomposition is defined by the single pair $(f, \{x_1\})$, as the only parametric derivatives are f and all its pure x_1-derivatives. Hence choosing an expansion point $(\hat{x}_1, \hat{x}_2)$, we recover the familiar initial condition $f(x_1, \hat{x}_2) = \rho(x_1)$ with an arbitrary function ρ.

Advancing to a second-order equation like the wave equation $(\partial_2^2 - \partial_1^2)f = 0$ with leading derivative $f_{x_2 x_2}$, we find that a simple complementary decomposition is defined by the two pairs $(f, \{x_1\})$ and $(f_{x_2}, \{x_1\})$. Hence our construction yields again the classical initial conditions $f(x_1, \hat{x}_2) = \rho(x_1)$ and $f_{x_2}(x_1, \hat{x}_2) = \sigma(x_1)$. This decomposition is shown in Fig. 1 on the left-hand side. The small black dots depict the terms ∂^μ of the underlying ring $\mathscr{D}$ or alternatively the derivatives f_μ. All dots lying in the blue two-dimensional cone correspond to principal derivatives contained in the ideal I and are multiples of the leading derivative $f_{x_2 x_2}$ shown as a large blue dot. All parametric derivatives are contained in the two red one-dimensional cones whose vertices are shown as large red dots. Obviously, the complementary decomposition provides a disjoint partitioning of the complement of the "blue" ideal.

If we consider the wave equation in characteristic coordinates $\partial_1 \partial_2 f = 0$, then we have two natural options for a complementary decomposition: $(f, \{x_1\})$, $(f_{x_2}, \{x_2\})$ (shown in Fig. 1 on the right-hand side) or $(f, \{x_2\})$, $(f_{x_1}, \{x_1\})$. Both correspond to the classical characteristic initial value problem (which is *not* an initial value problem in the strict sense) in which usually both $f(\hat{x}_1, x_2) = \rho(x_2)$ and $f(x_1, \hat{x}_2) = \sigma(x_1)$ are prescribed. However, this classical formulation is not formally well-posed, as the initial data must satisfy the consistency condition $\rho(0) = \sigma(0)$. The formulations induced by the above complementary decompositions avoid such a restriction by substituting in one of the initial conditions f by a first derivative. The quite different character of the standard and the characteristic initial value problem of the wave equation is here encoded in the fact that on the left-hand side of Fig. 1 the two red one-dimensional cones are pointing in the same direction, whereas on the right-hand side they have different directions.

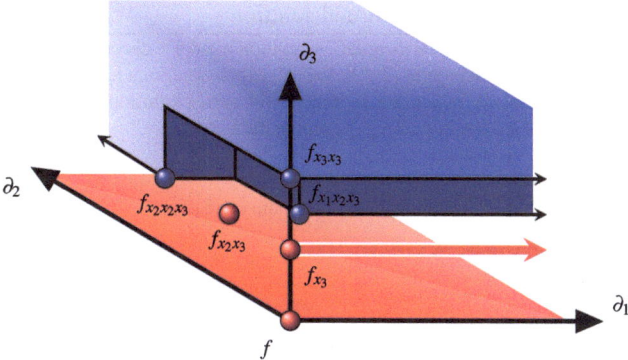

Fig. 2 Complementary decomposition for the monomial system (4.5)

The results become less obvious when we proceed to larger over-determined systems. As a simple example, we take the "monomial" system with

$$I = \langle \partial_3^2, \partial_2^2 \partial_3, \partial_1 \partial_2 \partial_3 \rangle \tag{4.5}$$

(any linear system with a Gröbner basis consisting of three operators with these generators as leading terms leads to the same initial conditions). Here a complementary decomposition (shown in Fig. 2 where again the blue colour marks the ideal I and the red colour the different cones of the complementary decomposition) is given by the three pairs $\left(f, \{x_1, x_2\} \right)$, $\left(f_{x_3}, \{x_1\} \right)$, and $\left(f_{x_2 x_3}, \emptyset \right)$. Hence a formally well-posed initial value problem is given by

$$f(x_1, x_2, \hat{x}_3) = \rho(x_1, x_2), \quad f_{x_3}(x_1, \hat{x}_2, \hat{x}_3) = \sigma(x_1), \quad f_{x_2 x_3}(\hat{x}_1, \hat{x}_2, \hat{x}_3) = \tau \tag{4.6}$$

where the initial data consist of two functions (of one and two arguments, resp.) and one constant. Note that this time the red cones are of different dimensions which is typical and unavoidable for over-determined systems. We are dealing here with a non-characteristic initial value problem, as the directions of the various cones form a flag: all directions defining a lower-dimensional cone are also contained in any higher-dimensional cone.

The considerations above also yield the simple Algorithm 1 for constructing any Taylor coefficient a_λ in the unique solution of the above constructed initial value problem. It computes the normal form (cf. Definition 7.4) of ∂^λ with respect to a Gröbner basis of I. If ∂^λ corresponds to a parametric derivative, it remains unchanged and we obtain in Line /4/ directly the value of a_λ from the appropriate initial condition. Otherwise the normal form computation expresses the principal derivative ∂^λ as a linear combination of parametric derivatives ∂^μ. We determine for each appearing ∂^μ in which unique cone (∂^ν, X_ν) of the given complementary

---

Algorithm 1 Taylor coefficient of formal solution

Input: Gröbner basis G of I for term order $\prec$, complementary decomposition $\mathscr{T}$ with corresponding initial data for each cone $(\partial^\nu, X_\nu) \in \mathscr{T}$, expansion point $\hat{\mathbf{x}}$, multi index $\lambda \in \mathbb{N}_0^n$

Output: Taylor coefficient a_λ of unique formal power series solution of initial value problem corresponding to given complementary decomposition

1: $h \leftarrow \mathrm{NormalForm}_\prec(\partial^\lambda, \mathscr{G})$
2: **for all** $\partial^\mu \in \operatorname{supp} h$ **do**
3:     find unique $(\partial^\nu, X_\nu) \in \mathscr{T}$ such that ∂^μ lies in the cone of ∂^ν
4:     $a_\mu \leftarrow \partial^{\mu-\nu} \rho_\nu(\hat{\mathbf{x}})$
5: **end for**
6: write $h = \sum_\mu c_\mu(\mathbf{x}) \partial^\mu$
7: **return** $\sum_\mu c_\mu(\hat{\mathbf{x}}) a_\mu$

---

decomposition it lies and then compute the required derivative of the corresponding initial data ρ_ν. Evaluating at the expansion point $\hat{\mathbf{x}}$, we obtain first the values of all required parametric coefficients a_μ and then compute the principal coefficient a_λ from these.

Remark 4.1 For notational simplicity, we restricted in the above discussion to the ideal case. The extension to the module case is straightforward; a separate complementary decomposition is now needed for each vector component (corresponding to one unknown function). Only in the application of Riquier's Theorem a slight technical complication arises. One must assume that the used module term order is not only degree compatible but also *Riquier* which means that if for one value of α the inequality $t\mathbf{e}_\alpha \prec s\mathbf{e}_\alpha$ with terms $s, t \in \mathbb{T}$ holds, then it holds for all values of α. This property does not appear in the usual theory of Gröbner bases, as it is only relevant in applications to differential equations. It has the following meaning. Using a classical trick due to Drach [10] (see [52] for an extensive discussion), any system of differential equations with m unknown functions can be transformed via the introduction of m additional independent variables into a system for only one unknown function (at the expense of raising the order by one). Algebraically, this trick transforms a submodule $M \subseteq \mathscr{D}^m$ defined over a skew polynomial ring $\mathscr{D}$ in n variables into an ideal I_M living in a skew polynomial ring $\mathscr{D}'$ in $n + m$ variables. Now the chosen module term order on $\mathscr{D}^m$ induces an ordering on the terms in $\mathscr{D}'$ which is guaranteed to be a term order only if the module term order is Riquier. Lemaire [28] constructed a concrete counter example to Riquier's Theorem where all of its assumptions are satisfied except that the chosen module term order is not Riquier and showed that the unique formal solution of the corresponding initial value problem diverges. Note that any TOP lift of a term order is automatically Riquier.

We have used above already several times the terms "under-" and "overdetermined" systems without actually having defined these notions. Somewhat surprisingly, it is not so easy to provide a rigorous definition for them (and this topic is not much considered in the literature). The linear algebra inspired approach of comparing the number of unknown functions and of equations fails in some instances, as we will show below with an explicit example. The main problem is that it is not clear what should be taken as the "right" number of equations. In linear algebra, one must count the number of linearly independent equations. In the case of linear differential equations, we are working with modules and ideals where almost never a linearly independent set of generators exists and where minimal generating sets may possess different cardinalities.

Example 4.4 Consider the following linear system for two unknown functions u, v of two independent variables x, t:

$$\begin{bmatrix} \partial_t^2 & -\partial_x \partial_t \\ \partial_x \partial_t & -\partial_x^2 \end{bmatrix} \begin{bmatrix} f \\ g \end{bmatrix} = 0 . \tag{4.7}$$

One may call it the *two-dimensional $U(1)$ Yang-Mills equations*, as it represents a gauge theoretic model of electromagnetism in one spatial and one temporal dimension. At first sight, one would call (4.7) a well-determined system, as it comprises as many equations as unknown functions. However, let $f = \phi(x,t)$ and $g = \psi(x,t)$ be a solution of (4.7) and choose an arbitrary function $\Lambda(x,t)$; then $f = \phi + \partial_x \Lambda$ and $g = \psi + \partial_t \Lambda$ is also a solution. Hence one of the unknown functions f, g may be chosen arbitrarily—the typical behaviour of an under-determined system! From a physical point of view, this phenomenon is a consequence of the invariance of (4.7) under gauge transformations of the form $f \rightarrow f + \partial_x \Lambda$ and $g \rightarrow g + \partial_t \Lambda$ and lies at the heart of modern theoretical physics.

Using our above considerations about the construction of formally well-posed initial value problems for a linear partial differential operator D with row module N, it is fairly straightforward to provide a rigorous definition of under- and overdeterminacy. At first sight, it is not clear that the definition presented below is independent of the chosen complementary decomposition. But this follows from the remarks after Definition 7.5, as it is only concerned with the cones of maximal dimension and their dimension and number are the same for any decomposition (in the Cartan–Kähler approach this maximal dimension is sometimes called the Cartan genus of the operator and the number of such cones the degree of arbitrariness; see [48] or [52, Chap. 8] for a more detailed discussion of these notions).

Definition 4.2 Let D be a linear differential operator of order q in n independent variables operating on m unknown functions and choose a complementary decomposition of the associated monomial module lt N (for some term order $\prec$). The operator D defines an *under-determined* system, if the decomposition contains at least one cone of dimension n. It defines a *well-determined* system, if a complementary decomposition exists consisting of mq cones of dimension $n - 1$ (and no other cones). In any other case, the system is *over-determined*.

The definition of underdeterminacy should be clear from the discussion above. If $k \geq 1$ cones of dimension n appear in the complementary decomposition, then $k \geq 1$ unknown functions can be chosen completely arbitrarily and thus are not constrained by the differential operator. This behaviour represents exactly what one intuitively expects from an under-determined system. Note that such cones will always appear, if there are less equations than unknown functions, as in this case there exists at least one unknown function to which no leading derivative belongs and for which thus all derivatives are parametric.

Example 4.5 It is less obvious that there exist under-determined systems comprising as many (or even more) equations as unknowns. If we go back to Example 4.4 and use the TOP extension of the reverse lexicographic order, then the two equations in (4.7) induce already a Gröbner basis of N with leading terms f_{tt} and f_{xt}. Thus there are no leading terms corresponding to derivatives of g and all g-derivatives are parametric. Hence the complementary decomposition consists for g of just one two-dimensional cone with vertex at g.

The well-determined case corresponds to what is called in the theory of partial differential equations a system in *Cauchy–Kovalevskaya form*. Note that for being in this form it is necessary but *not* sufficient that the system comprises as many equations as unknown functions. In addition, a distinguished independent variable t must exist such that each equation in the system can be solved for a pure t-derivative of order q (possibly after a coordinate transformation.[4]) We may assume that the αth equation is solved for $\partial^q f_\alpha / \partial t^q$ and a formally well-posed initial value problem is then given by prescribing $\partial^k f_\alpha / \partial t^k (t = \hat{t})$ for $0 \leq k < q$ and hence the initial data indeed consists of mq functions of $n - 1$ variables. The advection and the wave equation in Example 4.3 are of this type with $t = x_2$. If the wave equation is given in characteristic coordinates, then one must first transform to non-characteristic ones to find a suitable t.

Remark 4.2 There is a certain arbitrariness in Definition 4.2. We give precedence to under- and welldeterminacy; over-determined systems are the remaining ones. This could have been done the other way round. Consider the system

$$
\begin{bmatrix} \partial_1 & -\partial_2 & 0 \\ 0 & 0 & \partial_1 \\ 0 & 0 & \partial_2 \end{bmatrix} \begin{bmatrix} f \\ g \\ h \end{bmatrix} = 0. \tag{4.8}
$$

Obviously, it decouples into one under-determined equation for the two unknown functions f, g and an over-determined system for the unknown function h. According to our definition, the combined system is under-determined, although one could consider it equally well as an over-determined one. One reason for our choice is simply that underdeterminacy is the more intuitive notion (some unknown functions are unconstrained) compared to the rather technical concept of overdeterminacy.

---

[4]It is quite instructive to try to transform (4.7) into such a form: one will rapidly notice that this is not possible!

Another reason lies in the system theoretic interpretation of our results. As we will discuss in Sect. 6.3 in a more abstract setting, a system is *autonomous*, if and only if it is not under-determined. Those variables for which the complementary decomposition contains n-dimensional cones are *free* and thus represent the *inputs* of the system. Different complementary decompositions generally lead to different possible choices of the inputs. But as already remarked above, all complementary decompositions contain the same number of cones of dimension n; thus this number represents the *input dimension* of the system.

5 The Index

Both in the theory and in the numerical analysis of differential algebraic equations, the notion of an index plays an important role. One interpretation is that it provides a measure for the difficulties to expect in a numerical integration, as higher-index systems show a high sensitivity to perturbations (e.g., through numerical errors). It also shows up in the development of an existence and uniqueness theory.

Over the years, many different index concepts have been introduced. One may roughly distinguish two main types of indices. *Differentiation indices* basically count how many times some equations in the system must be differentiated, until a certain property of the system becomes apparent. For example, "the" differentiation index tells us when we can decide that the system is not under-determined (therefore it is called determinacy index in [49]). By contrast, *perturbation indices* are based on estimates for the difference of solutions of the given system and of a perturbed form of it and count how many derivatives of the perturbations have to be taken into account. Generally, differentiation indices are easier to compute, whereas the relevance of perturbation indices—e.g. for a numerical computation—is more obvious. In many cases, all the different approaches lead to the same index value. But it is not difficult to produce examples where the differences can become arbitrarily large. An overview of many index notions and their relationship is contained in [5]; some general references for differential algebraic equations are [2, 24, 27, 44].

We will now use Gröbner bases to introduce some index notions for linear differential equations. While these are strictly speaking differentiation indices, the main result about them is an estimate for a perturbation index. We will first consider the case of ordinary differential equations. Here the theory of Gröbner bases becomes of course rather trivial, as we are dealing with a univariate polynomial ring. Afterwards, we will discuss the extension to partial differential equations. From the point of view of mere definitions, this extension is straightforward and provides—in contrast to most results in the literature—an approach to introduce indices directly for partial differential equations without resorting to some sort of semi-discretization or other forms of reduction to the ordinary case. Using a more geometric approach to differential equations, one can also handle nonlinear equations in an analogous manner. For details we refer to [15, 49]. One could say that the here presented material represents a concrete algorithmic version of the ideas in [15,49] specialized to linear systems.

Let a homogeneous system $Df = 0$ be given where for the moment it does not matter whether D is an ordinary or a partial differential operator. We introduce for each equation in the system a new auxiliary unknown function δ_i and consider the inhomogeneous system $Df = \delta$. Depending on the context, there are different ways to interpret the new unknowns δ_i. One may consider them as perturbations of the original system or as residuals obtained by entering some approximate solution of the original system. For us, they are mainly a convenient way to keep track of what is happening during a Gröbner basis computation. Obviously, the following analysis will lead to the same results for an inhomogeneous system $Df = g$, as the right-hand side g can be absorbed into the perturbation δ.

We compute a Gröbner basis of the row module of the operator D. The most natural choice for the term order is the TOP lift of a degree compatible order. In the case of ordinary differential equations, this choice uniquely determines the term order. For partial differential equations, we will discuss further properties of the underlying order below. The outcome of this computation can be decomposed into two subsystems $\tilde{D}f = F\delta$ and $0 = G\delta$. Here the rows of $\tilde{D}$ form the Gröbner basis of the row module of D and we have the equality $\tilde{D} = FD$. Thus F tells us how the elements of the Gröbner basis have been constructed from the rows of D.

The rows of G form a generating set of the first syzygy module of D (see (7.6) for an explanation of syzygies). Indeed, the equations in the subsystem $0 = G\delta$ arise when some S-polynomial reduces to zero (that is the zero on the left-hand side!) and thus represent syzygies. The Schreyer Theorem 7.7 asserts that they actually generate the whole syzygy module. This second subsystem represents a necessary condition for the existence of solutions in the inhomogeneous case: only for right-hand sides δ satisfying it, the system can possess solutions. According to the fundamental principle discussed in the next section, it is also a sufficient condition (for "good" function spaces). Alternatively, we may consider δ as the residuals which arise when an approximate solution $\hat{f}$ is entered into the original system. Now the subsystem $0 = G\delta$ describes what is often called the *drift* which appears in the numerical integration of over-determined systems. In particular, a stability analysis of the trivial solution $\delta = 0$ provides valuable information about the behaviour of the drift, i.e. whether it is automatically damped or whether it accelerates itself.

Definition 5.1 The *first Gröbner index* γ_1 of $Df = 0$ is the order[5] of the operator F; the *second Gröbner index* γ_2 the order of G.

Example 5.1 For linear differential equations, most index concepts coincide (sometimes with a shift of 1). Hence it is not surprising that the first Gröbner index γ_1 yields often the same value as other approaches. Many applications lead to systems

---

[5]We define the order of an operator matrix as the maximal order of an entry.

in *Hessenberg form*. In our operator notation a (perturbed) Hessenberg system of index 3 has the form

$$
\begin{bmatrix}
\partial - a_{11} & -a_{12} & -a_{13} \\
-a_{21} & \partial - a_{22} & 0 \\
0 & -a_{32} & 0
\end{bmatrix}
\begin{bmatrix}
f_1 \\
f_2 \\
f_3
\end{bmatrix}
=
\begin{bmatrix}
\delta_1 \\
\delta_2 \\
\delta_3
\end{bmatrix}
$$

where all coefficients a_{ij} and the right-hand sides δ_i are functions of the independent variable t and where we assume that $a_{13}(t)a_{21}(t)a_{32}(t) \neq 0$ at every point t considered. According to [24, Theorem 4.23] such a system has strangeness index 2 and according to [2, Proposition 2.4.1] the classical differentiation index is 3.

Computing a Gröbner basis of the row module for the descending TOP lift corresponds exactly to the classical steps for computing the differentiation index. In the first two rows, the leading term is $\partial \mathbf{e}_1$ and $\partial \mathbf{e}_2$. In the last row we have only one non-vanishing term $-a_{32}\mathbf{e}_2$ so that it may be considered as an algebraic equation determining f_2 since our assumption implies that its coefficient $a_{32} \neq 0$.

Since the second and the third row have their leading term in the same component, we can compute an S-polynomial by adding ∂ times the third row to a_{32} times the second row—in other words: we differentiate the last equation and then simplify it with respect to the second one. This yields a new row $\begin{bmatrix} -a_{21}a_{32} & \dot{a}_{32} - a_{22}a_{32} & 0 \end{bmatrix}$ with right-hand side $\dot{\delta}_3 + a_{32}\delta_2$ (note how it "records" the performed operation). We could use the third equation for further simplifications (in fact, elimination of the second component). For our purposes, it suffices to observe that the new row yields an algebraic equation for f_1, since again our assumption implies that its coefficient $a_{21}a_{32}$ does not vanish.

As the leading term of the new row is $-a_{21}a_{32}\mathbf{e}_1$ for the descending TOP lift, we must now form its S-polynomial with the first row: we add ∂ times the new row to $a_{21}a_{32}$ times the first row. As the result still contains a term $\partial \mathbf{e}_2$, it must be reduced with respect to the second equation. We omit the explicit form of the result (which also should be simplified further with respect to the other algebraic equations) and only observe that it represents an algebraic equation for f_3 with nonvanishing coefficient $a_{13}a_{21}a_{32}$. The right-hand side of the new row is given by $\ddot{\delta}_3 + a_{32}\dot{\delta}_2 + 2\dot{a}_{32}\delta_2$.

Now we are done: a reduced Gröbner basis of the row module consists of the simplified form of the three algebraic equations. Their precise form is not relevant for us, but we see that we have $\gamma_1 = 2$, as the right-hand side for the last obtained row contains the second derivative of δ_3. Thus the first Gröbner index coincides here with the strangeness index. The classical differentiation index is one higher because for its determination one wants a *differential* equation for f_3 and hence multiplies the last row once more with ∂.

Example 5.2 We go back to the partial differential system (4.1) of Janet. If we add there a right-hand side with entries δ_1, δ_2, and redo the calculations described above in Example 4.1, then we obtain the following perturbed system:

$$
\begin{bmatrix} \partial_z^2 + y\partial_x^2 \\ \partial_y^2 \\ \partial_x^2\partial_y \\ \partial_x^4 \end{bmatrix} f = \begin{bmatrix} \delta_1 \\ \delta_2 \\ \frac{1}{2}\partial_y^2\delta_1 - \left(y\partial_x^2 + \partial_z^2\right)\delta_2 \\ \left(\frac{1}{2}y\partial_x^2\partial_y^2 - \frac{1}{2}\partial_y^2\partial_z^2 + \partial_x^2\partial_y\right)\delta_1 + \left(\frac{1}{2}y^2\partial_x^4 + y\partial_x^2\partial_y^2 + \frac{1}{2}\partial_z^4\right)\delta_2 \end{bmatrix}.
$$

Hence the first Gröbner index is $\gamma_1 = 4$ for this example. In order to obtain also the second Gröbner index, we must also record how the remaining four S-polynomials $(S(D_2, D_3), S(D_1, D_4), S(D_2, D_4), S(D_3, D_4))$ reduce to zero. This leads to the following four equations:

$$
0 = -\frac{1}{2}\partial_y^3\delta_1 + \left(\frac{1}{2}y\partial_x^2\partial_y + \frac{1}{2}\partial_y\partial_z^2 + \frac{3}{2}\partial_x^2\right)\delta_2,
$$

$$
0 = \left(\frac{1}{2}y^2\partial_x^4\partial_y^2 + y\partial_x^2\partial_y^2\partial_z^2 + \frac{1}{2}\partial_y^2\partial_z^4 - y\partial_x^4\partial_y - \partial_x^2\partial_y\partial_z^2 + \partial_x^4\right)\delta_1 -
$$

$$
\left(\frac{1}{2}y^3\partial_x^6 + \frac{3}{2}y^2\partial_x^4\partial_z^2 + \frac{3}{2}y\partial_x^2\partial_z^4 + \frac{1}{2}\partial_z^6\right)\delta_2,
$$

$$
0 = \left(\frac{1}{2}y\partial_x^2\partial_y^4 + \frac{1}{2}\partial_y^4\partial_z^2\right)\delta_1 -
$$

$$
\left(\frac{1}{2}y^2\partial_x^4\partial_y^2 + y\partial_x^2\partial_y^2\partial_z^2 + \frac{1}{2}\partial_y^2\partial_z^4 + 2y\partial_x^4 + \frac{3}{2}\partial_x^2\partial_z^2\right)\delta_2,
$$

$$
0 = \left(\frac{1}{2}y\partial_x^2\partial_y^3 + \frac{1}{2}\partial_x^3\partial_z^2\right)\delta_1 -
$$

$$
\left(\frac{1}{2}y^2\partial_x^4\partial_y + y\partial_x^2\partial_y\partial_z^2 + \frac{1}{2}\partial_y\partial_z^4 + \frac{3}{2}y\partial_x^4 + \frac{3}{2}\partial_x^2\partial_z^2\right)\delta_2.
$$

Thus we conclude that the second Gröbner index is 6 here.

For the definition of the Gröbner indices, it does not matter whether we are dealing with ordinary or partial differential equations. One should, however, note that in the case of partial differential equations the values γ_1 and γ_2 will generally depend on the used term order (in the above example the system was so simple and in addition homogeneous so that most term orders yield the same results). The whole interpretation of the indices is less obvious in this case and depends on certain properties of the considered system. For this reason, we discuss first the case of ordinary differential equations.

Definition 5.2 ([14]) Let $f(t)$ be a given solution of the linear system $Df = 0$ and $\hat{f}(t)$ an arbitrary solution of the perturbed system $D\hat{f} = \delta$ both defined on the interval $[0, T]$. The system $Df = 0$ has the *perturbation index* ν along the given solution $f(t)$, if ν is the smallest integer such that the following estimate holds for all $t \in [0, T]$ provided the right-hand side is sufficiently small:

$$|f(t) - \hat{f}(t)| \le C \left(|f(0) - \hat{f}(0)| + \|\delta\|_{\nu-1} \right). \tag{5.1}$$

Here the constant C may only depend on the operator D and the length T of the considered interval. $|\cdot|$ denotes some norm on $\mathbb{R}^m$ and $\|\delta\|_k = \sum_{i=0}^k \|\delta^{(i)}\|$ for some norm $\|\cdot\|$ on $C([0, T], \mathbb{R})$ (common choices are the L^1 or the L^∞ norm).

Note that the perturbation index is well-defined only for systems which are not under-determined, as an estimate of the form (5.1) can never hold, if there are free variables present. And obviously it is hard to determine, as it is firstly defined with respect to a specific solution and secondly, given an estimate of the form (5.1), it is very difficult to prove that no other estimate with a lower value of ν exists. On the other hand, Definition 5.2 shows why a solution with a perturbation index greater than one can be difficult to obtain with a standard numerical integration: it does not suffice to control only the size of the residuals δ—as automatically done by any reasonable numerical method—but one must also separately control the size of some derivatives $\delta^{(i)}$ (and thus for example ensure that the residuals do not oscillate with small amplitude but high frequency). The following result asserts that the first Gröbner index provides a uniform bound for the perturbation index along *any* solution of the considered linear system.

Proposition 5.3 *Let the linear system $Df = 0$ of ordinary differential equations be not under-determined. Along any solution $f(t)$ of it, the perturbation index ν and the first Gröbner index γ_1 satisfy the inequality $\nu \le \gamma_1 + 1$.*

Proof As discussed above, we compute for the perturbed system $Df = \delta$ a Gröbner basis with respect to the TOP lift of the ordering by degree and consider the subsystem $\tilde{D}f = F\delta$. We may assume without loss of generality that the operator $\tilde{D}$ is of first order, since a transformation to first order by introducing derivatives of f as additional unknown functions does not affect the value of γ_1 as the maximal order on the right-hand side. Because of the use of a TOP lift, the leading term tells us whether or not an equation in the system $\tilde{D}f = F\delta$ is algebraic. The algebraic equations are differentiated once in order to produce an underlying differential equation of the form $\dot{f} = Kf + \tilde{F}\delta$ where K is of order zero and $\tilde{F}$ of order $\gamma_1 + 1$.

This underlying equation leads immediately to an estimate of the form

$$|f(t) - \hat{f}(t)| \le |f(0) - \hat{f}(0)| + \left| \int_0^t K(\tau)\big(f(\tau) - \hat{f}(\tau)\big)d\tau \right| + \left| \int_0^t \tilde{F}\delta(\tau)d\tau \right|.$$

An application of the well-known Gronwall Lemma yields now the estimate

$$|f(t) - \hat{f}(t)| \le C \left(|f(0) - \hat{f}(0)| + \left| \int_0^t \tilde{F}\delta(\tau)d\tau \right| \right).$$

Since the integration kills one order of differentiation, our claim follows.   □

Example 5.4 Continuing Example 5.1, we see why one speaks of Hessenberg systems of index 3. The perturbation index of such a system is indeed 3, as we obtained above an algebraic equation for every component f_i and the one for f_3 contains a second derivative of δ_3. This observation leads immediately to an estimate of the form (5.1) with $\nu = 3$.

A similar result can be obtained in some cases of partial differential equations. We do not give a rigorous theorem, but merely describe a situation where the above approach works with only minor modifications. The main assumption is that the system is of an evolutionary character (so that it makes sense to consider an initial value problem), as eventually we want to consider our partial differential equation as an abstract ordinary differential equation. Thus we assume that the independent variable x_n plays the role of time $x_n = t$ and choose on our ring $\mathscr{D} = \Bbbk[\partial_1, \dots, \partial_n]$ of linear differential operators the TOP lift of a term order which gives precedence to terms containing the derivative $\partial_n = \partial_t$. Furthermore, we will always assume that the hyperplane $t = 0$ is not characteristic.

As a first step, we must generalize Definition 5.1 to partial differential equations. In principle, this is straightforward; however, one has a vast choice of possible function spaces and norms. We will restrict ourselves for simplicity to smooth functions; other choices should not lead to fundamentally different results. Let $C^\infty(\Omega, \mathbb{R}^m)$ denote the space of smooth $\mathbb{R}^m$-valued functions defined on some domain $\Omega \subseteq \mathbb{R}^n$. As in the ordinary case above, we introduce for functions $f \in C^\infty(\Omega, \mathbb{R}^m)$ and for any integer $\ell \in \mathbb{N}$ the Sobolev type norms

$$\|f\|_\ell = \sum_{0 \le |\mu| \le \ell} \left\| \frac{\partial^{|\mu|} f}{\partial x^\mu} \right\| \tag{5.2}$$

where $\| \cdot \|$ denotes some norm on $C^\infty(\Omega, \mathbb{R}^m)$.

Partial differential equations are usually accompanied by initial or boundary conditions which we accommodate by the following simple approach. Let $\Omega' \subset \Omega$ be a subdomain (typically of lower dimension) and introduce on $C^\infty(\Omega', \mathbb{R}^m)$ similar norms denoted by $\| \cdot \|'_\ell$ etc. The conditions are then written in the form $(Kf)|_{\Omega'} = 0$ for some linear differential operator K of order k. This setting comprises most kinds of initial or boundary conditions typically found in applications.

Definition 5.3 Let $Df = 0$ be a linear system of partial differential equations and let $f(\mathbf{x})$ be a smooth solution of it defined on the domain $\Omega \subseteq \mathbb{R}^m$ and satisfying some initial/boundary conditions $Kf = 0$ of order k on a subdomain $\Omega' \subset \Omega$. The system has the *perturbation index* ν along this solution, if ν is the smallest integer such that for any smooth solution $\hat{f}(\mathbf{x})$ of the perturbed system $D\hat{f} = \delta$ defined on Ω there exists an estimate

$$\left\| \hat{f}(\mathbf{x}) - f(\mathbf{x}) \right\| \le C \left(\|\hat{f} - f\|'_k + \|\delta\|_{\nu-1} \right), \tag{5.3}$$

whenever the right-hand side is sufficiently small. The constant C may depend only on the domains Ω, Ω' and on the operator D.

Given the assumed evolutionary character of our linear system, we consider for simplicity a pure initial value problem with initial conditions prescribed on the hyperplane $t = 0$. As in the proof above, we first compute a Gröbner basis for the perturbed system $D\hat{f} = \delta$ and assume without loss of generality that it is of first order. If the system is not under-determined, then our choice of the term order entails that (after differentiating potentially present algebraic equations) we find a subsystem of the form

$$\frac{\partial \hat{f}}{\partial t} = A\hat{f} + F\delta$$

where the linear differential operator A comprises only derivatives with respect to the remaining "spatial" variables $x_1, \ldots, x_{n-1}$. Our final assumption is that this operator A generates a strongly continuous semigroup $S(t)$ acting on an appropriately chosen Banach space of functions of $x_1, \ldots, x_{n-1}$ and we consider $\hat{f}$ and δ as functions of t alone taking values in this Banach space. Essentially, we have thus transformed our partial differential equation into an abstract ordinary one.

By an elementary result in the theory of semigroups (see, e.g. [39]), every solution of this subsystem can be written in the form

$$\hat{f}(t) = S(t)\hat{f}(0) + \int_0^t S(t - \tau)F\delta(\tau)d\tau$$

for an arbitrary element $\hat{f}(0)$ of the domain of A. Now a straightforward comparison with a solution $f(t)$ of the unperturbed system yields an estimate

$$\|f(t) - \hat{f}(t)\| \leq C\left(\|f(0) - \hat{f}(0)\| + \int_0^t \|F\delta(\tau)\|d\tau\right)$$

with a constant C, as the norms of elements $S(t)$ of a strongly continuous semigroup are bounded on any finite interval. Since the operator F is of order $\gamma_1 + 1$ with γ_1 the first Gröbner index, we finally obtain as bound for the perturbation index $\nu \leq \gamma_1 + 2$. We obtain a slightly worse result than in the case of an ordinary differential equation, as here we cannot necessarily argue that the integration kills one order of differentiation.[6]

Note that in both cases our basic strategy is the same: we derive a subsystem which can be taken as *underlying equation*, i.e. as a differential equation in Cauchy–

---

[6]In applications, it is actually quite rare that systems of partial differential equations contain algebraic equations. In this case, no differentiations are required and F is of order γ_1 so that we obtain the same estimate $\nu \leq \gamma_1 + 1$ as in the case of ordinary differential equations.

Kovalevskaya form (in the ordinary case this simply corresponds to an explicit equation $f' = Af + g$) such that every solution of our system is also a solution of it. For such equations many results and estimates are known; above we used the Gronwall lemma and the theory of semigroups, respectively. The crucial assumption for obtaining the above estimates on the perturbation index is that we may consider all other equations merely as constraints on the permitted initial data. In the ordinary case, this is a triviality. For over-determined systems of partial differential equations, one must be more careful. Under certain non-degenericity assumption on the used coordinates, one can show that this is indeed always the case (strictly speaking, we must assume that our Gröbner basis is in fact even a Pommaret basis). A detailed discussion of this point can be found in [52, Sect. 9.4].

6 Abstract Linear Systems

In his pioneering work, Kalman [19] developed an algebraic approach to control systems given by linear ordinary differential equations, using module theory and constructive methods. Independently, the school of Sato, Kashiwara et al. [21,22,34] was putting forward their algebraic analysis approach to linear systems of partial differential equations, using homological algebra and sheaf theory. This work highly inspired Malgrange [32], Oberst [36], and Pommaret [41–43], when they were developing their seminal contributions to linear partial differential and difference systems with constant or variable coefficients. One of the successes of the resulting new line of research in mathematical systems theory was to realize that the behavioural approach to dynamical systems developed by the school of Willems [53] provided just the right language to tackle also control theoretic questions within this framework. For a recent survey on these topics, see [46]. In this section, we present a unified algebraic approach to the classes of linear systems discussed so far, based on ideas and results from the papers mentioned above and others. Our aim is to extract the common features of these system classes on a general and abstract level.

6.1 Some Homological Algebra

Let $\mathscr{D}$ be a ring (with 1) and let M, N, P be left $\mathscr{D}$-modules. Let $f : M \rightarrow N$ and $g : N \rightarrow P$ be $\mathscr{D}$-linear maps. The sequence

$$M \xrightarrow{f} N \xrightarrow{g} P$$

is called *exact* if $\mathrm{im}(f) = \ker(g)$. Let $\mathscr{F}$ be a left $\mathscr{D}$-module. The contravariant functor $\mathrm{Hom}_{\mathscr{D}}(\cdot, \mathscr{F})$ assigns to each left $\mathscr{D}$-module M the Abelian group (w.r.t.

addition of functions) $\mathrm{Hom}_{\mathscr{D}}(M, \mathscr{F})$ consisting of all $\mathscr{D}$-linear maps from M to $\mathscr{F}$. Moreover, any $\mathscr{D}$-linear map $f : M \to N$ induces a group homomorphism

$$f' : \mathrm{Hom}_{\mathscr{D}}(N, \mathscr{F}) \to \mathrm{Hom}_{\mathscr{D}}(M, \mathscr{F}), \quad \varphi \mapsto \varphi \circ f. \qquad (6.1)$$

The left $\mathscr{D}$-module $\mathscr{F}$ is said to be a *cogenerator* if the functor $\mathrm{Hom}_{\mathscr{D}}(\cdot, \mathscr{F})$ is *faithful*, that is, if $f \neq 0$ implies $f' \neq 0$, where f' is as in (6.1). In other words, for any $f \neq 0$ there exists φ such that $\varphi \circ f \neq 0$.

The functor $\mathrm{Hom}_{\mathscr{D}}(\cdot, \mathscr{F})$ is *left exact*, that is, if

$$M \xrightarrow{f} N \xrightarrow{g} P \to 0$$

is exact (i.e. both $M \to N \to P$ and $N \to P \to 0$ are exact), then the induced sequence

$$\mathrm{Hom}_{\mathscr{D}}(M, \mathscr{F}) \xleftarrow{f'} \mathrm{Hom}_{\mathscr{D}}(N, \mathscr{F}) \xleftarrow{g'} \mathrm{Hom}_{\mathscr{D}}(P, \mathscr{F}) \leftarrow 0$$

is again exact.

The left $\mathscr{D}$-module $\mathscr{F}$ is said to be *injective* if the functor $\mathrm{Hom}_{\mathscr{D}}(\cdot, \mathscr{F})$ is *exact*, that is, if exactness of

$$M \xrightarrow{f} N \xrightarrow{g} P$$

implies exactness of

$$\mathrm{Hom}_{\mathscr{D}}(M, \mathscr{F}) \xleftarrow{f'} \mathrm{Hom}_{\mathscr{D}}(N, \mathscr{F}) \xleftarrow{g'} \mathrm{Hom}_{\mathscr{D}}(P, \mathscr{F}).$$

6.2 Application to Systems Theory

Let $\mathscr{D}$ be a ring (with 1) and let $\mathscr{F}$ be a left $\mathscr{D}$-module. One may think of $\mathscr{F}$ as a set of functions, and of $\mathscr{D}$ as a set of operators acting on them. Any operator $d \in \mathscr{D}$ can be applied to any function $f \in \mathscr{F}$ to yield another function $df \in \mathscr{F}$. Given a positive integer q, this action naturally extends to $d \in \mathscr{D}^{1 \times q}$ and $f \in \mathscr{F}^q$ via $df = \sum_{i=1}^{q} d_i f_i$. To any subset $D \subseteq \mathscr{D}^{1 \times q}$, one may associate

$$D^{\perp} := \{ f \in \mathscr{F}^q \mid \forall d \in D : df = 0 \}.$$

This is the solution set, in $\mathscr{F}^q$, of the homogeneous linear system of equations given in terms of the coefficient rows $d \in D$. Note that $D^{\perp} = \langle D \rangle^{\perp}$, where $\langle D \rangle$ denotes the submodule of $\mathscr{D}^{1 \times q}$ generated by the set D. In general, the set $D^{\perp} \subseteq \mathscr{F}^q$

has no particular algebraic structure besides being an Abelian group with respect to addition. Conversely, given a set $F \subseteq \mathscr{F}^q$, one considers

$$^\perp F := \{d \in \mathscr{D}^{1 \times q} \mid \forall f \in F : df = 0\}$$

which formalizes the set of all equations (given by their coefficient rows) satisfied by the given solutions $f \in F$. It is easy to check that $^\perp F$ is a left submodule of $\mathscr{D}^{1 \times q}$. Thus we have a *Galois correspondence* between the left submodules of $\mathscr{D}^{1 \times q}$ on the one hand, and the Abelian subgroups of $\mathscr{F}^q$ on the other. This means that $(\cdot)^\perp$ and $^\perp(\cdot)$ are both inclusion-reversing, and that

$$D \subseteq {}^\perp(D^\perp) \quad \text{and} \quad F \subseteq ({}^\perp F)^\perp$$

hold for all D and F as above. This implies that

$$(^\perp(D^\perp))^\perp = D^\perp \quad \text{and} \quad {}^\perp(({}^\perp F)^\perp) = {}^\perp F.$$

A *linear system* $S \subseteq \mathscr{F}^q$ takes the form $S := E^\perp$ for some finite subset $E \subseteq \mathscr{D}^{1 \times q}$. We may identify E with a matrix in $\mathscr{D}^{g \times q}$ and write $S = \{f \in \mathscr{F}^q \mid Ef = 0\}$. One calls E a *(kernel) representation matrix* of S. Consider the exact sequence

$$\mathscr{D}^{1 \times g} \overset{\cdot E}{\to} \mathscr{D}^{1 \times q} \to M := \mathscr{D}^{1 \times q} / \mathscr{D}^{1 \times g} E \to 0.$$

Applying the contravariant functor $\mathrm{Hom}_{\mathscr{D}}(\cdot, \mathscr{F})$, and using the standard isomorphism $\mathrm{Hom}_{\mathscr{D}}(\mathscr{D}^{1 \times k}, \mathscr{F}) \cong \mathscr{F}^k$ which relates a linear map defined on the free module $\mathscr{D}^{1 \times k}$ to the image of the standard basis, one obtains an exact sequence

$$\mathscr{F}^g \overset{E}{\leftarrow} \mathscr{F}^q \leftarrow \mathrm{Hom}_{\mathscr{D}}(M, \mathscr{F}) \leftarrow 0.$$

This proves the *Malgrange isomorphism* [32]

$$S \cong \mathrm{Hom}_{\mathscr{D}}(M, \mathscr{F}),$$

which is an isomorphism of Abelian groups. One calls M the *system module* of S.

From now on, we shall assume that $\mathscr{D}$ is a left Noetherian ring. Then every submodule of $\mathscr{D}^{1 \times q}$ is finitely generated, and thus, $D^\perp$ is a linear system for every subset $D \subseteq \mathscr{D}^{1 \times q}$, because $D^\perp = \langle D \rangle^\perp = E^\perp$ for some finite set E.

For any linear system S, we have $(^\perp S)^\perp = S$. If the left $\mathscr{D}$-module $\mathscr{F}$ is a cogenerator, then we also have $^\perp(D^\perp) = D$ for every submodule $D \subseteq \mathscr{D}^{1 \times q}$. To see this, we show that $d \notin D$ implies $d \notin {}^\perp(D^\perp)$. If $d \notin D$, then $0 \neq [d] \in M = \mathscr{D}^{1 \times q} / D$ and thus the $\mathscr{D}$-linear map $g : \mathscr{D} \to M$ with $g(1) = [d]$ is nonzero. By the cogenerator property, there exists $\varphi \in \mathrm{Hom}_{\mathscr{D}}(M, \mathscr{F})$ such that $\varphi \circ g$ is nonzero, that is, $\varphi([d]) \neq 0$. Using the Malgrange isomorphism, this φ corresponds to $f \in S := D^\perp$ with $df \neq 0$. Thus $d \notin {}^\perp S = {}^\perp(D^\perp)$.

Summing up: If $\mathcal{D}$ is left Noetherian and $\mathcal{F}$ a cogenerator, we have a duality between linear systems in $\mathcal{F}^q$ and submodules of $\mathcal{D}^{1 \times q}$, that is, $(\cdot)^\perp$ and ${}^\perp(\cdot)$ are bijections and inverse to each other. More concretely, since $\mathcal{D}$ is left Noetherian, any submodule of $\mathcal{D}^{1 \times q}$ can be written in the form $D = \mathcal{D}^{1 \times g} E$ for some matrix $E \in \mathcal{D}^{g \times q}$. Then its associated system is $S = \{f \in \mathcal{F}^q \mid Ef = 0\}$, and we have both $S = D^\perp$ and $D = {}^\perp S$. For instance, let $E_i \in \mathcal{D}^{g_i \times q}$ be representation matrices of two systems $S_i \subseteq \mathcal{F}^q$. Then we have $S_1 \subseteq S_2$ if and only if $\mathcal{D}^{1 \times g_1} E_1 \supseteq \mathcal{D}^{1 \times g_2} E_2$, that is, if $E_2 = XE_1$ holds for some $X \in \mathcal{D}^{g_2 \times g_1}$. In particular, $S_1 = \{0\}$ holds if and only if E_1 is left invertible.

If $\mathcal{F}$ is injective, then the exactness of

$$\mathcal{D}^{1 \times m} \overset{\cdot E}{\to} \mathcal{D}^{1 \times n} \overset{\cdot G}{\to} \mathcal{D}^{1 \times p} \tag{6.2}$$

implies the exactness of

$$\mathcal{F}^m \overset{E}{\leftarrow} \mathcal{F}^n \overset{G}{\leftarrow} \mathcal{F}^p. \tag{6.3}$$

Thus the inhomogeneous system of equations $Gg = f$ (where $G \in \mathcal{D}^{n \times p}$ and $f \in \mathcal{F}^n$ are given) has a solution $g \in \mathcal{F}^p$ if and only if $f \in \text{im}(G) = \ker(E)$, that is, if and only if $Ef = 0$. Since $EG = 0$ by construction, it is clear that $Gg = f$ implies $Ef = 0$. The crucial aspect of injectivity is that $Ef = 0$ is also sufficient for the existence of g. The resulting solvability condition for $Gg = f$ is often called the "fundamental principle": One computes a matrix E whose rows generate the left kernel $\ker(\cdot G) \subseteq \mathcal{D}^{1 \times n}$ of G. In other words, one computes the "syzygies" of the rows of G as in Eq. (7.6). Then the inhomogeneous system $Gg = f$ has a solution g if and only if the right-hand side f satisfies the "compatibility condition" $Ef = 0$. If $\mathcal{F}$ is an injective cogenerator, then the exactness of (6.2) is in fact equivalent to the exactness of (6.3).

6.3   Autonomy

Let $\mathcal{D}$ be left Noetherian and let $\mathcal{F}$ be an injective cogenerator. Let q be a positive integer and let $S \subseteq \mathcal{F}^q$ be a linear system. The system S is said to be autonomous if it has no free variables, that is, if none of the projection maps

$$\pi_i : S \to \mathcal{F}, \quad f \mapsto f_i$$

is surjective for $1 \leq i \leq q$. Let E be a representation matrix of S and let $M = \mathcal{D}^{1 \times q} / \mathcal{D}^{1 \times g} E$ be the system module of S.

Theorem 6.1 *The following are equivalent:*

1. *S is autonomous.*
2. *For* $1 \leq i \leq q$, *there exist* $0 \neq d_i \in \mathscr{D}$ *and* $X \in \mathscr{D}^{q \times g}$ *such that* $\mathrm{diag}(d_1, \ldots, d_q) = XE$.

If $\mathscr{D}$ *is a domain, then these conditions are also equivalent to:*

3. *M is a torsion module, that is, for all* $m \in M$ *there exists* $0 \neq d \in \mathscr{D}$ *such that* $dm = 0$.
4. *E has full column rank.*

Note that the rank of E is well-defined, since a left Noetherian domain $\mathscr{D}$ possesses a quotient (skew) field $\mathscr{K} = \{d^{-1}n \mid d, n \in \mathscr{D}, d \neq 0\}$. Recall that the number $m := q - \mathrm{rank}(E)$ is known as the input dimension of S. Thus, the theorem above describes the situation where $m = 0$. In other words, S is not under-determined.

Proof Via the injective cogenerator property, a surjection $\pi_i : S \to \mathscr{F}$ exists if and only if there is an injection $j_i : \mathscr{D} \to M$ with $j_i(1) = [\mathbf{e}_i]$, where $\mathbf{e}_i \in \mathscr{D}^{1 \times q}$ denotes the ith standard basis row. Hence autonomy is equivalent to j_i being noninjective for all $1 \leq i \leq q$, that is, $j_i(d_i) = [d_i \mathbf{e}_i] = 0 \in M$ for some $0 \neq d_i \in \mathscr{D}$, and then $d_i \mathbf{e}_i \in \mathscr{D}^{1 \times g} E$.

Now let $\mathscr{D}$ be a domain. The following implications are straightforward: "3 $\Rightarrow$ 2 $\Rightarrow$ 4". We show "4 $\Rightarrow$ 2". Assume that $\mathrm{rank}(E) = q \leq g$. Let E_1 denote the matrix E after deleting the first column. Then $\mathrm{rank}(E_1) = q - 1 < g$ and thus there exists $0 \neq x \in \mathscr{K}^{1 \times g}$ with $xE_1 = 0$. This x can be chosen such that $xE = [k, 0, \ldots, 0]$ for some $0 \neq k \in \mathscr{K}$. Writing $x = d^{-1}n_1$ for some $n_1 \in \mathscr{D}^{1 \times g}$, we get $n_1 E = [d_1, 0, \ldots, 0]$ with $0 \neq d_1 \in \mathscr{D}$. Proceeding like this with the remaining columns, we get $XE = \mathrm{diag}(d_1, \ldots, d_q)$ as desired. Finally, "2 $\Rightarrow$ 3" follows from the fact that in a left Noetherian domain, any two nonzero elements have a nonzero common left multiple, see [13, Example 4N] or [25, Corollary 10.23]. □

To test whether condition 2 is satisfied, define the augmented matrix

$$E_i' := \begin{bmatrix} \mathbf{e}_i \\ E \end{bmatrix} \in \mathscr{D}^{(1+g) \times q}.$$

Consider the left $\mathscr{D}$-module $\ker(\cdot E_i') := \{x \in \mathscr{D}^{1 \times (1+g)} \mid xE_i' = 0\}$. Let

$$\pi : \mathscr{D}^{1 \times (1+g)} \to \mathscr{D}, \quad x \mapsto x_1$$

denote the projection on the first component. Then S is autonomous if and only if $\pi(\ker(\cdot E_i')) \neq 0$ holds for all $1 \leq i \leq q$. Thus autonomy can be tested constructively if $\mathscr{D}$ admits the computation of generating sets of kernels of matrices (or, in other words, "syzygies" as described in Eq. (7.6)).

Algorithm 2 Test for autonomy of system S

Input: kernel representation matrix E of S
Output: message "autonomous" or "not autonomous"
1: $q \leftarrow$ number of columns of E
2: **for all** $1 \leq i \leq q$ **do**
3:     $E_i' \leftarrow \begin{bmatrix} \mathbf{e}_i \\ E \end{bmatrix}$
4:     $F \leftarrow$ matrix whose rows generate $\ker(\cdot E_i')$
5:     $F_1 \leftarrow$ set of first column entries of F
6:     **if** $F_1 = \{0\}$ **then**
7:         **return** "not autonomous" and **stop**
8:     **end if**
9: **end for**
10: **return** "autonomous"

For $\mathscr{D} = \Bbbk[s_1, \dots, s_r]$, that is, for partial differential or difference equations with constant coefficients, the notion of autonomy can be refined as follows. The *autonomy degree* of S [54] is defined to be $r - d$, where d denotes the maximal dimension of a cone in the complementary decomposition of the row module of E (note that the number d coincides with the Krull dimension of the system module M). Thus, nonautonomous systems have autonomy degree zero, and nonzero autonomous systems have autonomy degrees between 1 and r. The value r corresponds to systems which are finite-dimensional as $\Bbbk$-vector spaces, which is the strongest autonomy notion. Systems whose autonomy degree is at least 2 are always over-determined. Analytic characterizations of this property in terms of the system trajectories are given in [47, 59].

6.4 Controllability

Let $\mathscr{D}$ be left and right Noetherian and let $\mathscr{F}$ be an injective cogenerator. Let q be a positive integer and let $S \subseteq \mathscr{F}^q$ be a linear system. The system S is said to be controllable if it has an image representation, that is, if there exists a matrix $L \in \mathscr{D}^{q \times l}$ such that

$$S = \{Lv \mid v \in \mathscr{F}^l\}.$$

The motivation for this definition comes from behavioural systems theory. There, controllability corresponds to concatenability of trajectories. Roughly speaking, this amounts to being able to join any given "past" trajectory with any desired "future" trajectory by a connecting trajectory. For many relevant system classes, a properly defined version of this concatenability property has been shown to be equivalent to the existence of an image representation [40, 47, 53, 57].

Theorem 6.2 *Let E be a kernel representation matrix of S and let M be the system module of S. The following are equivalent:*

1. *S has an image representation.*
2. *E is a left syzygy matrix, that is, for some matrix $L \in \mathscr{D}^{q \times l}$, we have*

$$\mathrm{im}(\cdot E) = \{ xE \mid x \in \mathscr{D}^{1 \times g} \} = \ker(\cdot L) := \{ y \in \mathscr{D}^{1 \times q} \mid yL = 0 \}.$$

3. *M is torsionless, that is, for every $0 \neq m \in M$ there exists $\varphi \in \mathrm{Hom}_{\mathscr{D}}(M, \mathscr{D})$ with $\varphi(m) \neq 0$.*

If $\mathscr{D}$ is a domain, then these conditions are also equivalent to:

4. *M is torsionfree, that is, $dm = 0$ with $d \in \mathscr{D}$ and $m \in M$ implies that $d = 0$ or $m = 0$.*

Proof The sequence $\mathscr{D}^{1 \times g} \xrightarrow{\cdot E} \mathscr{D}^{1 \times q} \xrightarrow{\cdot L} \mathscr{D}^{1 \times l}$ is exact if and only if $\mathscr{F}^g \xleftarrow{E} \mathscr{F}^q \xleftarrow{L} \mathscr{F}^l$ is exact. This proves "1 $\Leftrightarrow$ 2". For "2 $\Rightarrow$ 3", we use that $M = \mathscr{D}^{1 \times q}/\mathrm{im}(\cdot E) = \mathscr{D}^{1 \times q}/\ker(\cdot L) \cong \mathrm{im}(\cdot L) \subseteq \mathscr{D}^{1 \times l}$ by the homomorphism theorem, and submodules of free modules are torsionless. For "3 $\Rightarrow$ 2", let K be a $\mathscr{D}$-matrix such that

$$\ker(E) = \{ x \in \mathscr{D}^q \mid Ex = 0 \} = \mathrm{im}(K) = \{ Ky \mid y \in \mathscr{D}^k \}$$

and let $\bar{E}$ be a $\mathscr{D}$-matrix such that

$$\ker(\cdot K) = \{ y \in \mathscr{D}^{1 \times q} \mid yK = 0 \} = \mathrm{im}(\cdot \bar{E}) = \{ z\bar{E} \mid z \in \mathscr{D}^{1 \times \bar{g}} \}.$$

Since $EK = 0$, we have $\mathrm{im}(\cdot E) \subseteq \mathrm{im}(\cdot \bar{E})$. The proof is finished if we can show that this inclusion is in fact an equality, because then $\mathrm{im}(\cdot E) = \mathrm{im}(\cdot \bar{E}) = \ker(\cdot K)$ and we may take $L = K$ in condition 2. Assume conversely that $d \in \mathrm{im}(\cdot \bar{E}) \setminus \mathrm{im}(\cdot E)$. Then $0 \neq [d] \in M$. Any homomorphism $\phi : M \to \mathscr{D}$ takes the form $\phi([d]) = dx$ for some $x \in \mathscr{D}^q$ which must satisfy $Ex = 0$ for well-definedness, and this implies $x = Ky$. Since $d = z\bar{E}$, we have $\phi([d]) = dx = z\bar{E}Ky = 0$. This contradicts the assumption of torsionlessness. Now let $\mathscr{D}$ be a domain. For "3 $\Rightarrow$ 4", suppose that $dm = 0$ and $0 \neq d$. Then any $\varphi \in \mathrm{Hom}_{\mathscr{D}}(M, \mathscr{D})$ satisfies $d\varphi(m) = 0$ and thus $\varphi(m) = 0$. By condition 3, this implies that $m = 0$. The converse implication holds as well, since M is finitely generated [13, Proposition 7.19]. $\square$

To test whether condition 2 is satisfied, one proceeds as in the proof. One computes the $\mathscr{D}$-matrices K and $\bar{E}$ as described above. This requires that generating sets of kernels of matrices (that is, "syzygies" as in Eq. (7.6)) can be computed over $\mathscr{D}$. The assumption that $\mathscr{D}$ is left and right Noetherian guarantees that left and right kernels are finitely generated. Then E is a left syzygy matrix if and only if

$$\mathrm{im}(\cdot E) = \mathrm{im}(\cdot \bar{E}).$$

Algorithm 3 Test for controllability of system S

Input: kernel representation matrix E of S
Output: message "controllable" or "not controllable"
1: $\mathscr{G} \leftarrow$ Gröbner basis of row module of E
2: $K \leftarrow$ matrix whose columns generate $\ker(E)$
3: $\bar{E} \leftarrow$ matrix whose rows generate $\ker(\cdot K)$
4: **for all** rows d of $\bar{E}$ **do**
5:     **if** $\mathrm{NormalForm}(d, \mathscr{G}) \neq 0$ **then**
6:         **return** "not controllable" and **stop**
7:     **end if**
8: **end for**
9: **return** "controllable"

This can be shown similarly as in the proof. The condition can be tested if $\mathscr{D}$ allows to decide module membership: It holds if and only if each row of $\bar{E}$ is contained in the row module of E. If $\mathscr{D}$ admits a Gröbner basis theory, then a row $d \in \mathscr{D}^{1 \times q}$ belongs to a submodule $D \subseteq \mathscr{D}^{1 \times q}$ if and only if its normal form with respect to a Gröbner basis of D is zero.

The data K and $\bar{E}$ computed by this algorithm are useful on their own: Due to $\mathrm{im}(\cdot \bar{E}) = \ker(\cdot K)$, we have

$$S_c := \{ f \in \mathscr{F}^q \mid \bar{E} f = 0 \} = \{ K v \mid v \in \mathscr{F}^k \}$$

because of the injectivity of $\mathscr{F}$. It turns out that S_c is the largest controllable subsystem of S, and that $S = S_c$ holds if and only if S is controllable. Thus, the factor group S/S_c (sometimes called the obstruction to controllability [56]) measures how far S is from being controllable.

The controllability test described above has been developed, for certain operator domains $\mathscr{D}$, in a series of papers [41–43] by Pommaret and Quadrat, see also [6]. These authors also introduced a concept of controllability degrees, similar to the autonomy degrees. However, the definition of the controllability degrees is more involved and uses extension functors (for an introduction in the systems theoretic setting, see e.g. [55]). The controllability test has been generalized to a large class of noncommutative rings with zero-divisors in [60].

7 Appendix: Gröbner Bases

Gröbner bases are a fundamental tool in constructive algebra, as they permit to perform many basic algebraic constructions algorithmically. They were formally introduced for ideals in a polynomial ring in the Ph.D. thesis of Buchberger [4] (written under the supervision of Gröbner); modern textbook presentations can be found, e.g. in [1, 8]. Most general purpose computer algebra systems like Maple or Mathematica provide an implementation. However, the computation of a Gröbner basis can be quite demanding (with respect to both time and space) and for larger examples the use of specialised systems like CoCoA,[7] Macaulay 2,[8] Magma[9] or Singular[10] is recommended.

Gröbner bases were originally introduced for ideals in the standard commutative polynomial ring $\mathscr{P} = \Bbbk[x_1, \ldots, x_n]$ where $\Bbbk$ is any field (e.g. $\Bbbk = \mathbb{R}$ or $\Bbbk = \mathbb{C}$). Since then, several generalizations to non-commutative rings have been studied. We will present here one such extension covering all rings appearing in this survey: *polynomial rings of solvable type*. This class of algebras was first introduced in [20]; further studies can be found in [23, 29, 50, 52]. Furthermore, we will also discuss besides ideals the case of submodules of a free module $\mathscr{P}^m$. A general introduction to modules over certain non-commutative rings covering also algorithmic aspects and Gröbner bases was recently given by Gómez–Torrecillas [12].

Gröbner bases are always defined with respect to a *term order*. In the polynomial ring $\mathscr{P}$ there does not exist a natural ordering on the set $\mathbb{T}$ of all terms x^μ with an exponent vector $\mu \in \mathbb{N}_0^n$ (only in the univariate case the degree provides a canonical ordering). However, the use of such an ordering is crucial for many purposes like extending the familiar polynomial division to the multivariate case. Elements of the free module $\mathscr{P}^m$ can be represented as m-dimensional column (or row) vectors where each component is a polynomial. Here a "term" $\mathbf{t}$ is a vector where all components except one vanish and the non-zero component consists of a term $t \in \mathbb{T}$ in the usual sense and thus can be written as $\mathbf{t} = t\mathbf{e}_i$ where $\mathbf{e}_i$ denotes the ith vector in the standard basis of $\mathscr{P}^m$. We denote the set of all such vector terms by $\mathbb{T}^m$. We say that $\mathbf{t}$ *divides* another term $\mathbf{s} = s\mathbf{e}_j$, written $\mathbf{t} \mid \mathbf{s}$, if $i = j$ and $t \mid s$, i.e. only terms living in the same component can divide each other.

Definition 7.1 A total order $\prec$ on $\mathbb{T}$ is a *term order*, if it satisfies: (i) given three terms $r, s, t \in \mathbb{T}$ such that $s \prec t$, we also have $rs \prec rt$ (monotonicity), and (ii) any term $t \in \mathbb{T}$ different from 1 is greater than 1. A term order for which additionally terms of higher degree are automatically greater than terms of lower degree is called *degree compatible*.

A total order $\prec$ on $\mathbb{T}^m$ is a *module term order*, if it satisfies: (i) for two vector terms $\mathbf{s}, \mathbf{t} \in \mathbb{T}^m$ with $\mathbf{s} \prec \mathbf{t}$ and an ordinary term $r \in \mathbb{T}$, we also have $r\mathbf{s} \prec r\mathbf{t}$ and (ii) for any $\mathbf{t} \in \mathbb{T}^m$ and $s \in \mathbb{T}$, we have $\mathbf{t} \prec s\mathbf{t}$.

Given a (non-zero) polynomial $f \in \mathscr{P}$ and a term order $\prec$, we can sort the finitely many terms actually appearing in f according to $\prec$. We call the largest one the *leading term* lt f of f and its coefficient is the *leading coefficient* lc f; finally,

---

[7]http://cocoa.dima.unige.it.

[8]http://www.math.uiuc.edu/Macaulay2.

[9]http://magma.usyd.edu.au.

[10]http://www.singular.uni-kl.de.

the *leading monomial*[11] of f is then the product $\operatorname{lm} f = \operatorname{lc}(f)\operatorname{lt}(f)$. The same notations are also used for polynomial vectors $\mathbf{f} \in \mathscr{P}^m$.

Example 7.1 Usually, term orders are introduced via the exponent vectors. The *lexicographic order* is defined as follows: $x^\mu \prec_{\text{lex}} x^\nu$, if the first non-vanishing entry of $\mu - \nu$ is negative. Thus it implements the ordering used for words in a dictionary: if we take $x_1 = a$, $x_2 = b$ etc, then $ab^2 \prec_{\text{lex}} a^2$ and a^2 is ordered before ab^2, although it is of lower degree. The lexicographic order is very useful in elimination problems and for solving polynomial equations. Unfortunately, it tends to be rather inefficient in Gröbner bases computations.

An example of a degree compatible order is the *reverse lexicographic order*: $x^\mu \prec_{\text{revlex}} x^\nu$, if $\deg x^\mu < \deg x^\nu$ or $\deg x^\mu = \deg x^\nu$ and the last non-vanishing entry of $\mu - \nu$ is positive. Now we find $a^2 \prec_{\text{revlex}} ab^2$ because of the different degrees and $a^3 \prec_{\text{revlex}} ab^2$ because only the latter term contains b. Usually, the reverse lexicographic order is the most efficient order for Gröbner bases computations.

Given a term order $\prec$ on $\mathbb{T}$, there exist two natural ways to lift it to a module term order on $\mathbb{T}^m$. In the *(ascending) TOP lift*, we put *t*erm *o*ver *p*osition and define $s\mathbf{e}_i \prec_{\text{TOP}} t\mathbf{e}_j$, if $s \prec t$ or $s = t$ and $i < j$ (in the descending TOP lift one uses $i > j$). The *(ascending) POT lift* works the other way round: $s\mathbf{e}_i \prec_{\text{POT}} t\mathbf{e}_j$, if $i < j$ or $i = j$ and $s \prec t$ (again we use $i > j$ for the descending version). Such lifts are the most commonly used module term orders.

Finally, assume that $F = \{\mathbf{f}_1, \ldots, \mathbf{f}_r\} \subset \mathscr{P}^m$ is a finite set of r (non-zero) vectors and $\prec$ a module term order on $\mathbb{T}^m$. Then the set F induces a module term order $\prec_F$ on $\mathbb{T}^r \subset \mathscr{P}^r$ as follows: $s\mathbf{e}_i \prec_F t\mathbf{e}_j$, if $\operatorname{lt}(s\mathbf{f}_i) \prec \operatorname{lt}(t\mathbf{f}_j)$ or $\operatorname{lt}(s\mathbf{f}_i) = \operatorname{lt}(t\mathbf{f}_j)$ and $i > j$ (no, the direction of this relation is not a typo!). As we will see later, this induced order is very important for computing the syzygies of the set F.

Let $\star$ be a non-commutative multiplication on $\mathscr{P}$. We allow both that our variables x_i do no longer commute, i.e. $x_i \star x_j \neq x_j \star x_i$, and that the variables act on the coefficients $c \in \mathbb{k}$, i.e. $x_i \star c \neq c \star x_i$. However, we do not permit that the coefficients act on the variables, i.e. $c \star x_i = cx_i$ where on the right-hand side the usual product is used. A prototypical example are linear differential operators where we may choose $\mathbb{k} = \mathbb{R}(t_1, \ldots, t_n)$, the fields of real rational functions in some unknowns $t_1, \ldots, t_n$, and take as "variables" for the polynomials the partial derivatives with respect to these unknowns, $x_i = \partial/\partial t_i$. Here we still find $x_i \star x_j = x_j \star x_i$, as for smooth functions partial derivatives commute, but $x_i \star c = cx_i + \partial c/\partial t_i$ for any rational function $c \in \mathbb{k}$. Non-commutative variables occur, e.g. in the Weyl algebra or in "quantized algebras" like q-difference operators.

For the definition of Gröbner bases in such non-commutative polynomial rings, it is important that the product does not interfere with the chosen term order. This motivates the following definition of a special class of polynomial rings.

---

[11]For us a *term* is a pure power product x^μ whereas a *monomial* is of the form cx^μ with a coefficient $c \in \mathbb{k}$; beware that some text books on Gröbner bases use the words term and monomial with exactly the opposite meaning.

Definition 7.2 Let $\prec$ be a term order and $\star$ a non-commutative product on the polynomial ring $\mathscr{P} = \Bbbk[x_1, \ldots, x_n]$. The triple $(\mathscr{P}, \star, \prec)$ defines a *solvable polynomial ring*, if the following three conditions are satisfied:

(i) $(\mathscr{P}, \star)$ is a ring;
(ii) $c \star f = cf$ for all coefficients $c \in \Bbbk$ and polynomials $f \in \mathscr{P}$;
(iii) $\mathrm{lt}\,(f \star g) = \mathrm{lt}\,(f) \cdot \mathrm{lt}\,(g)$ for all polynomials $f, g \in \mathscr{P} \setminus \{0\}$ (note the use of the ordinary commutative product on the right-hand side!).

The first condition ensures that the arithmetics in $\mathscr{P}$ satisfies all the usual rules like associativity or distributivity. The second condition implies that $\mathscr{P}$ is still a $\Bbbk$-linear space (as long as we multiply with field elements only from the left). The third condition enforces the compatibility of the new product $\star$ with the chosen term order: the non-commutativity does not affect the leading terms. If such a compatibility holds, then the usual commutative Gröbner bases theory remains valid in our more general setting without any changes.

Example 7.2 All non-commutative rings appearing in this article belong to a subclass of the solvable polynomial rings, namely the Ore algebras which may be considered as generalizations of the ring of linear differential operators. This class was first considered by Noether and Schmeidler [35] and then more extensively by Ore [38]; our presentation follows [3].

Let $\sigma : \Bbbk \to \Bbbk$ be an automorphism of the field $\Bbbk$. A *pseudo-derivation* with respect to σ is a map $\delta : \Bbbk \to \Bbbk$ such that $\delta(c + d) = \delta(c) + \delta(d)$ and $\delta(cd) = \sigma(c)\delta(d) + \delta(c)d$ for all $c, d \in \Bbbk$. If $\sigma = \mathrm{id}$, the identity map, the second condition is the standard Leibniz rule for derivations. $\sigma(c)$ is called the *conjugate* and $\delta(c)$ the *derivative* of $c \in \Bbbk$.

Given the maps σ and δ, the ring $\Bbbk[\partial; \sigma, \delta]$ of univariate Ore polynomials consists of all polynomials $\sum_{i=0}^{d} c_i \partial^i$ in ∂ with coefficients $c_i \in \Bbbk$. The addition is defined as usual. The "variable" ∂ operates on an element $c \in \Bbbk$ according to $\partial \star c = \sigma(c)\partial + \delta(c)$. Note that we may interpret this equation as a rewrite rule which tells us how to bring a ∂ from the left to the right of a coefficient. This rewriting can be used to define the multiplication $\star$ on the whole ring $\Bbbk[\partial; \sigma, \delta]$: given two elements $\theta_1, \theta_2 \in \Bbbk[\partial; \sigma, \delta]$, we can transform the product $\theta_1 \star \theta_2$ to the normal form of a polynomial (coefficients to the left of the variable) by repeatedly applying our rewrite rule. The product of two linear polynomials evaluates then to

$$(f_1 + f_2\partial) \star (g_1 + g_2\partial) = f_1g_1 + f_2\delta(g_1) +$$
$$\left[f_1g_2 + f_2\sigma(g_1) + f_2\delta(g_2)\right]\partial + f_2\sigma(g_2)\partial^2 .$$
$$(7.1)$$

The fact that σ is an automorphism ensures that $\deg(\theta_1 \star \theta_2) = \deg \theta_1 + \deg \theta_2$. We call $\Bbbk[\partial; \sigma, \delta]$ the *Ore algebra* generated by σ and δ.

A simple familiar example is given by $\Bbbk = \mathbb{Q}(x)$, $\delta = \frac{d}{dx}$, and $\sigma = \mathrm{id}$ yielding linear ordinary differential operators with rational functions as coefficients.

Similarly, we can obtain recurrence and difference operators. We set $\Bbbk = \mathbb{C}(n)$, the field of sequences $(s_n)_{n \in \mathbb{Z}}$ with complex elements $s_n \in \mathbb{C}$, and take for σ the shift operator, i.e. the automorphism mapping s_n to s_{n+1}. Then $\Delta = \sigma - \mathrm{id}$ is a pseudo-derivation. $\Bbbk[E; \sigma, 0]$ consists of linear ordinary recurrence operators, $\Bbbk[E; \sigma, \Delta]$ of linear ordinary difference operators.

For multivariate Ore polynomials, we take a set $\Sigma = \{\sigma_1, \ldots, \sigma_n\}$ of automorphisms and a set $\Delta = \{\delta_1, \ldots, \delta_n\}$ where each δ_i is a pseudo-derivation with respect to σ_i. For each pair (σ_i, δ_i) we introduce a "variable" ∂_i satisfying a commutation rule as in the univariate case. If we require that all the maps σ_i, δ_j commute with each other, one easily checks that $\partial_i \star \partial_j = \partial_j \star \partial_i$, i.e. the "variables" ∂_i also commute. Setting $D = \{\partial_1, \ldots, \partial_n\}$, we denote by $\Bbbk[D; \Sigma, \Delta]$ the ring of multivariate Ore polynomials. Because of the commutativity of the variables ∂_i, we may write the terms as ∂^μ with multi indices $\mu \in \mathbb{N}_0^n$, so that it indeed makes sense to speak of a polynomial ring. The proof that $(\Bbbk[D; \Sigma, \Delta], \star, \prec)$ is a solvable polynomial ring for any term order $\prec$ is trivial.

From now on, we always assume that we have chosen a fixed solvable polynomial algebra $(\mathscr{P}, \star, \prec)$. All references to a leading term etc. are then meant with respect to the term order $\prec$ contained in this choice. In a non-commutative ring we must distinguish between left, right and two-sided ideals. In this appendix, we exclusively deal with left ideals: if $F = \{f_1, \ldots, f_r\} \subset \mathscr{P}$ is some finite set of polynomials, then the left ideal generated by the basis F is the set

$$\langle F \rangle = \left\{ \sum_{\alpha=1}^{r} g_\alpha \star f_\alpha \mid g_\alpha \in \mathscr{P} \right\} \tag{7.2}$$

of all left linear combinations of the elements of F and it satisfies $g \star f \in \langle F \rangle$ for any $f \in \langle F \rangle$ and any $g \in \mathscr{P}$. Right ideals are defined correspondingly[12] and a two-sided ideal is simultaneously a left and a right ideal.

A for computational purposes highly relevant property of the commutative polynomial ring is that it is *Noetherian*, i.e. any ideal in it possesses a finite basis (Hilbert's Basis Theorem). For solvable polynomial rings, the situation is more complicated. The book [52, Sect. 3.3] collects a number of possible approaches to prove for large classes of such rings that they are Noetherian, too. In particular, this is the case for all Ore algebras. In [52, Proposition 3.2.10], it is also shown that all solvable algebras (over a coefficient field) satisfy the left Ore condition so that one can define a left quotient skew field [25].

Remark 7.1 A complication in the treatment of non-commutative polynomial rings is given by the fact that in general the product of two terms is no longer a term. Hence the notion of a monomial ideal makes no longer sense. In the sequel, we will use the convention that when we speak about the divisibility of terms $s, t \in \mathbb{T}$,

---

[12]Beware that the left and the right ideal generated by a set F are generally different.

this is always to be understood within the *commutative* polynomial ring, i.e. $s \mid t$, if and only if a further term $r \in \mathbb{T}$ exists such that $r \cdot s = s \cdot r = t$. In other words, we consider $(\mathbb{T}, \cdot)$ as an Abelian monoid. Given a left ideal $I \lhd \mathscr{P}$ in a solvable polynomial ring, we then introduce within this monoid the *leading ideal* $\operatorname{lt} I = \langle \operatorname{lt} f \mid f \in I \setminus \{0\} \rangle$. Thus $\operatorname{lt} I$ is always to be understood as a monomial ideal in the *commutative* polynomial ring, even if the considered polynomial ring $\mathscr{P}$ has a non-commutative product $\star$.

Definition 7.3 Let $I \lhd \mathscr{P}$ be a left ideal in the polynomial ring $\mathscr{P}$. A finite set $G \subset I$ is a *Gröbner basis* of I, if for every non-zero polynomial $f \in I$ in the ideal a generator $g \in G$ exists such that $\operatorname{lt} g \mid \operatorname{lt} f$ and thus $\operatorname{lt} I = \langle \operatorname{lt} G \rangle$.

Above definition is rather technical and it is neither evident that Gröbner bases exist nor that they are useful for anything. We will now show that they allow us to solve effectively the *ideal membership problem*: given an ideal $I = \langle F \rangle \lhd \mathscr{P}$ and a polynomial $f \in \mathscr{P}$, decide whether or not $f \in I$. A solution of this problem is for example mandatory for an effective arithmetics in the factor ring $\mathscr{P}/I$. As a by-product, we will see that a Gröbner basis is indeed a basis, i.e. $I = \langle G \rangle$.

Definition 7.4 Let $G = \{g_1, \dots, g_r\} \subset \mathscr{P} \setminus \{0\}$ be a finite set of non-zero polynomials. A further polynomial $f \in \mathscr{P} \setminus \{0\}$ is called *reducible* with respect to G, if f contains a term $t \in \mathbb{T}$ for which a polynomial $g_i \in G$ exists such that $\operatorname{lt} g_i \mid t$. If this is the case, then a term $s \in \mathbb{T}$ exists such that $\operatorname{lt}(s \star g_i) = t$ and we can perform a *reduction step*: $f \to_{g_i} \tilde{f} = f - (c/\operatorname{lc}(s \star g_i))s \star g_i$ where $c \in \Bbbk$ denotes the coefficient of t in f.[13] A polynomial $h \in \mathscr{P}$ is a *normal form* of f modulo the set G, if we can find a sequence of reduction steps

$$f \to_{g_{i_1}} h_1 \to_{g_{i_2}} h_2 \to_{g_{i_3}} \cdots \to_{g_{i_s}} h_s = h \tag{7.3}$$

and h is not reducible with respect to G. In this case, we also write shortly $f \to_G^+ h$ for (7.3) or $h = \operatorname{NF}(f, G)$.

It should be noted that the notation $h = \operatorname{NF}(f, G)$ is somewhat misleading. A polynomial f may have many different normal forms with respect to some set G, as generally different sequences of reduction steps lead to different results. If h is some normal form of the polynomial f, then (non-unique) *quotients* $q_1, \dots, q_r \in \mathscr{P}$ exist such that $f = \sum_{i=1}^{r} q_i \star g_i + h$. Thus $h = 0$ immediately implies $f \in \langle G \rangle$; however, the converse is not necessarily true: even if $f \in \langle G \rangle$, it may possess non-vanishing normal forms. In the univariate case (and for $r = 1$), we recover here the familiar polynomial division from high school where both the normal form (or *remainder*) h and the quotient q are unique. A concrete multivariate *division*

---

[13]The term "reduction" refers to the fact that the monomial ct in f is replaced by a linear combination of terms which are all smaller than t with respect to the used term order. It does *not* imply that $\tilde{f}$ is simpler in the sense that it has less terms than f. In fact, quite often the opposite is the case!

Algorithm 4 Multivariate polynomial division

Input: finite set $G = \{g_1, \ldots, g_r\} \subset \mathscr{P} \setminus \{0\}$, polynomial $f \in \mathscr{P}$
Output: a normal form $h = \mathrm{NF}(f, G)$, quotients $q_1, \ldots, q_r$
 1: $h \leftarrow 0; \quad q_1 \leftarrow 0; \quad \ldots \quad q_r \leftarrow 0$
 2: **while** $f \neq 0$ **do**
 3:     **if** $\exists\, i : \mathrm{lt}\, g_i \mid \mathrm{lt}\, f$ **then**
 4:        choose smallest index i with this property and $s \in \mathbb{T}$ such that $\mathrm{lt}\,(s \star g_i) = t$
 5:        $m \leftarrow \frac{\mathrm{lc}\, f}{\mathrm{lc}\,(s\star g_i)} s; \quad f \leftarrow f - m \star g_i; \quad q_i \leftarrow q_i + m$
 6:     **else**
 7:        $h \leftarrow h + \mathrm{lm}\, f; \quad f \leftarrow f - \mathrm{lm}\, f$
 8:     **end if**
 9: **end while**
10: **return** $h, q_1, \ldots, q_r$

---

algorithm for computing a normal form together with some quotients is shown in Algorithm 4.

In this article, we are always assuming that our polynomials are defined over a field $\Bbbk$. This assumption entails that the divisibility of two monomials cx^μ and dx^ν is decided exclusively by the contained terms x^μ and x^ν. Over a coefficient ring $\mathscr{R}$, this is no longer the case: note that in Line /5/ of Algorithm 4 we must perform the division $\mathrm{lc}\, f / \mathrm{lc}\,(s \star g_i)$ which may not be possible in a coefficient ring. Under certain technical assumptions on the ring $\mathscr{R}$, it is still possible to set up a theory of Gröbner bases which, however, becomes more complicated. For details on this extension, we refer to the literature, see e.g. [1].

The following fundamental characterization theorem collects a number of equivalent definitions for Gröbner bases. It explains their distinguished position among all possible bases of a given ideal. In particular, (ii) already solves the ideal membership problem. Furthermore, (iii) implies that for a Gröbner basis G the notation $NF(f, G)$ is well-defined, as in this case any sequence of reduction steps leads to same final result.

Theorem 7.3 *Let $0 \neq I \lhd \mathscr{P}$ be an ideal and $G \subset I$ a finite subset. Then the following statements are equivalent.*

(i) G is a Gröbner basis of I.
(ii) Given a polynomial $f \in \mathscr{P}$, ideal membership $f \in I$ is equivalent to $f \to_G^+ 0$.
(iii) $I = \langle G \rangle$ and every $f \in \mathscr{P}$ has a unique normal form with respect to G.
(iv) A polynomial $f \in \mathscr{P}$ is contained in the ideal I, if and only if it possesses a
      standard representation with respect to G, i.e. there are coefficients $h_g \in \mathscr{P}$
      such that $f = \sum_{g \in G} h_g \star g$ and $\mathrm{lt}\,(h_g \star g) \preceq \mathrm{lt}\, f$ whenever $h_g \neq 0$.

Obviously, we may also consider the ring $\mathscr{P}$, any ideal $I \lhd \mathscr{P}$ and the corresponding factor space $\mathscr{P}/I$ as $\Bbbk$-linear spaces. The following observation shows why it is of interest to know the leading ideal $\mathrm{lt}\, I$.

Theorem 7.4 (Macaulay) *Let $I \lhd \mathcal{P}$ be an ideal and $\prec$ an arbitrary term order. Then $\mathcal{P}/I$ and $\mathcal{P}/\operatorname{lt} I$ are isomorphic as $\Bbbk$-linear spaces.*

Proof (Sketch) Denote by $\mathcal{B} = \mathbb{T} \setminus \operatorname{lt} I$ the set of all terms *not* contained in the leading ideal. One now shows that the respective equivalence classes of the elements of $\mathcal{B}$ define a $\Bbbk$-linear basis of $\mathcal{P}/I$ and $\mathcal{P}/\operatorname{lt} I$, respectively. The linear independence is fairly obvious and $\mathcal{B}$ induces generating sets, as the normal form of any polynomial $f \in \mathcal{P}$ with respect to a Gröbner basis is a $\Bbbk$-linear combination of elements of $\mathcal{B}$.   □

The use of the term "basis" in commutative algebra is a bit misleading, as one does not require linear independence. Opposed to the situation in linear algebra, elements of an ideal generally do not possess a unique representation as linear combination of the basis. For homogeneous ideals, which may also be considered as graded vector spaces, the following concept of a combinatorial composition leads again to unique representations.

Definition 7.5 Let $I \lhd \mathcal{P}$ be a homogeneous ideal. A *Stanley decomposition* of I is an isomorphism as graded vector spaces

$$I \cong \bigoplus_{t \in \mathcal{T}} \Bbbk[X_t] \cdot t \tag{7.4}$$

with a finite set $\mathcal{T} \subset \mathbb{T}$ of terms and subsets $X_t \subseteq \{x_1, \ldots, x_n\}$. The elements of X_t are called the *multiplicative variables* of the generator t. One speaks of a *Rees decomposition*, if all sets of multiplicative variables are of the form $X_t = \{x_1, x_2, \ldots, x_{k_t}\}$ where $0 \leq k_t \leq n$ is called the *class* of t. A *complementary (Stanley) decomposition* is an analogous isomorphism for the factor space $\mathcal{P}/I$.

Vector spaces of the form $\Bbbk[X_t] \cdot t$ are called *cones* and the number of multiplicative variables is the *dimension* of such a cone. While Stanley decompositions are anything but unique and different decompositions may consist of differently many cones, one can show that the highest appearing dimension of a cone is always the same (the dimension of the ideal I) and also the number of cones with this particular dimension (algebraically it is given by the multiplicity or degree of the ideal) is an invariant. This observation is a simple consequence of the connection between complementary decompositions and Hilbert polynomials (or functions) which, however, cannot be discussed here (see e.g. [52] and references given there). Complementary decomposition got their name from the simple observation that in the monomial case they are equivalent to expressing the complement $\mathcal{P} \setminus I$ as a direct sum of cones. Concrete examples are shown in Figs. 1 and 2 on pp. 300 and 301.

Because of Macaulay's Theorem 7.4, it indeed suffices for complementary decompositions to consider $\mathcal{P}/\operatorname{lt} I$ and thus monomial ideals. This observation reduces the task of their construction to a purely combinatorial problem. A simple solution is provided by the recursive Algorithm 5. It takes as input the minimal basis of I and returns a set of pairs (t, X_t) consisting of a generator t and its multiplicative

---

Algorithm 5 Complementary decomposition of monomial ideal

Input: minimal basis $\mathscr{B}$ of monomial ideal $I \subset \mathscr{P}$
Output: finite complementary decomposition $\mathscr{T}$ of $\overline{I}$
1: **if** $n = 1$ **then** {in this case $\mathscr{B} = \{x^\nu\}$ with $\nu \in \mathbb{N}$}
2: $\quad q_0 \leftarrow \nu; \quad \mathscr{T} \leftarrow \{([x^0], \emptyset), \ldots, ([x^{q_0-1}], \emptyset)\}$
3: **else**
4: $\quad q_0 \leftarrow \max\{\nu_n \mid x^\nu \in \mathscr{B}\}; \quad \mathscr{T} \leftarrow \emptyset$
5: $\quad$ **for** q from 0 to q_0 **do**
6: $\quad\quad \mathscr{B}'_q \leftarrow \{x^{\nu'} \in \mathbb{N}_0^{n-1} \mid x^\nu \in \mathscr{B}, \, \nu_n \leq q\}$
7: $\quad\quad \mathscr{T}'_q \leftarrow \texttt{ComplementaryDecomposition}(\mathscr{B}'_q)$
8: $\quad\quad$ **if** $q < q_0$ **then**
9: $\quad\quad\quad \mathscr{T} \leftarrow \mathscr{T} \cup \{(x^{[\nu',q]}, X_{\nu'}) \mid (x^{\nu'}, X_{\nu'}) \in \mathscr{T}'_q\}$
10: $\quad\quad$ **else**
11: $\quad\quad\quad \mathscr{T} \leftarrow \mathscr{T} \cup \{(x^{[\nu',q]}, X_{\nu'} \cup \{n\}) \mid (x^{\nu'}, X_{\nu'}) \in \mathscr{T}'_q\}$
12: $\quad\quad$ **end if**
13: $\quad$ **end for**
14: **end if**
15: **return** $\mathscr{T}$

---

variables. The recursion is on the number n of variables in the polynomial ring $\mathscr{P}$. If $\nu = (\nu_1, \ldots, \nu_n) \in \mathbb{N}_0^n$ is an exponent vector, we denote by $\nu' = [\nu_1, \ldots, \nu_{n-1}]$ its truncation to the first $n - 1$ entries and write $\nu = [\nu', \nu_n]$. We remark that a special type of Gröbner bases, the *involutive bases* (first introduced by Gerdt and Blinkov [11]), is particularly adapted to this problem [52, Sect. 5.1]. We refer to [50, 51] and [52, Chap. 3] for an extensive treatment of these bases and further references.

Despite its great theoretical importance, Theorem 7.3 still does not settle the question of the existence of Gröbner bases, as none of the given characterisations is effective. For a constructive approach, we need syzygies. The fundamental tool is the *S-polynomial* of two polynomials $f, g \in \mathscr{P}$. Let $x^\mu = \text{lt } f$, $x^\nu = \text{lt } g$ be the corresponding leading terms and $x^\rho = \text{lcm}(\dot{x}^\mu, x^\nu)$ their least common multiple. If $x^\rho = x^{\bar{\mu}} x^\mu = x^{\bar{\nu}} x^\nu$ (in the commutative sense, cf. Remark 7.1), then the *S*-polynomial of f and g is defined as the difference

$$S(f, g) = \frac{x^{\bar{\mu}} \star f}{\text{lc}(x^{\bar{\mu}} \star f)} - \frac{x^{\bar{\nu}} \star g}{\text{lc}(x^{\bar{\nu}} \star g)}. \tag{7.5}$$

Note that the coefficients are chosen in such a way that the leading monomials cancel in the subtraction. With the help of this construction, one can provide an effective criterion for a set to be a Gröbner basis of an ideal.

Theorem 7.5 (Buchberger) *A finite set $G \subset \mathscr{P}$ of polynomials is a Gröbner basis of the left ideal $I = \langle G \rangle$ generated by it, if and only if for every pair $f, g \in G$ the S-polynomial $S(f, g)$ reduces to zero with respect to G.*

This theorem translates immediately into a simple algorithm for the effective construction of Gröbner bases, the *Buchberger Algorithm 6*, and also ensures its correctness. The termination is guaranteed, if the solvable polynomial ring $\mathscr{P}$ is

---

Algorithm 6 Gröbner basis (Buchberger)

Input: finite set $F \subset \mathscr{P}$, term order $\prec$
Output: Gröbner basis G of the ideal $\langle F \rangle$
1: $G \leftarrow F$
2: $\mathscr{S} \leftarrow \{\{g_1, g_2\} \mid g_1, g_2 \in G, g_1 \neq g_2\}$
3: **while** $\mathscr{S} \neq \emptyset$ **do**
4:    choose $\{g_1, g_2\} \in \mathscr{S}$
5:    $\mathscr{S} \leftarrow \mathscr{S} \setminus \{\{g_1, g_2\}\}$; $\bar{g} \leftarrow \mathrm{NF}(S(g_1, g_2), G)$
6:    **if** $\bar{g} \neq 0$ **then**
7:       $\mathscr{S} \leftarrow \mathscr{S} \cup \{\{\bar{g}, g\} \mid g \in G\}$; $G \leftarrow G \cup \{\bar{g}\}$
8:    **end if**
9: **end while**
10: **return** G

Noetherian (which is the case for all rings appearing in this article). It should be noted that the basic form of the Buchberger algorithm shown here can handle only very small examples. A version able to handle substantial problems requires many optimizations and the development and improvement of efficient implementations is still an active field of research.

Gröbner bases are anything but unique: if G is a Gröbner basis of the ideal I for some term order $\prec$, then we may extend G by arbitrary elements of I and still have a Gröbner basis. For obtaining uniqueness, one must impose further conditions on the basis. It is easy to show that a monomial ideal I (in the commutative polynomial ring) always possesses a unique *minimal* basis $\mathscr{B}$ consisting entirely of monomials. Minimial means here that no element of $\mathscr{B}$ divides another element.

Definition 7.6 A Gröbner basis G of an ideal $I \lhd \mathscr{P}$ is called *minimal*, if the set lt G is the minimal basis of the monomial ideal lt I. We call G a *reduced Gröbner basis*, if every generator $g \in G$ is in normal form with respect to $G \setminus \{g\}$ and every leading coefficient lc g equals 1.

It is not difficult to show that augmenting Algorithm 6 by autoreductions of the set G (i.e. every element of G is reduced with respect to all other elements) leads to an algorithm that always returns a reduced Gröbner basis. With a little bit more effort, one obtains in addition the following uniqueness result which allows for effectively deciding whether two ideals are equal.

Proposition 7.6 ([1, Theorem 1.8.7]) *Every ideal $I \lhd \mathscr{P}$ possesses for any term order $\prec$ a unique reduced Gröbner basis.*

Although there exist infinitely many different term orders, one can show that any given ideal I has only finitely many different reduced Gröbner bases [33, Lemma 2.6].

All the presented material on Gröbner bases is readily translated to left submodules $\mathscr{M} \subseteq \mathscr{P}^m$ of a free polynomial module using the module term orders introduced in Definition 7.1 and all results remain true in this more general situation. The only slight difference concerns the definition of the S-polynomial. In the case of two elements $\mathbf{f}, \mathbf{g} \in \mathscr{P}^m$, their S-"polynomial" (which now is of course also a vector in $\mathscr{P}^m$) is set to zero, if $\operatorname{lt} \mathbf{f}$ and $\operatorname{lt} \mathbf{g}$ live in different components, as then our construction of the S-"polynomial" (7.5) makes no sense (recall that in a free module terms are only divisible, if they are in the same component, and thus we can speak of a least common multiple only in this case).

Remark 7.2 The Buchberger algorithm may be considered as a simultaneous generalization of the Gauß algorithm for linear systems of equations and of the Euclidean algorithm for determining the greatest common divisor of two univariate polynomials. One can easily verify that the S-polynomial of two polynomials with relatively prime leading terms always reduces to zero. Hence, in the case of linear polynomials it suffices to consider pairs of polynomials with the same leading term (variable) for which the construction of the S-polynomial amounts to a simple Gaußian elimination step.

In the case of two univariate polynomials, the construction of their S-polynomial and its subsequent reduction with respect to the two polynomials is equivalent to the familiar polynomial division. Hence computing the Gröbner basis of a set F amounts simply to determine the greatest common divisor of the elements of F and any reduced Gröbner basis consists of a single polynomial (this observation may be considered as an alternative proof that univariate polynomials define a principal ideal domain). By the same reasoning, we conclude that a reduced Gröbner basis of a submodule of a free module $\mathscr{P}^m$ over a univariate polynomial ring $\mathscr{P}$ may have at most m elements, the leading terms of which are all in different components.

The terminology "S-polynomial" is actually an abbreviation of "syzygy polynomial." Recall that a *syzygy* of a finite set $F = \{\mathbf{f}_1, \ldots, \mathbf{f}_r\} \subset \mathscr{P}^m$ is a vector $\mathbf{S} \in \mathscr{P}^r$ with components $S_i \in \mathscr{P}$ such that

$$S_1 \star \mathbf{f}_1 + \cdots + S_r \star \mathbf{f}_r = 0. \tag{7.6}$$

All syzygies of F together form again a left submodule $\operatorname{Syz}(F) \subseteq \mathscr{P}^r$. Note that this submodule may be understood as the solution set of a linear system of equations over the ring $\mathscr{P}$ or—in a more abstract terminology—as the kernel of a linear map. Thus the effective determination of syzygy modules represents a natural and important problem, if one wants to do linear algebra over a ring.

The Schreyer Theorem shows that, by retaining information that is automatically computed during the determination of a Gröbner basis G with the Buchberger Algorithm 6, one obtains for free a Gröbner basis of the syzygy module $\operatorname{Syz}(G)$. More precisely, assume that $G = \{\mathbf{g}_1, \ldots, \mathbf{g}_r\} \subset \mathscr{P}^m$ is a Gröbner basis and let $\mathbf{g}_i, \mathbf{g}_j \in G$ be two generators with leading terms in the same component. According to (7.5), their S-"polynomial" can be written in the form $S(\mathbf{g}_i, \mathbf{g}_j) = m_i \star \mathbf{g}_i - m_j \star \mathbf{g}_j$ for suitable monomials m_i, m_j, and Theorem 7.3 implies the existence of coefficients $h_{ijk} \in \mathscr{P}$ such that $\sum_{k=1}^r h_{ijk} \star \mathbf{g}_k$ is a standard representation of $S(\mathbf{g}_i, \mathbf{g}_j)$. Combining these two representations, we obtain a syzygy

$$\mathbf{S}_{ij} = m_i \mathbf{e}_i - m_j \mathbf{e}_j - \sum_{k=1}^{r} h_{ijk} \mathbf{e}_k \qquad (7.7)$$

where $\mathbf{e}_k$ denote the vectors of the standard basis of $\mathscr{P}^r$. Recalling the module term order introduced at the end of Example 7.1, we obtain now the following fundamental result on the syzygy module of a Gröbner basis.

Theorem 7.7 (Schreyer [9, Theorem 3.3]) *Let $G \subset \mathscr{P}^m$ be a Gröbner basis for the term order $\prec$ of the submodule generated by it. Then the set of all the syzygies $\mathbf{S}_{ij}$ defined by (7.7) is a Gröbner basis of* $\mathrm{Syz}(G) \subseteq \mathscr{P}^r$ *for the induced term order $\prec_G$.*

For a general finite set $F \subset \mathscr{P}^m$, one can determine $\mathrm{Syz}(F)$ by first computing a Gröbner basis G of $\langle F \rangle$, then using Theorem 7.7 to obtain a generating set $\mathscr{S}$ of $\mathrm{Syz}(G)$ and finally transforming $\mathscr{S}$ into a generating set of $\mathrm{Syz}(F)$ essentially by linear algebra. Details can be found, e.g., in [1].

By iterating Schreyer's Theorem 7.7, one obtains a *free resolution* of the submodule $\langle G \rangle \subseteq \mathscr{P}^m$ (although this is not necessarily the most efficient way to do this), i.e. an exact sequence

$$0 \longrightarrow \mathscr{P}^{r_n} \longrightarrow \cdots \longrightarrow \mathscr{P}^{r_1} \longrightarrow \mathscr{P}^{r_0} \longrightarrow \langle G \rangle \longrightarrow 0 \qquad (7.8)$$

(Hilbert's Syzygy Theorem guarantees that the length of the resolution is at most the number n of variables in the polynomial ring $\mathscr{P}$). The *minimal free resolution* which can be constructed from any free resolution via some linear algebra gives access to many important invariants of the submodule $\langle G \rangle$ like Betti numbers. However, it is beyond the scope of this article to discuss this application of Gröbner bases.

References

1. Adams, W., Loustaunau, P.: An Introduction to Gröbner Bases. Graduate Studies in Mathematics, vol. 3. American Mathematical Society, Providence (1994)
2. Brenan, K., Campbell, S., Petzold, L.: Numerical Solution of Initial-Value Problems in Differential-Algebraic Equations. Classics in Applied Mathematics, vol. 14. SIAM, Philadelphia (1996)
3. Bronstein, M., Petkovšek, M.: An introduction to pseudo-linear algebra. Theor. Comput. Sci. **157**, 3–33 (1996)
4. Buchberger, B.: Ein algorithmus zum auffinden der basiselemente des restklassenringes nach einem nulldimensionalen polynomideal. Ph.D. thesis, Universität Innsbruck (1965) [English translation: J. Symb. Comput. **41**, 475–511 (2006)]
5. Campbell, S., Gear, C.: The index of general nonlinear DAEs. Numer. Math. **72**, 173–196 (1995)

6. Chyzak, F., Quadrat, A., Robertz, D.: Effective algorithms for parametrizing linear control systems over Ore algebras. Appl. Algebra Eng. Commun. Comput. **16**, 319–376 (2005)
7. Cohn, P.: Free Rings and Their Relations. Academic, New York (1971)
8. Cox, D., Little, J., O'Shea, D.: Ideals, Varieties, and Algorithms. Undergraduate Texts in Mathematics. Springer, New York (1992)
9. Cox, D., Little, J., O'Shea, D.: Using Algebraic Geometry. Graduate Texts in Mathematics, vol. 185. Springer, New York (1998)
10. Drach, J.: Sur les systèmes complètement orthogonaux dans l'espace à n dimensions et sur la réduction des systèmes différentielles les plus généraux. C. R. Acad. Sci. **125**, 598–601 (1897)
11. Gerdt, V., Blinkov, Y.: Involutive bases of polynomial ideals. Math. Comp. Simul. **45**, 519–542 (1998)
12. Gómez-Torrecillas, J.: Basic module theory over non-commutative rings with computational aspects of operator algebras. In: Barkatou, M., Cluzeau, T., Regensburger, G., Rosenkranz, M. (eds.) Algebraic and Algorithmic Aspects of Differential and Integral Operators. Lecture Notes in Computer Science, vol. 8372, pp. 23–82. Springer, Heidelberg (2014)
13. Goodearl, K., Warfield, R.: An Introduction to Noncommutative Noetherian Rings. London Mathematical Society Student Texts, vol. 61, 2nd edn. Cambridge University Press, Cambridge (2004)
14. Hairer, E., Lubich, C., Roche, M.: The Numerical Solution of Differential-Algebraic Equations by Runge-Kutta Methods. Lecture Notes in Mathematics, vol. 1409. Springer, Berlin (1989)
15. Hausdorf, M., Seiler, W.: Perturbation versus differentiation indices. In: Ghanza, V., Mayr, E., Vorozhtsov, E. (eds.) Computer Algebra in Scientific Computing—CASC 2001, pp. 323–337. Springer, Berlin (2001)
16. Hausdorf, M., Seiler, W.: An efficient algebraic algorithm for the geometric completion to involution. Appl. Algebra Eng. Commun. Comput. **13**, 163–207 (2002)
17. Jacobson, N.: The Theory of Rings. Americal Mathematical Society, Providence (1943)
18. Janet, M.: Leçons sur les Systèmes d'Équations aux Dérivées Partielles. Cahiers Scientifiques, Fascicule IV. Gauthier-Villars, Paris (1929)
19. Kalman, R.: Algebraic structure of linear dynamical systems. Proc. Natl. Acad. Sci. USA **54**, 1503–1508 (1965)
20. Kandry-Rody, A., Weispfenning, V.: Non-commutative Gröbner bases in algebras of solvable type. J. Symb. Comput. **9**, 1–26 (1990)
21. Kashiwara, M., Kawai, T., Kimura, T.: Foundations of Algebraic Analysis. Princeton University Press, Princeton (1986)
22. Kato, G., Struppa, D.: Fundamentals of Algebraic Microlocal Analysis. Pure and Applied Mathematics, vol. 217. Dekker, New York (1999)
23. Kredel, H.: Solvable Polynomial Rings. Verlag Shaker, Aachen (1993)
24. Kunkel, P., Mehrmann, V.: Differential-Algebraic Equations: Analysis and Numerical Solution. EMS Textbooks in Mathematics. EMS Publishing House, Zürich (2006)
25. Lam, T.: Lectures on Modules and Rings. Graduate Texts in Mathematics, vol. 189. Springer, New York (1999)
26. Lam, T.: On the equality of row rank and column rank. Expo. Math. **18**, 161–163 (2000)
27. Lamour, R., März, R., Tischendorf, C.: Differential-Algebraic Equations: A Projector Based Analysis. Differential-Algebraic Equations Forum. Springer, Berlin/Heidelberg (2013)
28. Lemaire, F.: An orderly linear PDE with analytic initial conditions with a non-analytic solution. J. Symb. Comput. **35**, 487–498 (2003)
29. Levandovskyy, V.: Non-commutative computer algebra for polynomial algebras: Gröbner bases, applications and implementation. Ph.D. thesis, Fachbereich Mathematik, Universität Kaiserslautern (2005)
30. Levandovskyy, V., Schindelar, K.: Computing diagonal form and Jacobson normal form of a matrix using Gröbner bases. J. Symb. Comput. **46**, 595–608 (2011)
31. Li, P., Liu, M., Oberst, U.: Linear recurring arrays, linear systems and multidimensional cyclic codes over quasi-Frobenius rings. Acta Appl. Math. **80**, 175–198 (2004)

32. Malgrange, B.: Systemes différentiels à coefficients constants. Semin. Bourbaki **15**, 1–11 (1964)
33. Mora, T., Robbiano, L.: The Gröbner fan of an ideal. J. Symb. Comput. **6**, 183–208 (1988)
34. Morimoto, M.: An Introduction to Sato's Hyperfunctions. Translation of Mathematical Monographs, vol. 129. American Mathematical Society, Providence (1993)
35. Noether, E., Schmeidler, W.: Moduln in nichtkommutativen Bereichen, insbesondere aus Differential- und Differenzausdrücken. Math. Z. **8**, 1–35 (1920)
36. Oberst, U.: Multidimensional constant linear systems. Acta Appl. Math. **20**, 1–175 (1990)
37. Oberst, U., Pauer, F.: The constructive solution of linear systems of partial difference and differential equations with constant coefficients. Multidim. Syst. Sign. Process. **12**, 253–308 (2001)
38. Ore, O.: Theory of non-commutative polynomials. Ann. Math. **34**, 480–508 (1933)
39. Pazy, A.: Semigroups of Linear Operators and Applications to Partial Differential Equations. Applied Mathematical Sciences, vol. 44. Springer, New York (1983)
40. Pillai, H., Shankar, S.: A behavioral approach to control of distributed systems. SIAM J. Control Optim. **37**, 388–408 (1999)
41. Pommaret, J., Quadrat, A.: Generalized Bezout identity. Appl. Algebra Eng. Commun. Comput. **9**, 91–116 (1998)
42. Pommaret, J., Quadrat, A.: Algebraic analysis of linear multidimensional control systems. IMA J. Math. Control Inf. **16**, 275–297 (1999)
43. Pommaret, J., Quadrat, A.: Localization and parametrization of linear multidimensional control systems. Syst. Control Lett. **37**, 247–260 (1999)
44. Rabier, P., Rheinboldt, W.: Theoretical and numerical analysis of differential-algebraic equations. In: Ciarlet, P., Lions, J. (eds.) Handbook of Numerical Analysis, vol. VIII, pp. 183–540. North-Holland, Amsterdam (2002)
45. Riquier, C.: Les Systèmes d'Équations aux Derivées Partielles. Gauthier-Villars, Paris (1910)
46. Robertz, D.: Recent progress in an algebraic analysis approach to linear systems. Multidimensional System Signal Processing (2015, to appear). doi: 10.007/s11045-014-0280-9
47. Rocha, P., Zerz, E.: Strong controllability and extendibility of discrete multidimensional behaviors. Syst. Control Lett. **54**, 375–380 (2005)
48. Seiler, W.: On the arbitrariness of the general solution of an involutive partial differential equation. J. Math. Phys. **35**, 486–498 (1994)
49. Seiler, W.: Indices and solvability for general systems of differential equations. In: Ghanza, V., Mayr, E., Vorozhtsov, E. (eds.) Computer Algebra in Scientific Computing—CASC '99, pp. 365–385. Springer, Berlin (1999)
50. Seiler, W.: A combinatorial approach to involution and δ-regularity I: involutive bases in polynomial algebras of solvable type. Appl. Algebra Eng. Commun. Comput. **20**, 207–259 (2009)
51. Seiler, W.: A combinatorial approach to involution and δ-regularity II: structure analysis of polynomial modules with Pommaret bases. Appl. Algebra Eng. Commun. Comput. **20**, 261–338 (2009)
52. Seiler, W.: Involution: The Formal Theory of Differential Equations and Its Applications in Computer Algebra. Algorithms and Computation in Mathematics, vol. 24. Springer, Berlin (2009)
53. Willems, J.: Paradigms and puzzles in the theory of dynamical systems. IEEE Trans. Autom. Control **36**, 259–294 (1991)
54. Wood, J., Rogers, E., Owens, D.: A formal theory of matrix primeness. Math. Control Signals Syst. **11**, 40–78 (1998)
55. Zerz, E.: Extension modules in behavioral linear systems theory. Multidim. Syst. Sign. Process. **12**, 309–327 (2001)
56. Zerz, E.: Multidimensional behaviours: an algebraic approach to control theory for PDE. Int. J. Control **77**, 812–820 (2004)
57. Zerz, E.: An algebraic analysis approach to linear time-varying systems. IMA J. Math. Control Inf. **23**, 113–126 (2006)

58. Zerz, E.: Discrete multidimensional systems over $\mathbb{Z}_n$. Syst. Control Lett. **56**, 702–708 (2007)
59. Zerz, E., Rocha, P.: Controllability and extendibility of continuous multidimensional behaviors. Multidim. Syst. Sign. Process. **17**, 97–106 (2006)
60. Zerz, E., Seiler, W., Hausdorf, M.: On the inverse syzygy problem. Commun. Algebra **38**, 2037–2047 (2010)

Index

© Springer International Publishing Switzerland 2015

335

A. Ilchmann, T. Reis (eds.), *Surveys in Differential-Algebraic Equations II*,
Differential-Algebraic Equations Forum, DOI 10.1007/978-3-319-11050-9